Lecture Notes in Computer Science 12629

More information about this subseries at http://www.springer.com/series/7409

Éric Renault · Selma Boumerdassi ·
Paul Mühlethaler (Eds.)

Machine Learning for Networking

Third International Conference, MLN 2020
Paris, France, November 24–26, 2020
Revised Selected Papers

 Springer

Editors
Éric Renault ⓘ
Laboratoire LIGM UMR 8049 CNRS
ESIEE Paris
Noisy-le-Grand, France

Selma Boumerdassi
CNAM/CEDRIC
Paris, France

Paul Mühlethaler
Inria/EVA Project
Paris, France

ISSN 0302-9743   ISSN 1611-3349 (electronic)
Lecture Notes in Computer Science
ISBN 978-3-030-70865-8   ISBN 978-3-030-70866-5 (eBook)
https://doi.org/10.1007/978-3-030-70866-5

LNCS Sublibrary: SL3 – Information Systems and Applications, incl. Internet/Web, and HCI

This Springer imprint is published by the registered company Springer Nature Switzerland AG
The registered company address is: Gewerbestrasse 11, 6330 Cham, Switzerland

Preface

The rapid development of new network infrastructures and services has led to the generation of huge amounts of data, and machine learning now appears to be the best solution to process these data and make the right decisions for network management.

The International Conference on Machine Learning for Networking (MLN) aimed at providing a top forum for researchers and practitioners to present and discuss new trends in deep and reinforcement learning, pattern recognition and classification for networks, machine learning for network slicing optimization, 5G systems, user behavior prediction, multimedia, IoT, security and protection, optimization and new innovative machine learning methods, performance analysis of machine learning algorithms, experimental evaluations of machine learning, data mining in heterogeneous networks, distributed and decentralized machine learning algorithms, intelligent cloud-support communications, resource allocation, energy-aware communications, software-defined networks, cooperative networks, positioning and navigation systems, wireless communications, wireless sensor networks, and underwater sensor networks. Due to the special health situation all around the world, all presentations at MLN 2020 were done remotely.

The call for papers resulted in a total of 50 submissions from all around the world: Algeria, Brazil, Canada, France, Greece, India, Italy, Morocco, The Netherlands, Norway, Pakistan, Russia, South Africa, Sweden, Taiwan, Tunisia, Turkey, the United Kingdom, and the United States. All submissions were assigned to at least three members of the Program Committee for review. The Program Committee decided to accept 22 papers. Three intriguing keynotes from Mérouane Debbah, Huawei, France, Philippe Jacquet, Inria, France, and Fakhri Karray, University of Waterloo, Canada, completed the technical program.

We would like to thank all who contributed to the success of this conference, in particular the members of the Program Committee and the reviewers for carefully reviewing the contributions and selecting a high-quality program. Our special thanks go to the members of the Organizing Committee for their great help.

Thursday afternoon was dedicated to the Third International Workshop on Networking for Smart Living (NSL). The technical program of NSL included two presentations and an invited paper by Nirmalya Thakur, University of Cincinnati, USA.

We hope that all participants enjoyed this successful conference.

December 2020

Paul Mühlethaler
Éric Renault

Organization

MLN 2020 was jointly organized by the EVA Project of Inria Paris, the Laboratoire d'Informatique Gaspard-Monge (LIGM) and ESIEE Paris of Université Gustave Eiffel, and CNAM Paris.

General Chairs

Paul Mühlethaler   Inria, France
Éric Renault   ESIEE Paris, France

Steering Committee

Selma Boumerdassi   CNAM, France
Éric Renault   ESIEE Paris, France

Publicity Chair

Christophe Maudoux   CNAM, France

Organization Committee

Lamia Essalhi   ADDA, France
Nour El Houda Yellas   CNAM, France

Technical Program Committee

Claudio A. Ardagna	Università degli Studi di Milano, Italy
Maxim Bakaev	NSTU, Russia
Mohamed Belaoued	University of Constantine 1, Algeria
Aissa Belmeguenai	University of Skikda, Algeria
Nicola Bena	Università degli Studi di Milano, Italy
Indayara Bertoldi Martins	Ekinops, France
Luiz Bittencourt	University of Campinas, Brazil
Naïla Bouchemal	ECE, France
Nardjes Bouchemal	University Center of Mila, Algeria
Selma Boumerdassi	CNAM, France
Alberto Ceselli	Università degli Studi di Milano, Italy
Hervé Chabanne	Télécom Paris, France
Nirbhay Chaubey	Ganpat University, India
Antonio Cianfrani	University of Rome Sapienza, Italy
Domenico Ciuonzo	University of Naples Federico II, Italy

Alberto Conte	Nokia Bell Labs, France
Vincenzo Eramo	University of Rome "La Sapienza", Italy
Flavio Esposito	Saint Louis University, USA
Smain Femmam	Université de Haute-Alsace, France
Hacène Fouchal	Université de Reims Champagne-Ardenne, France
Aravinthan Gopalasingham	Nokia Bell Labs, France
Róża Goścień	Wrocław University of Science and Technology, Poland
Jean-Charles Grégoire	INRS, Canada
Viet Hai Ha	University of Education, Hue University, Vietnam
Gloria Elena Jaramillo	DFKI, Germany
Müge Fesci-Sayit	Ege University, Turkey
Mariam Kiran	Lawrence Berkeley National Laboratory, USA
Emmanuel Lavinal	Toulouse III University, France
Piotr Lechowicz	Wrocław University of Science and Technology, Poland
Cherkaoui Leghris	Hassan II University, Morocco
Feng Liu	Huawei, Germany
Olaf Maennel	Tallinn University of Technology, Estonia
Ruben Milocco	Universidad Nacional del Comahue, Argentina
Paul Mühlethaler	Inria, France
Gianfranco Nencioni	University of Stavanger, Norway
Van Khang Nguyen	Hue University, Vietnam
Thakur Nirmalya	University of Cincinnati, USA
Kenichi Ogaki	KDDI Corporation, Japan
Satoshi Ohzahata	The University of Electro-Communications, Japan
Paulo Pinto	Universidade Nova de Lisboa, Portugal
Sabine Randriamasy	Nokia Bell Labs, France
Éric Renault	ESIEE Paris, France
Patrick Sondi	Université du Littoral Côte d'Opale, France
Mounir Tahar Abbes	University of Chlef, Algeria
Van Long Tran	Hue Industrial College, Vietnam
Vinod Kumar Verma	SLIET, India
Martine Wahl	IFSTTAR, France
Haibo Wu	Chinese Academy of Sciences, P.R. China
Kui Wu	University of Victoria, Canada
Miki Yamamoto	Kansai University, Japan
Sherali Zeadally	University of Kentucky, USA
Jin Zhao	Fudan University, P.R. China
Jun Zhao	Nanyang Technological University (NTU), Singapore

Sponsoring Institutions

CNAM, Paris, France
ESIEE, Paris, France
Inria, Paris, France
Université Gustave Eiffel, France

Contents

Better Anomaly Detection for Access Attacks Using Deep Bidirectional LSTMs

Henry Clausen[1(✉)], Gudmund Grov[2], Marc Sabate[1], and David Aspinall[1,3]

[1] University of Edinburgh, Edinburgh, UK
{henry.clausen,m.sabate,david.aspinall}@ed.ac.uk
[2] Norwegian Defence Research Establishment (FFI), Oslo, Norway
Gudmund.Grov@ffi.no
[3] The Alan Turing Institute, London, UK

Abstract. Recent evaluations show that the current anomaly-based network intrusion detection methods fail to detect remote access attacks reliably [10]. Here, we present a deep bidirectional LSTM approach that is designed specifically to detect such attacks as contextual network anomalies. The model efficiently learns short-term sequential patterns in network flows as conditional event probabilities to identify contextual anomalies. To verify our improvements on current detection rates, we re-implemented and evaluated three state-of-the-art methods in the field. We compared results on an assembly of datasets that provides both representative network access attacks as well as real normal traffic over a long timespan, which we contend is closer to a potential deployment environment than current NIDS benchmark datasets. We show that by building a deep model, we are able to reduce the false positive rate to 0.16% while detecting effectively, which is significantly lower than the operational range of other methods. Furthermore, we reduce overall misclassification by more than 100% from the next best method.

1 Introduction

We present a short-term contextual model of network flows that aims at improving detection rates of remote access attacks. Remote access attacks are used to gain control or access information on remote devices by exploiting vulnerabilities in network services, and are involved in many of today's data breaches [1]. A recent survey [10] showed that these attacks are detected at significantly lower rates than more high-volume probing or DoS attacks. We present the construction of a short-term contextual model of network flows, and show how this model handles suspicious behaviour. Our idea is to capture probability distributions over sequences of network flows that quantify their overall likelihood, much like a language model. We hypothesise that this improves the detection of low-rate access attacks. Our model is based on deep bidirectional LSTM networks.

Recently, deep learning models such as LSTMs have been a popular tool in network intrusion detection [2,7,12]. However, persistent failings in evaluations have made it difficult to assess the performance and real-world applicability of

© Springer Nature Switzerland AG 2021
E. Renault et al. (Eds.): MLN 2020, LNCS 12629, pp. 1–18, 2021.
https://doi.org/10.1007/978-3-030-70866-5_1

currently proposed methods to access attack detection, and have lead to a chaotic and convoluted NIDS landscape [10].

To avoid these pitfalls and demonstrate that our approach delivers a significant improvement in detection rates and real-world applicability, we evaluated our model carefully on two modern network intrusion detection datasets. Furthermore, we reimplemented and evaluated three state-of-the-art methods on these datasets and compared their performance against ours. By carefully selecting input parameters based on their sequential interdependence as well as increasing model complexity in terms of depth and efficient input embedding compared to preceding models, we are able to detect remote access attacks at a false positive rate of 0.16%, a rate at which none of the comparison models are able to detect any attacks reliably.

This work provides the following novel contributions:

1. We present a new and efficient contextual network flow model based on a deep bidirectional LSTM model. It is specifically designed to detect low-volume network access attacks, and significantly improves on current results through selected input parameters as well as increased model depth and efficient input embedding which enables us to detect attacks at a low false positive rate of around 0.16%.
2. We perform a careful evaluation to avoid common failings and demonstrate that our model is capable of both detecting attacks while remaining stable and consistent over time, using two modern datasets.
3. We reimplement and evaluate three prominent anomaly-based intrusion detection models as benchmarks as well as including a smaller and more shallow version of our model. We perform an appropriate, discerning, and comparative evaluation of their performance and conclude that none of these models are able to detect remote access attacks reliably at the false positive rate we achieve.

1.1   Overview

In verbal or written speech, we expect the words "I will arrive by ..." to be followed by a word from a smaller set such as "car" or "bike" or "5 pm". Similarly, on an average machine we may expect DNS lookups to be followed by outgoing HTTP/HTTPS connections. These short-term structures in network traffic are a reflection of the computational order of information exchange. Attacks that exploit vulnerabilities in network communication protocols often achieve their target by deviating from the regular computational exchange of a service, which should be reflected in the generated network pattern.

Table 1(a) depicts a flow sequence from an XSS-attack. Initial larger flows are followed by a long sequence of very small flows which are likely generated by the embedded attack script trying to download multiple inaccessible locations. Flows of this size are normally immediately followed by larger flows, as depicted in Fig. 1, which makes the repeated occurrence of small HTTP flows in this sequence very unusual.

Table 1. The left side depicts a flow sequence from an XSS-attack. The right side depicts a benign SMB-sequence (top), and a sequence from a *Pass-the-hash* attack via the same SMB service.

Src	Dst	DPort	bytes	# packets
A	B	80	247956	315
A	B	80	7544	13
A	B	80	328	6
A	B	80	2601	10
A	B	80	328	6
A	B	80	328	6
A	B	80	380	7
A	B	80	328	6
⋮				

(a) XSS-attack, A = 192.168.10.50, B = 172.16.0.1

Src	Dst	DPort	bytes	# packets
D	C	N33	600	5
C	D	445	77934	1482
D	C	N33	600	5
C	D	445	5202	10

(b) Benign SMB, C = C6267, D = C754

Src	Dst	DPort	bytes	# packets
C	D	445	4106275	2830
C	D	445	358305611	242847

(c) *Pass-the-hash* attack via SMB

Table 1(b) depicts a regular SMB service sequence while Table 1(c) depicts a *Pass-the-hash* attack via the same SMB service. As shown, the flows to port N33 necessary to trigger the communication on the SMB port are missing while the second flow is significantly larger than any regular SMB flows due to it being misused for exfiltration purposes.

The underlying idea of our model is to predict probabilities of connections in a host's traffic stream conditional on adjacent connections. The probabilities are assigned based on the connection's protocol, network port, direction, and size, and the model is trained to maximise the overall predicted probabilities.

To assign probabilities, we map each connection event to two discrete sets of states, called vocabularies, according to the protocol, the network port, and the direction of the connection for the first, and according to number of transmitted bytes for the second. The size of the vocabulary is chosen large enough to capture meaningful structures without capturing rare events that can deteriorate prediction quality. We then designed a deep bidirectional LSTM (long short-term memory) network that takes bivariate sequences of mapped events as input to efficiently capture the conditional probabilities for each event.

1.2   Outline

The remainder of the paper is organised as follows. Section 2 discusses the current state of network intrusion detection. Section 3 explains the methodology and architecture of our model as well as the data preprocessing. Section 4 explains the advantages of the datasets used in this work as well as current state of the art models that we compare our results with. Section 5 discusses our detection rates on attack traffic, the false positive rate on benign traffic, and compares our results with those of the implemented comparison models. Section 6 concludes our results.

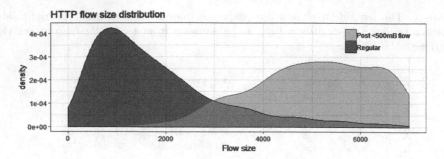

Fig. 1. HTTP flow size distribution overall, and if preceded by an HTTP flow smaller than 500 bytes.

2   Related Work and Evaluation Pitfalls

LSTM-models for web attack detection, such as by Yu et al. [19], improve detection rates of simpler preceding models such as Song et al. [16]. They rely on deep packet inspection, and are often targeted at protecting selected web-servers rather than network-wide, due to a lack of computational scalability and increasing traffic encryption. Methodologically, vocabularies are created from string sequences with well-known NLP methods, while our work provides a new vocabulary-construction method suitable for traffic metadata.

The majority of LSTM-based metadata approaches rely on labelled attack data for classification, and do not have the scope of anomaly-based models to detect previously unseen attacks. A prominent example of this comes from Kim et al. [7], who classify flow sequences based on 41 numeric input features. Anomaly-based approaches, such as ours, mostly rely on iterative one-step ahead forecasts, with the forecasting error acting as the anomaly indicator. This is for instance done in GAMPAL by Wakui et al. [18], who use flow data aggregation as numerical input features, which are computationally easier to process, but cannot encapsulate high-level information such as the used protocol, port, or direction. These models are best used for detecting high-volume attacks. Apart from our work, only Radford et al. [12] create event vocabularies from flow pro-tocols and sizes. We use a more sophisticated model in terms of stacked recurrent layers and embeddings for more input features, which results in higher detection rates, as demonstrated in see Sect. 2.1.

Notable work outside of network traffic includes Tiresias [15] and DeepLog [5]. The design of Tiresias has similarities to ours, but the scope of the model is attack forecasting rather than intrusion detection, and relies on both different input data in the form of IDS logs as well as different evaluation metrics.

DeepLog is combined with a novel log parser to create a sequence of symbolic log keys, which is then also modelled using one-step forecasting. The authors achieve good detection results in regulated environments such as Hadoop with limited variety of events (e.g., 29 events in Hadoop). Here, our model goes further by being applied to a much more heterogeneous data source and creating a more than 30 times larger vocabulary.

2.1 Evaluation Pitfalls

According to Nisioti et al. [10], the trustworthiness of published low volume access attack detection rates is debatable due to evaluation shortcomings. We designed our evaluation to avoid four common pitfalls that are regularly seen:

Outdated Datasets. Two datasets and their derivatives, DARPA-98 and KDD-99, have been extensively used to benchmark network intrusion detection models [11]. However, both datasets are now more than 20 years old and have been pointed out as significantly flawed and prone to give overoptimistic results [17].

Lack of Attack Class Distinction. Most intrusion datasets include attack events from both low volume access attack classes such as R2L (Remote-to-Local) and U2R (User-to-Root) as well as attacks like DoS or port scans which generate a large number of events. If reported detection rates do not distinguish between different attacks or attack classes, performance metrics will be dominated and potentially inflated by DoS and probing attacks.

Arbitrary False Positive Rates. There is no agreed upon value for a suitable false positive rate in network intrusion detection. This leads many authors to report very high detection rates at the expense of having unrealistically high false positive rates, often around 5% and above. In our evaluation, we report overall AUC scores, which describe the separation of benign and anomalous traffic.

Lack of Long-Term Evaluation. To be effective, an intrusion detection system has to produce a consistently low false positive rate in the presence of concept drift. A crucial aspect when assessing the deployability of an intrusion detection system is the long-term stability of a trained model [8], which is often neglected in the literature. We include a dataset focused on long-term traffic evolution in our evaluation to demonstrate the stability and deployability of our model.

3 Experiment Setting

3.1 Session Construction

To order the raw network flows, we first gather all outgoing and incoming flows for each of the hosts selected for examination according to their IP address.

The traffic a host generates is often seen as a series of *session*, which are intervals of time during which the host is engaging in the same, continued, activity

[13]. In our context, flows that occur during the same session can be seen as having strong short-term dependencies. We therefore group flows going from or to the same host to sessions using an established statistical approach [13]:

If a network flow starts less than α seconds after the previous flow for that host, then it belongs to the same session; otherwise a new session is started. If a session exceeds β events, a new session is started.

We chose the number of $\alpha = 8$ s as we have found that on average around 90% of flows on a host start less than 8 s after the previous flow, a suitable threshold to create cohesive sessions according to Rubin-Delanchy et al. [13]. We introduced the β parameter in order to break up long sessions that potentially contain a small amount of malicious flows, and estimated $\beta = 25$ to be a suitable parameter, detection rates do not seem to be very sensitive to the exact choice of β though.

3.2   Contextual Modelling

Each session is now a sequence of flows that are assumed to be interdependent. We observed in an initial traffic analysis that the protocol, port, and direction of a flow as well as its size are highly dependent on the surrounding flows, which motivates their use in the modelling process. We treat flows as symbolic events that can take different states, much like words in a language model. The state of a flow is defined as the tuple consisting the protocol, network port, and the direction of the flow. We consider only the server port numbers, which indicate the used service, in the state-building process. We introduce the following notation:

M:	number of states	N_j:	the length of session j
C:	number of host groups	$x^{i,j}$:	the state of flow i in session j
S:	number of size groups	c^j:	the host group
N^i_{embed}:	embedding dimension	$s^{i,j}$:	size group of flow i in
N^i_{hidden}:	LSTM layers dimension		session j

The collection of all states is called a *vocabulary*. For prediction, the total size of a vocabulary directly correlates with the number of weights needed to be inferred in an LSTM network, thus influencing the time and data volume needed for training. Too large vocabularies also lead to decreased predictive performance by including rare events that are hard to predict [4]. We therefore bound the total number of states and only distinguish between the $M - 2$ tuples of protocol, port, and direction most commonly seen on a machine, with less popular combinations being grouped as "Other". Furthermore, the end of a session is treated as an additional artificial event with its own state. The total vocabulary size is then given by M.

Our experimentation has shown that detection rates improve when including the size as an additional variable, especially for brute force web attacks. Rather than making a point estimate of the size, we want to produce a probability distribution for different size intervals. This provides better accuracy for situations in which both small and large flows have a similar occurency likelihood. We group flows into S different size quantile intervals, with the set of all size intervals forming a third vocabulary.

Hosts are grouped according to their functionality (Windows, Ubuntu, servers, etc.) a distinction that can easily be performed using signals in the traffic. The group is provided to the model as an additional input parameter and forms a third vocabulary.

3.3 Trained Architecture

We use a deep bidirectional LSTM network which process a sequence in both forward and reverse direction to predict the state and size group of individual flows. The architecture of the network we trained is depicted in Fig. 2. The increased model complexity we present has not been explored in previous LSTM applications to network intrusion detection, and enables us to boost detection rates while lowering false positive rates, which we demonstrate in Sect. 2.1.

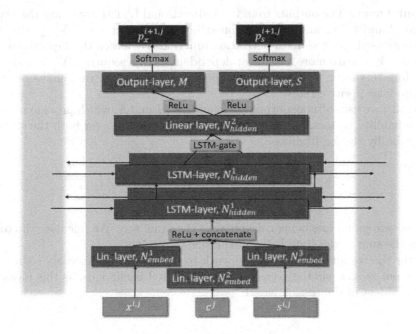

Fig. 2. Architecture of the trained bidirectional LSTM network.

Embedding. First, each of the three inputs of the three vectors is fed through an embedding layer, which assigns them a vector of size N^i_{embed}, $i \in \{1, 2, 3\}$. This embedding allows the network to project the data into a space with easier temporal dynamics. This step significantly extends existing designs of LSTM models for anomaly detection and allows us to project multiple input vocabularies simultaneously without a large increase in the model size. By treating the state, the size group, and the host group as separate dictionaries, we avoid the creation of one large vocabulary of size $M \times C \times S$, which makes training faster and avoids the creation of rare states [4].

LSTM-Layer. In the second step, the vectors are concatenated and fed to a stacked bidirectional LSTM layer with N^1_{hidden} hidden cells. This layer is responsible for the transport of sequential information in both directions. The usage of bidirectional LSTM layers compared to unidirectional ones significantly improved the prediction of events at the beginning of a session and consequently boosted detection rates within short sessions. Increasing the number of LSTM layers from one to two decreases false positive rates in longer sessions while maintaining similar detection rates.

Output Layer. The outputs from the bidirectional LSTM layers are then concatenated and fed to an additional linear hidden layer of size N^2_{hidden} with the commonly used rectified linear activation function. We added this layer to enable the network to learn more non-linear dependencies in a sequence. We found that by adding this layer, we are able to capture complex and rare behaviours and decrease false positive rates.

Finally, we use softmax output layers of size M and S, which provide us with two probability vectors, $p_x^{i,j,k}$ and $p_s^{i,j,l}$. The prediction loss for both the state is then given by the log-likelihood;

$$\text{lh}_x^j = \sum_{i=1}^{N_j} \text{lh}_x^{i,j} = \sum_{k=1,i=1}^{M,(N_j)} x_k^{(i),j} \cdot \log(p_x^{i,j,k})$$

with the size group loss being calculated in the same way. We calculate the total loss as the sum of the state loss and the size group loss.

After the training, we use the network to determine the anomaly score of a given input session via the average of the predicted likelihoods, as this measure is independent of the session length:

$$\text{AS}^j = 1 - \sum_{i=1}^{N_j} \Big(\exp(\text{lh}_x^{i,j}) + \exp(\text{lh}_s^{i,j}) \Big)/N_j$$

An anomaly score close to 0 corresponds to a benign session with a very high likelihood while a score close to 1 corresponds to an anomalous session with events which the network would not predict in the context of previous events.

As we mentioned above, too large vocabularies can cause problems both for model training and event prediction. We achieved the best results for $M = 200$ for the available data and computational resources. The size of the size group was chosen smaller with $S = 7$. Host groups were determined for each dataset individually. We trained on a quad-core CPU with 3.2 GHz, 16 GB RAM, a single NVIDIA Tesla V100 GPU. Training each model could be achieved in under three hours. We chose $N_{embed}^1 = 10$, $N_{embed}^2 = N_{embed}^3 = 5$ for the embedding layers, and $N_{hidden}^1 = N_{hidden}^2 = 50$ for the hidden layers, which achieve the best results without overfitting. We trained each model for 500 epochs. The parameters of the network are optimised to minimise the total loss in minibatches of size 30 using the ADAM optimiser. The optimal value for the learning rate was found to be 0.0003, and decays by a factor of 2 after each ten subsequent epochs the training set. We use a parameter weight decay of 5×10^{-4} per epoch to avoid overfitting, and a drop-out rate of 0.5. Our implementation uses PyTorch.

4 Datasets and Benchmark Models

4.1 Dataset Assembly

We selected two publicly available datasets that complement each other: CICIDS-17 [14], and UGR-16 [9]. The CICIDS-17 dataset contains traffic from a variety of modern attacks, while the UGR-16 dataset's length is suitable for long-term evaluation. We train models with the same hyperparameters on each dataset to demonstrate the capability of our model to detect various attacks and perform well in a realistic environment.

CICIDS-17 [14]. This dataset, released by the Canadian Institute for Cybersecurity (CIC), contains 5 days of network traffic collected from 12 computers with attacks that were conducted in a laboratory setting. The attack data of this dataset is one of the most diverse among NID datasets and contains SQL-injections, Heartbleed attacks, brute-forcing, various download infiltrations, and cross-site scripting (XSS) attacks.

UGR-16 Dataset [9]. Released by the University of Grenada in 2016, this dataset contains real-world network flows from a Spanish ISP. It contains both clients' access to the Internet and traffic from servers hosting a number of services. The data contains a wide variety of real-world traffic patterns. The main advantage of this dataset over previous ones is its usefulness for evaluating IDSs that consider long-term evolution and traffic periodicity, as it spans from March to August of 2016.

4.2 Detection Method

We identify an individual session as malicious if it contains labelled attack traffic from the seven remote access attacks in the CICIDs-17 dataset. We use the 99.9% anomaly-score quantile as a simple threshold T for our model to distinguish between malicious and benign. By determining T from the training data, we control the expected false positive rate in the test data. Finding an appropriate threshold value is a compromise between higher detection rates and lower false positive rates. Our chosen threshold would translate to about one false alert every three days for host with an activity similar to the CICIDS-17 data, and about one false alert every seven days for hosts in the UGR-16 dataset, which is about 20–50 times lower than the false-positive rates that our benchmark models were evaluated on. We additionally provide AUC-scores, a performance measure independent of a particular threshold choice (see next Section).

4.3   Dataset Split

To evaluate detection rates, we split the CICIDS-17 data into a test set and a training set. We selected the four hosts in the data that are subject to remote access attacks, two web servers and two personal computers. We choose our test set to contain the known attack data while the training data should only contain the benign data. Due to the short timespan of the dataset, we have to train on traffic from all five days, with the test data intervals being placed around the attack. For this reason, we evaluate temporal model consistency only on the UGR-16 data. In total, the test set contains 14 h of traffic for each host while the training set contains 31 h of traffic. We chose our training data to contain about 10 000 sessions per host if possible. Overall, we included for the data 24128 sessions in the training and 32414 sessions in the test set for the CICIDS-17, whereas we included 50010 sessions in the training and 100018 sessions in the test set for the UGR-16 data.

To test long-term stability and robustness of our model against concept drift, we split the UGR-16 data into one training set interval and two test set intervals, for which we can compare model performance. The training set interval stretches over the first month, with the first test set interval containing the sessions from the following two months, and the second test set interval containing the last two months. We then isolated traffic from five web-servers that provide a variety of services that show behavioural evolution. Figure 3 depicts the changes of these servers in terms of protocol and port usage over the different intervals.

4.4   Benchmark Comparison Models

We compare our detection and false positive rates against three network anomaly models that we have re-implemented and re-evaluated.

A recent and well cited survey by Nisioti et al. [10] identified the **UNIDS model by Casas et al.** [3] as achieving the highest detection rates of remote access attacks on the KDD-99 dataset, so we chose to include this method as our first benchmark. The authors rely on subspace-projection and density-based clustering (DBSCAN) for outlier detection, and achieve detection rates of access attacks at around 80–85% on the KDD-99 dataset with a false positive rate of 3.5%.

Niyaz et al. [6] present a more recent deep-learning model combines anomaly detection and classification by using sparse autoencoders and detection through

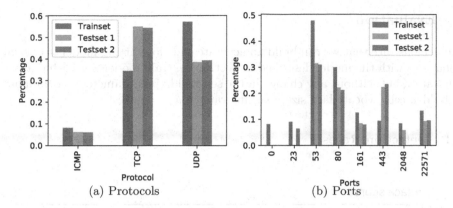

Fig. 3. Temporal change in protocol and port usage over the different train and test intervals across selected servers in the UGR-16 dataset.

reconstruction error. The authors classify individual flows and claim a detection precision of 99% with a recall of 97.5%, even higher than the UNIDS model.

Finally, **Radford et al.** [12] predict sequences of individual flows between pairs of hosts using a two-layer LSTM network. Flows are tokenised according to their protocol and size, and the model detects 60% of the attacks at a false positive rate of about 2% on the CICIDS-2010 dataset. This model is closest to ours in terms of contextual anomaly detection from flow metadata, and achieves the best results out of the three bechmark models during our evaluation. We include it to highlight the improvements our design choices yield over other contextual LSTM-models.

Lastly, we include a more **shallow version of our model**, depicted in Fig. 4, to highlight the benefits of a deeper structure. This model only contains one LSTM-layer, and no linear layer before the output layer.

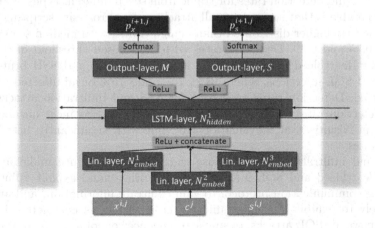

Fig. 4. Architecture of the shallow LSTM-model version we use as a benchmark.

5   Evaluation

We demonstrate that we can build an accurate and close-fitting model of normal behaviour with the model described in Sect. 3. We train models for each dataset separately, but without any change in the selected hyperparameters, i.e. number of hidden cells, vocabulary size, learning rate etc.

5.1   Detection Rates

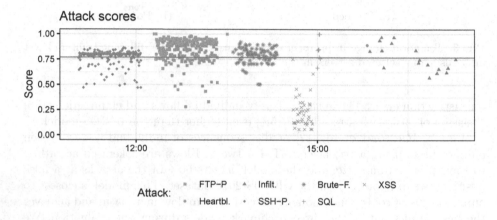

Fig. 5. Score distribution for access attacks contained in the CICIDS-17 dataset.

As described above, we estimate detection rates using traffic of various remote access attacks in the CICIDS-17 dataset. Table 3 and Fig. 5 depict anomaly score distributions and detection rates for traffic from seven different types of attacks.

Most notable is that scores from all attacks except cross-site scripting (XSS) are significantly higher distributed than benign traffic, with median scores lying between 0.75 and 0.89. Detection rates with our chosen threshold of 0.77 are highest for Heartbleed attacks (100%), followed by FTP and SSH brute-force attacks and SQL-injections, where 91%, 74%, and 75% of all affected sessions are detected. Detection rates are lowest for XSS and Infiltration attacks. The overall detection rates we achieve are in a similar range as most unsupervised methods in Nisioti et al.'s evaluation [10], but with significantly better false positive rates.

XSS and infiltration attacks cause the victim to execute malicious code locally. Heartbleed and SQL injections on the other hand exploit vulnerabilities in the communication protocol to gain exfiltrate information, and are thus more likely to exhibit unusual traffic patterns, visible as completely isolated TCP-80 flows for SQL attacks, or unusual sequences of connections initiated by the attacked server during Heartbleed attacks (Table 2).

Table 2. Anomaly score distributions and detection rates for malicious sessions in the CICIDS-17 dataset, as well as detection rates for comparison models on the right. Depicted are the minimum, maximum, and median score for sessions in each attack class along with the rate of sessions exceeding the detection threshold (99.9% quantile).

	Scores and det. rates				Comparison det. rates			
	min	max	median	det. rate	UNIDS	Radford	Niyaz	shallow m.
Brute force web	0.50	0.92	0.80	0.66	0.0	0.0	0.07	0.28
FTP-Patator	0.28	1.00	0.82	0.91	0.07	0.0	0.03	0.38
Heartbleed	0.89	0.89	0.89	1.00	0.0	0.0	0.0	0.0
Infiltration	0.57	0.97	0.75	0.41	0.0	0.166	0.0	0.0
SQL-injection	0.67	1.00	0.84	0.75	0.0	0.0	0.02	0.21
SSH-Patator	0.47	0.86	0.80	0.74	0.0	0.0	0.0	0.67
XSS	0.06	0.75	0.20	0.00	0.0	0.0	0.20	0.0

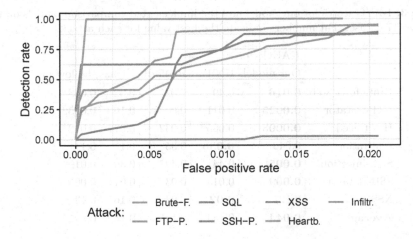

Fig. 6. ROC-curves for different attack types in the CICIDS-17 dataset.

Brute-force attacks on the other hand cause longer sequences of incoming connections to the same port of a server, in this case to port 21 for FTP, 22 for SSH, and 80 for web brute-force. Especially for port 80, such sequences are not necessarily unusual, which explains the difference in detection rates between web brute-force, which our model does not detect reliably, and FTP and SSH brute-force, which are detected at a higher rate. Depending on how much benign traffic the particular sessions are overlayed, the estimated anomaly scores can vary. Brute-Force attacks are not low in volume, and spread over many sessions since we introduced a maximum session length. For these types of attack, our model therefore only has to flag a smaller percentage of malicious sessions the attack generates to detect anomalous behaviour.

Figure 6 provides *ROC* (Receiver operating characteristic) curves for each attack type. As seen, for Heartbleed, FTP brute-force, SQL injection, and infiltration attacks, our model starts detecting attacks with close to zero false positives.

Comparison Models. We now compare detection rates between our model and the three models described in Sect. 4.4 that we chose as benchmarks.

All three models ultimately detect anomalies when an anomaly score exceeds a threshold, which controls the balance between low false positive rates and high detection rates and usually depends on the given data. To create a fair comparison, we chose threshold providing similar false positive rates of 0.01%, e.g. the 99.9% anomaly score quantile of the training data, which is necessary for assessing suitability for real-world deployment as we have argued in Sect. 4.2. Table 3 depicts detection rates for each model. As seen, none of the other models are achieving any meaningful detection rates at a false positive rate this low.

Table 3. 1-AUC scores for our model and the implemented comparison models on the CICIDS-17 dataset. Fat numbers indicate the best value for each attack.

	1-AUC scores				
	Our model	UNIDS	Radford	Niyaz	shallow m.
Brute force web	**0.016**	0.49	0.027	0.32	0.048
FTP-Patator	**0.0025**	0.011	0.0048	0.16	0.0052
Heartbleed	**0.0003**	0.0057	0.032	0.077	0.012
Infiltration	0.046	**0.033**	0.35	0.15	0.11
SQL-injection	**0.005**	0.44	0.497	0.39	0.019
SSH-Patator	**0.009**	0.013	0.035	0.011	0.005
XSS	0.127	**0.02**	0.03	0.16	0.13
Average	**0.044**	0.144	0.135	0.18	0.091

To assess the overall separation between benign and malicious traffic for each model, we calculated $1-AUC$ *(Area under ROC curve)* scores by varying the thresholds for each model, and calculating the area under the ROC-curve, depicted in Fig. 6 and Table 3.

In comparison to the other benchmark models, our shallow model is capable of making some detection at the chosen false positive rate, but cannot reach the levels of our deeper model. While brute-force attacks are still detected, more nimble attacks such as Heartbleed or XSS are less distinctive from benign traffic for the shallow model. It is remarkable that by adding the described additional layers, we are able to more than double the overall detection power, as indicated by the 1-AUC-scores.

5.2   Benign Traffic and Long-Term Stability

We assess the performance of our model on benign traffic by looking at the quantiles and visual distribution of sessions scores. The plots and tables in Table

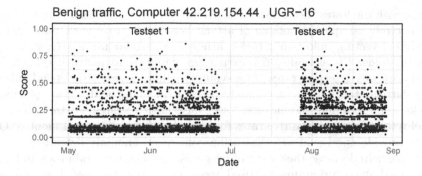

Fig. 7. Anomaly score distribution for benign sessions in the UGR-16 dataset

4 as well as Fig. 7 depict the score distribution of benign sessions for each dataset in the corresponding test sets.

Table 4. Anomaly score quantiles for benign sessions in each testset in the UGR-16 data, along with the maximum session score and the percentage of sessions exceeding the detection threshold (false alerts), averaged over all hosts.

	50%q	90%q	99.9%q	Max score	Pr(>T)
UGR-16, testset 1	0.14	0.33	0.76	0.9	0.0018
UGR-16, testset 2	0.17	0.37	0.72	0.86	0.0014

As shown, the centres of the distributions are concentrated very well in the lower region of the $[0, 1]$ interval, with about 50% of all sessions receiving scores in the region between 0.1 and 0.25. High scores are rare, with only very small percentages exceeding our chosen detection threshold of T. If we look at the hosts in the UGR-16 data, on average less than 0.15% of all assumed-benign sessions exceed the threshold, which would translate to fewer than ten false-alerts over the span of four months on a host with similar activity rates.

A clear banding structure is visible in the plotted distributions, with most session scores being very concentrated on narrow intervals. These scores represent frequently reoccurring activities that generate similar traffic sequences. Figure 7 shows that these banding structures remain virtually unchanged over several months. Similarly, Score distributions remain stable over several months, as depicted in Table 4. Although more investigations are required for definitive conclusions, these results indicate that the identified contextual structures in network traffic remain relatively stable over time.

6   Conclusion

Our proposed model presents a new and promising angle to anomaly-based intrusion detection that significantly improves detection rates on the types of network attacks with the lowest detection rates. We use an anomaly-based approach that does not rely on specific notions of attack behaviours, and is therefore better suited at detecting unknown attacks rather than regular misuse- or signature-based systems. By assigning contextual probabilities to network events, our model improves detection rates of low-volume remote access attacks and outperforms current state-of-the-art anomaly-based models in the detection of several attacks while retaining significantly lower false positive rates. Furthermore, our model retains low false positive rates for periods stretching several months. Our results provide good evidence that using contextual anomaly detection may in the future help decrease the threat of previously unseen vulnerabilities and malware aimed at acquiring unauthorised access on a host. We specifically focused on short-term anomalies as they are often an unavoidable byproduct of an attack thus very difficult for an attacker to avoid without preexisting control over the victim device or other network devices.

6.1   Resilience and Evasion

Evasion tactics and corresponding model resilience against them have been a concern in the development of NIDS. We specifically focused on short-term sequential anomalies as they are often an unavoidable byproduct of attack sequences, and it is thus very difficult for an attacker to pertube attack sequences that rely on a specific exploit without preexisting control over the victim device or other network devices. We therefore believe that our model is relatively robust against evasion, however we identified potential improvements for future work.

A specific evasion tactic that has been discussed extensively in the context of machine learning is model poisoning in the training/retraining phase. As our model uses symbolic features instead of numerical ones, there is little possibility to introduce gradual shifts, and attempts to introduce new sequences would likely exceed the anomaly threshold. Additionally, without control over the victim device, the attacker can only introduce alterations in one direction. Lastly, we showed in Sect. 5.2 that short-term contextual traffic patterns remain stable over several months, which means that retraining of our model is only necessary at a low rate and attackers will have to wait for a long time to execute successful model poisoning.

An issue we encountered is the overlay of malicious and benign traffic. Currently, the existence of known traffic patterns in a session can deplete the overall anomaly score of a session. A potential evasion tactic could therefore conceal an attack behind benign communication with the victim device, an already common approach for C&C communication. Possible improvements for this issue are a refined notion of a session that groups related traffic better, and a better scoring method that identifies smaller anomalous sequences in an otherwise normal sequence of flows.

Acknowledgements. We are grateful for our ongoing collaboration with our industry partners (blinded) on this topic area, who provided both support and guidance to this work. Discussions with them have helped reinforce the need for a better evaluation and understanding of the possibilities that new intelligent tools can provide.

Full funding sources are currently blinded.

References

1. M-trends 2015: a view from the front lines. Technical report (2015). https://www2.fireeye.com/rs/fireye/images/rpt-m-trends-2015.pdf
2. Bontemps, L., Cao, V.L., McDermott, J., Le-Khac, N.-A.: Collective anomaly detection based on long short-term memory recurrent neural networks. In: Dang, T.K., Wagner, R., Küng, J., Thoai, N., Takizawa, M., Neuhold, E. (eds.) FDSE 2016. LNCS, vol. 10018, pp. 141–152. Springer, Cham (2016). https://doi.org/10.1007/978-3-319-48057-2_9
3. Casas, P., Mazel, J., Owezarski, P.: Unsupervised network intrusion detection systems: Detecting the unknown without knowledge. Comput. Commun. **35**(7), 772–783 (2012)
4. Chen, W., Grangier, D., Auli, M.: Strategies for training large vocabulary neural language models. arXiv preprint arXiv:1512.04906 (2015)
5. Du, M., Li, F., Zheng, G., Srikumar, V.: DeepLog: anomaly detection and diagnosis from system logs through deep learning. In: Proceedings of the 2017 ACM SIGSAC Conference on Computer and Communications Security, pp. 1285–1298. ACM (2017)
6. Javaid, A., Niyaz, Q., Sun, W., Alam, M.: A deep learning approach for network intrusion detection system. In: Proceedings of the 9th EAI International Conference on Bio-inspired Information and Communications Technologies (formerly BIONETICS), pp. 21–26. ICST (Institute for Computer Sciences, Social-Informatics and Telecommunications Engineering) (2016)
7. Kim, J., Kim, J., Thu, H.L.T., Kim, H.: Long short term memory recurrent neural network classifier for intrusion detection. In: 2016 International Conference on Platform Technology and Service (PlatCon), pp. 1–5. IEEE (2016)
8. Koutrouki, E.: Mitigating concept drift in data mining applications for intrusion detection systems. arXiv preprint arXiv:1010.4784 (2018)
9. Maciá-Fernández, G., Camacho, J., Magán-Carrión, R., García-Teodoro, P., Therón, R.: UGR '16: a new dataset for the evaluation of cyclostationarity-based network IDSs. Comput. Secur. **73**, 411–424 (2018)
10. Nisioti, A., Mylonas, A., Yoo, P.D., Katos, V.: From intrusion detection to attacker attribution: a comprehensive survey of unsupervised methods. IEEE Commun. Surve. Tutorials **20**(4), 3369–3388 (2018)
11. Özgür, A., Erdem, H.: A review of KDD99 dataset usage in intrusion detection and machine learning between 2010 and 2015. PeerJ Preprints **4**, e1954v1 (2016)
12. Radford, B.J., Apolonio, L.M., Trias, A.J., Simpson, J.A.: Network traffic anomaly detection using recurrent neural networks. arXiv preprint arXiv:1803.10769 (2018)
13. Rubin-Delanchy, P., Lawson, D.J., Turcotte, M.J., Heard, N., Adams, N.M.: Three statistical approaches to sessionizing network flow data. In: 2014 IEEE Joint Intelligence and Security Informatics Conference, pp. 244–247. IEEE (2014)
14. Sharafaldin, I., Lashkari, A.H., Ghorbani, A.A.: Toward generating a new intrusion detection dataset and intrusion traffic characterization. In: ICISSP, pp. 108–116 (2018)

15. Shen, Y., Mariconti, E., Vervier, P.A., Stringhini, G.: Tiresias: predicting security events through deep learning. In: Proceedings of the 2018 ACM SIGSAC Conference on Computer and Communications Security, pp. 592–605. ACM (2018)
16. Song, Y., Keromytis, A.D., Swtolfo, S.: Spectrogram: a mixture-of-Markov-chains model for anomaly detection in web traffic (2009)
17. Tavallaee, M., Bagheri, E., Lu, W., Ghorbani, A.A.: A detailed analysis of the KDD CUP 99 data set. In: 2009 IEEE Symposium on Computational Intelligence for Security and Defense Applications, pp. 1–6. IEEE (2009)
18. Wakui, T., Kondo, T., Teraoka, F.: GAMPAL: anomaly detection for internet backbone traffic by flow prediction with LSTM-RNN. In: Boumerdassi, S., Renault, É., Mühlethaler, P. (eds.) MLN 2019. LNCS, vol. 12081, pp. 196–211. Springer, Cham (2020). https://doi.org/10.1007/978-3-030-45778-5_13
19. Yu, Y., Liu, G., Yan, H., Li, H., Guan, H.: Attention-based Bi-LSTM model for anomalous HTTP traffic detection. In: 2018 15th International Conference on Service Systems and Service Management (ICSSSM), pp. 1–6. IEEE (2018)

Using Machine Learning to Quantify
the Robustness of Network Controllability

Ashish Dhiman, Peng Sun[✉], and Robert Kooij

Delft University of Technology, Delft, The Netherlands
ashish06.dhiman@gmail.com, P.Sun-1@tudelft.nl

Abstract. This paper presents machine learning based approximations for the minimum number of driver nodes needed for structural controllability of networks under link-based random and targeted attacks. We compare our approximations with existing analytical approximations and show that our machine learning based approximations significantly outperform the existing closed-form analytical approximations in case of both synthetic and real-world networks. Apart from targeted attacks based upon the removal of so-called critical links, we also propose analytical approximations for out-in degree-based attacks.

Keywords: Network controllability · Network robustness · Driver nodes · Machine learning

1 Introduction

In the modern world, we see networks everywhere such as the Internet, transportation networks, and communication networks [18]. It is important that these networks perform their desired functions properly. Naturally, we need to control these networks to ensure their proper functioning and maintenance. Network science offers a way to study and analyze these networks using graph theory. The entities in a network are represented by the nodes and interconnections between the nodes are represented by links. For example, in an air-transportation network, the nodes represent different airports and the links represent the flight paths that connect these airports. Network controllability is the ability to drive a system from an initial state to any other state in a finite time by application of external inputs on certain nodes [3]. For directed networks, Liu *et al.* [2] showed that the minimum number of nodes required to control a network can be identified through the maximum matching of the network. However, Cowan *et al.* [5] pointed out that the results of Liu *et al.* [2] are based on the assumption of no self-links. In other words, a state of a node can only be changed through interacting with its adjacent nodes. Recently, Sun *et al.* [1] derived closed-form analytical approximations for the minimum number of driver nodes as a function of the fraction of removed links for both random and targeted attacks. However, the approximations sometimes do not fit well with the simulations, especially

ⓒ Springer Nature Switzerland AG 2021
E. Renault et al. (Eds.): MLN 2020, LNCS 12629, pp. 19–39, 2021.
https://doi.org/10.1007/978-3-030-70866-5_2

when the fraction of removed links is not small. Figure 1 shows the performance of Sun's approximation as compared to simulation for a Erdős-Rényi network under targeted attack. We will discuss the analytical approximations by Sun *et al.* [1] for both random and targeted attacks in Sect. 3 of this paper.

The objective of this work is to improve the analytical approximations for both random and targeted attacks using machine learning methods. We will compare our machine learning based approximations with the existing analytical approximations and simulations. Furthermore, we will also derive an analytical approximation for out-in degree-based attacks and evaluate its performance on both synthetic and real-world networks.

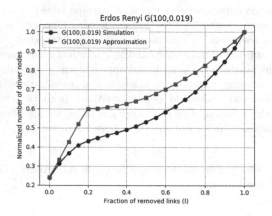

Fig. 1. Performance comparison of Sun's approximation for the normalized minimum number of driver nodes as a function of the fraction of removed links in a Erdős-Rényi network under targeted attack.

In the remainder of this paper, in Sect. 2 we describe the concept of network robustness. In Sect. 3, network controllability is discussed. In Sect. 4, we discuss the closed-form analytical approximations for the minimum number of driver nodes given by Sun *et al.* [1]. Machine learning methods and information related to training and testing data are discussed in Sect. 5. Machine learning based approximations for both random and targeted attacks are presented in Sect. 6. An analytical approximation for out-in degree-based attacks is also derived in this section. Additionally, we also analyze and compare our machine learning based approximations with Sun's approximations and simulations. Finally, in Sect. 7 we conclude this paper.

2 Network Robustness

Network robustness is the ability of a network to deal with failures and errors. In real-world networks, we encounter various failures such as power transmission line failures in an electrical network and network disruption due to natural disasters. It is important to make networks robust to deal with such failures. A generic quantitative definition of network robustness does not exist but there are various metrics to assess network robustness depending on the type of network and its purpose. In this work, we assess network robustness in terms of controllability. Network robustness under perturbations has been studied extensively. Socievole et al. [6] studied network robustness in case of epidemic spreads. They investigated Susceptible-Infected-Susceptible (SIS) spreads with N-Intertwined Mean-Field Approximation (NIMFA) epidemic threshold as the robustness metric. Trajanovski et al. [7] considered node removals in both random and targeted attacks to study network robustness. They used two metrics to evaluate the network robustness, the size of the giant component and efficiency. Wang et al. [8] considered effective graph resistance as the robustness metric to investigate network robustness in case of both synthetic and real-world networks. Koç et al. [9] studied the robustness of networks in terms of cascading failures that lead to blackouts in electrical power grids.

Real-world networks are often challenged by perturbations in the form of random and targeted attacks [29]. In this work, we simulate these attacks by removing links. We do not consider node removals. Random attacks are the unintentional failures such as disruption of networks due to natural disasters and failures due to exhausted mechanical parts [24]. Targeted attacks are carried out by people with malicious intent to maximize the damage [25–28]. In targeted attacks, it is assumed that the attacker has the information related to network topology, functions and vulnerabilities.

3   Network Controllability

Network controllability is the ability to drive a system from an initial state to any other state in a finite time by application of external inputs on certain nodes [3]. It is classified as state controllability and structural controllability.

3.1   State Controllability

State controllability, also known as complete controllability, was introduced by Kalman in the 1960s [3]. Even though non-linear processes govern most of the real-world systems, a linearized counterpart offers a way to study the controllability of non-linear systems [2]. In this work, we consider directed networks with linear time-invariant (LTI) dynamics which are described by:

$$\frac{dx(t)}{dt} = Ax(t) + Bu(t), \tag{1}$$

where $x(t) = [x_1(t), x_2(t), ..., x_N(t)]^T$ indicates the state vector of the system at time t. $x_i(t)$ represents the state that could be the amount of traffic that passes through node i in a communication network. The $N \times N$ adjacency matrix A represents the interconnections of a network [19]. The input $N \times M$ ($M \leq N$) matrix B represents the nodes that are directly controlled.

$u(t) = [u_1(t), u_2(t), ..., u_M(t)]^T$ is the input vector. According to Kalman's controllability condition, the system described in Eq. (1) is said to be controllable if the controllability matrix $C = (B, AB, A^2B, ..., A^{N-1}B)$ has full rank. In other words, $Rank(C) = N$. However, Kalman's rank condition for network controllability has some limitations. It is computationally expensive to check Kalman's rank condition for larger networks that consist of thousands of nodes. The rank condition also requires exact weights of the parameters of A and B but in reality, often the link weights are not known. To account for such limitations, the concept of structural controllability was introduced.

3.2 Structural Controllability

Structural controllability was introduced by Lin in 1974 [10]. The system described in Eq. (1) is said to be structurally controllable if we can fix some weights to the non-zero parameters in A and B so that the system becomes controllable in Kalman's controllability condition. To ensure full rank condition, we have to appropriately choose B which consists of a minimum number of driver nodes. A structurally controllable system is also controllable for different possible parametric realizations except for some pathological cases [2]. One of the advantages of studying structural controllability is that the controllability of a network can still be determined even if we lack information about some or all the link weights.

Liu *et al.* [2] developed the minimum input theory to achieve structural controllability of directed networks. According to Liu *et al.* [2], the minimum number of driver nodes to which external inputs needs to be applied to achieve structural controllability is determined by the maximum matching of the network. They found that the minimum number of driver nodes required to fully control a network depends on the degree distribution. Furthermore, they observed that sparse and homogeneous networks are difficult to control as compared to dense and heterogeneous networks. Liu's work is based on the assumption that there are no self-links in the networks. In this work, we also follow this assumption. Next, we discuss the concept of maximum matching to determine the minimum number of driver nodes required to fully control a network. Next, we find the matching links i.e. the links that do not have common start or end nodes. The nodes at which these links terminate are the matched nodes. The remaining nodes are the unmatched nodes or driver nodes. We will apply external input to these unmatched or driver nodes to fully control the network. In a network, a matching of maximum size is known as maximum matching. There could be multiple maximum matchings in a network but the number of driver nodes remains the same [1]. The Hopcroft-Karp algorithm [11] provides a method to find the maximum matching of a network from its bipartite equivalent.

Now we discuss the robustness of network controllability under perturbations. On removal of a critical link, the number of driver nodes increases by one [2]. In other words, we need more driver nodes to fully control the network when a critical link is removed. It means that there is a decrease in network controllability or the network becomes less robust. Nie *et al.* [12] studied the robustness of network controllability of Erdős-Rényi and Barabási-Albert networks and observed that a Barabási-Albert network with a modest power-law exponent is more robust than an Erdős-Rényi network with a modest average degree. Pu *et al.* [13] studied network controllability and found that degree based attacks are more efficient than random attacks in affecting network controllability. Sun *et al.* [1] quantified the robustness of network controllability for two types of attacks based on the removal of links, random attacks and targeted attacks. They derived closed-form analytical approximations for the minimum number of driver nodes for both random and targeted attacks. While the results of Sun *et al.* [1] fit well with the simulations for small fractions of remove links, there is still room for improvement. In the next section, we will use machine learning to construct more accurate approximations and analyze their performance.

4 Analytical Approximations

The analytical approximations for random and targeted link removals by Sun *et al.* [1] are based on the concept of critical links. If the number of driver nodes required to control a network increases when removing a specific link, then that link is called a critical link. A link that does not belong to any maximum matching is dubbed a redundant link. A link that is neither critical nor redundant is an ordinary link. The initial number of driver nodes N_{DO} i.e. the number of driver nodes before any attack, is calculated using the Hopcroft-Karp algorithm [11]. To find the number of critical links, each link in a network is removed one by one and the Hopcroft-Karp algorithm [11] is applied simultaneously. If the current number of driver nodes N_D exceeds the initial number of driver nodes N_{DO}, then the removed link is a critical link. In a network with N nodes and L links, the Hopcroft-Karp algorithm [11] is applied L times to identify all the critical links.

4.1 Number of Driver Nodes Under Random Attacks

According to Sun *et al.* [1], for random attacks, the normalized minimum number of driver nodes is expressed as,

$$n_{D,rand} = \begin{cases} \frac{N_{DO}+lL_C}{N}, & l \leq l_C \\ al^2 + bl + c, & l \geq l_C \end{cases} \tag{2}$$

where $n_{D,rand}$ represents the normalized value of the minimum number of driver nodes required to fully control a network, L_C represents the number of critical links, l represents the fraction of removed links and $l_C = \frac{L_C}{L}$ represents the fraction of critical links. The values of a, b and c are derived from the boundary conditions described in [1] such that $a = \frac{N-N_{DO}-L_C}{N(l_C-1)^2}$, $b = \frac{L_C}{N} - 2al_C$ and $c = 1 - \frac{L_C}{N} + a(2l_C - 1)$.

4.2 Number of Driver Nodes Under Targeted Attacks

In targeted attacks, first we randomly remove all the critical links and then the remaining links. Sun *et al.* [1] derived the following analytical approximation for targeted attacks.

$$n_{D,crit} = \begin{cases} \frac{N_{DO}+lL}{N}, & l \leq l_C \\ dl^2 + el + f, & l \geq l_C \end{cases} \tag{3}$$

where d, e and f are derived from the boundary conditions described in [1] such that $d = \frac{N-N_{DO}-l_C L}{N(l_C-1)^2}$, $e = -2dl_C$ and $f = 1 + d(2l_C - 1)$.

5 Machine Learning

Machine learning is a technique to predict the outcome of a certain event by learning from data. The data could already be available from experiments, data centers or it can be generated through proper simulations. There are numerous applications of machine learning such as predicting customer's buying habits based on historical data in e-Commerce, weather forecasts and Virtual Personal Assistants such as Siri and Alexa. In broader terms, machine learning is classified as supervised learning, unsupervised learning and reinforcement learning. Furthermore, supervised machine learning is divided into classification and regression problems. In this work, we use various supervised learning methods for regression problems to predict the number of driver nodes under various attacks. Specifically, we use Linear Regression, Random Forest and Artificial Neural Networks. Recently, Lou *et al.* [30] also investigated the use of neural networks for network controllability. However, they used another type of neural networks, Convolution Neural Networks.

To develop our machine learning models, various hyper-parameters are used. Table 1 and Table 2 shows the number of hidden layers and other hyper-parameters that are used to develop our ANN models. For our linear regression model, we use the least-squares to minimize the errors. Additionally, we also use k-fold cross-validation with $k = 10$ to check for over-fitting. In our Random-Forest model, we select the number of trees as 50. Moreover, we also use feature importance scores to determine the features that contribute more to the output. A detailed explanation of the choice of hyper-parameters is presented in the master thesis report [4].

Table 1. Selection of ANN size for different networks under targeted, random and out-in degree-based attacks.

Attack	Number of hidden layers		
	Real-world	Erdős-Rényi	Barabási-Albert
Targeted critical link attack	512/512/512	128	512/512/512
Random attack	512/512/512	128/512/512/512	128/512/512/512
Out-in degree based attack	512/512/512	128	512/512/512

Table 2. ANN hyper-parameters selection.

Hyper-parameters	Activation function	Loss function	Dropout rate	Early stopping	Patience	Epochs	Batch size
Selection	ReLU	MSE	0.2	Yes	50	300	32

Table 3. Properties of 10 real-world networks used for testing our models.

Network	N	L	L_C	N_{DO}
Colt	153	177	38	81
Surfnet	50	68	23	15
EliBackbone	20	30	12	5
Garr200912	54	68	9	30
GtsPoland	33	37	12	14
Ibm	18	24	6	6
Arpanet19706	9	10	6	2
GtsHungary	30	31	8	18
BellCanada	48	64	17	16
Uninet	69	96	19	4

5.1 Dataset for Real-World Networks

Now we discuss the real-world dataset that we consider to construct our models. For synthetic networks, we generate data through simulations. We use the dataset available at The Internet Topology Zoo [14] for real-world networks. It is a collection of a publicly accessible dataset provided by different network operators. As the networks evolve and change, the dataset is updated and in this sense, it is not fixed. Network operators provide maps of their networks and this dataset is interpreted from those maps. However, there are various ambiguities in the dataset as the interpretations are not accurate for some networks. The dataset is available in Graph Markup Language (GML) [15] and GraphML [16] formats. In this work, we consider the dataset that is available in GraphML format as it is easy to parse using python's NetworkX library [17]. We pre-process the data to remove any disconnected networks and multigraphs. After pre-processing of the dataset, we have 232 networks out of which we use 192 networks for training and the remaining 40 networks for testing. The networks in the dataset are not directed, however, we use the information available in two attributes of the GraphML format, edge source and target, to make these networks directed.

The networks in the dataset have small average degrees. The smallest network is the Arpanet196912 network with 4 nodes and 4 links. Cogentco network is the largest network with 197 nodes and 243 links. Additionally, there are some networks that have zero critical links. We conclude that the networks in this dataset vary a lot and machine learning models might have difficulties in learning from such a varying dataset. Table 3 lists the properties of some of the real-world networks we use for testing.

5.2   Datasets for Synthetic Networks

We generate data for synthetic networks using simulations. We consider two types of synthetic networks, Erdős-Rényi and Barabási-Albert networks. These networks come under the class of random graphs [20]. In Erdős-Rényi (ER) random graphs $G(N, p)$ [21], N denotes the number of nodes and p denotes the probability of an outbound link from a node to another node. For Erdős-Rényi networks, we generate networks with different values of N and p. For each such network, we generate 100 corresponding networks and determine the average values of network characteristics such as the average degree, the average number of links, the number of critical links and graph metrics such as diameter and clustering coefficient.

In the Barabási-Albert (BA) scale-free model $G(N, M)$ [22,23], N indicates the number of nodes and M indicates the number of links of a new node that attaches itself to the original network. To generate a BA network, we assume a complete digraph of M_O nodes where M_O equals M. Then we add new nodes one by one with a probability proportional to the number of links of the existing nodes. We generate BA networks with different values of N and M using simulations. For each BA network, we also generate 100 corresponding networks to get the average values of the network characteristics such as the average degree, the average number of links, the average number of critical links and graph metrics such as diameter and clustering coefficient. Moreover, it is to be noted that in a targeted critical link attack, first, the critical links are removed randomly and then the remaining links. For such random removal of links, we use 10,000 simulations. Furthermore, in random attacks, all the links are removed uniformly at random and we also use 10,000 simulations to get the average values of the minimum number of driver nodes.

6   Measuring the Robustness of Network Controllability Using Machine Learning

6.1   Targeted Critical Link Attack

To develop a machine learning based approximation for targeted critical link attack, we predict the difference in the normalized minimum number of driver nodes between the simulation value and the analytical approximation Eq. (3) at $l = l_C$. We use various input features such as the number of nodes N, number of links L, number of critical links L_C, clustering coefficient, average degree and diameter. We choose to estimate the difference at l_C as the original approximation fits well with the simulation for $l \ll l_C$ [1], while the difference can be significant at $l = l_C$, see also Fig. 1 , where $l_c = 0.2$. We subtract this predicted difference to get a new value n_{DX} that is closer to the simulation. We assume a linear relationship similar to the analytical approximation Eq. (3) for $l \leq l_C$. The value of the normalized minimum number of driver nodes at $l = 0$ is n_{DO} where, $n_{DO} = \frac{N_{DO}}{N}$ and at $l = l_C$, the value is assumed to be n_{DX}. From these two conditions we get,

$$n_{D,crit,ML} = n_{DO} + \frac{n_{DX} - n_{DO}}{l_C}l, \tag{4}$$

where $n_{D,crit,ML}$ gives us the new machine learning based normalized minimum number of driver nodes for $l \leq l_C$. When the fraction of removed links l is greater than or equal to the fraction of critical links l_C i.e. for $l \geq l_C$, we estimate the normalized minimum number of driver nodes using a parabolic approximation of the form,

$$n_{D,crit,ML} = d_{ML}l^2 + e_{ML}l + f_{ML}, \tag{5}$$

where d_{ML}, e_{ML} and f_{ML} are derived from the boundary conditions. For the first boundary condition, $n_{D,crit,ML}$ equals n_{DX} at $l = l_C$. When all the links are removed, we need to control all the nodes. Hence, at $l = 1$, $n_{D,crit,ML}$ equals one. Finally, for the third boundary condition, we assume the derivative of the parabola is zero at $l = l_C$. Using these boundary conditions, we get $d_{ML} = \frac{1-n_{DX}}{l_C^2 - 2l_C + 1}$, $e_{ML} = -2d_{ML}l_C$ and $f_{ML} = 1 + d_{ML}(2l_C - 1)$. Finally, the machine learning based approximation for targeted attacks can be expressed as,

$$n_{D,crit\_ML} = \begin{cases} n_{DO} + \frac{n_{DX} - n_{DO}}{l_C}l, & l \leq l_C \\ d_{ML}l^2 + e_{ML}l + f_{ML}, & l \geq l_C \end{cases} \tag{6}$$

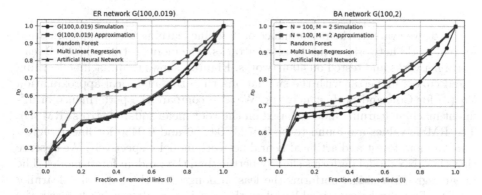

Fig. 2. Comparison of different methods to get the normalized values of minimum number of driver nodes n_D needed to control the network as a function of the fraction of removed links in synthetic networks under targeted attacks. Simulations are based on 10,000 realizations of attacks.

In Fig. 2, we compare the performance of linear regression, random forest and artificial neural network models with simulation and analytical approximation Eq. (3) for synthetic networks under targeted attacks. We notice that the machine learning based approximation fits better with the simulations than the analytical approximation Eq. (3). To further quantify the performance, we use mean absolute errors and mean relative errors to compare the performance of different approximations. Table 4 compares the performance of ANN with the analytical approximation Eq. (3) for a few synthetic networks. We observe that the mean relative error decreases from 19.07% to 2.13% using the ANN-based approximation for ER network with $N = 100$ and $p = 0.019$. For BA network with $N = 100$ and $M = 2$, we see an improvement from 7.04% to 4.67%. Furthermore, the mean relative errors are larger for Barabási-Albert networks as compared to Erdős-Rényi networks. This is because, in BA networks, there are a few nodes with high degrees, so even after removal of some links, the minimum number of driver nodes does not change significantly and hence, the curve is less steep in BA networks as compared to ER networks as also evident from Fig. 2.

Table 4. Performance indicators for synthetic networks under targeted attacks.

Network	Mean absolute error		Mean relative error	
	Approximation	ANN	Approximation	ANN
ER (100, 0.019)	0.1000	0.0124	0.1907	0.0213
ER (200, 0.0063)	0.0663	0.0115	0.1008	0.0175
ER (400, 0.0026)	0.0472	0.0046	0.0659	0.0071
BA (50, 2)	0.0590	0.426	0.0821	0.0582
BA (100, 2)	0.051	0.0351	0.0704	0.0467

Next, we evaluate the performance of machine learning based approximation for real-world networks under targeted attacks. The model is trained on 192 real-world networks and tested on 40 networks. Figure 3 shows that machine learning based curves fit better with the simulations than the analytical approximation Eq. (3) for Colt and Surfnet network. We also compare the performance of different machine learning models based on the root mean squared errors (RMSE). The RMSE values are found to be 0.0723, 0.0550 and 0.0430 for linear regression, random forest and artificial neural network model respectively. We observe that the ANN model performs slightly better than the random forest model. The linear regression model performs the least amongst the three machine learning models. This can be explained based on the non-linear relationship between the input features and the difference that we predict.

In Table 5, we compare the performance of the ANN-based approximation and the analytical approximation Eq. (3) for 10 real-world networks. We notice that machine learning based approximation performs the best in the case of the Colt network with a mean relative error of 1.46% and the worse in Ibm network with a mean relative error of 8.3%. Furthermore, we observe that 9 out of 10 networks have mean relative errors of less than 5%. Among the 40 test networks, the machine learning based approximation performs better than the analytical approximation Eq. (3) in 30 networks. For the remaining 10 networks, the analytical approximation performs only slightly better with a difference of less than 2%. The results of the remaining test networks are available in the master thesis report [4].

6.2   Random Attack

In this section, we develop a machine learning based approximation for the normalized minimum number of driver nodes as a function of the fraction of removed links for random attacks. Furthermore, we compare our approximation with the analytical approximation Eq. (2) and simulations. We also evaluate the performance of different machine learning algorithms. For real-world networks, the RMSE comes out to be 0.0165 for the ANN model and 0.0192 for the random forest model. Again, the ANN model performs slightly better in terms of RMSE.

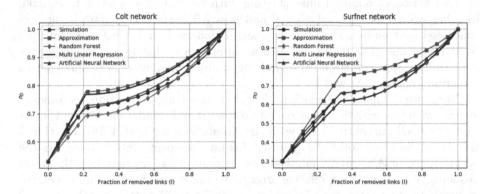

Fig. 3. Comparison of different methods to get the normalized values of minimum number of driver nodes n_D needed to control the network as a function of the fraction of removed links in real-world under targeted attacks. Simulations are based on 10,000 realizations of attacks.

Table 5. Performance indicators for real-world networks under targeted attacks.

Network	Mean absolute error		Mean relative error	
	Approximation	ANN	Approximation	ANN
Colt	0.0393	0.0116	0.0512	0.0146
Surfnet	0.0597	0.0095	0.0866	0.0151
EliBackbone	0.1468	0.0201	0.2471	0.0376
Garr200912	0.0223	0.0202	0.0277	0.0251
GtsPoland	0.0266	0.0171	0.0335	0.0235
Ibm	0.0595	0.0519	0.0956	0.0832
Arpanet19706	0.0440	0.0255	0.0588	0.0434
GtsHungary	0.0269	0.0321	0.0311	0.0373
BellCanada	0.0502	0.0135	0.0757	0.0230
Uninet	0.1195	0.0309	0.184	0.0485

Table 6. Performance indicators for synthetic networks under random attacks.

Network	Mean absolute error		Mean relative error	
	Approximation	ANN	Approximation	ANN
ER (50, 0.082)	0.0712	0.0105	0.3080	0.0675
ER (100, 0.016)	0.0085	0.0024	0.0137	0.0044
BA (50, 2)	0.035	0.0032	0.0517	0.0051
BA (100, 2)	0.032	0.0030	0.0455	0.0049

In the remainder of this section, we will only consider ANN. For random attacks, we predict the normalized minimum number of driver nodes for different values of the fraction of removed links starting with $l = 0$ to $l = 1$ in steps of 0.05. In other words, for each value of N and p in ER networks, 21 data points are generated for training. The same approach is followed for BA networks for each N and M value. The reason for such an approach is that at l_C, the difference between the approximation value and the simulation value is not significant as the approximation fits well for $l \leq l_C$ [1].

Next, we compare our machine learning based approximation for random attacks with the analytical approximation Eq. (2) and simulation. Figure 4 shows that the ANN curves fit better with the simulations for both Erdős-Rényi and Barabási-Albert networks. To quantify this improvement, Table 6 compares the performance of ANN and analytical approximation Eq. (2) based on the mean absolute errors and mean relative errors. We notice a significant improvement in mean relative error from 30.80% to 6.75% for ER network with $N = 50$ and $p = 0.082$ using ANN. Similarly, we see an improvement from 13.70% to 0.44% in the mean relative error in ER network with $N = 100$ and $p = 0.016$. Furthermore, for BA network with $N = 100$ and $M = 2$, the mean relative error improves from 4.55% to 0.49%.

Specifically for ER networks under random attacks, Liu *et al.* [2] also derived an approximation based on generating functions. According to Liu *et al.* [2], the normalized minimum number of driver nodes is given by,

$$n_D = w_1 - w_2 + k(1 - l)w_1(1 - w_2), \tag{7}$$

Table 7. Performance indicators for all three approximations for ER networks under random attacks.

Network	Mean relative error		
	Approximation by Sun *et al.* Eq. (2)	ANN	Approximation by Liu *et al.* Eq. (7)
ER (100, 0.015)	0.0162	0.0084	0.0045
ER (100, 0.017)	0.0156	0.0097	0.0020
ER (200, 0.006)	0.0117	0.0059	0.0018

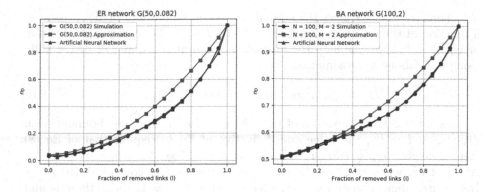

Fig. 4. Comparison of different methods to get the normalized values of minimum number of driver nodes n_D needed to control the network as a function of the fraction of removed links in synthetic networks under random attacks. Simulations are based on 10,000 realizations of attacks.

where k is the average out-degree of an ER network expressed as $k = p(N-1)$. The solution of the implicit equation $w_1 = e^{-k(1-l)}e^{-k(1-l)w_1}$ gives us the value of w_1 and w_2 is given by, $w_2 = 1 - e^{-k(1-l)w_1}$.

Now we will compare our ANN-based approximation with Sun's approximation Eq. (2), Liu's approximation Eq. (7) and simulations. From Table 7, it is evident that Liu's approximation Eq. (7) outperforms both ANN based approximation and Sun's approximation Eq. (2). In $ER(100, 0.015)$ network, the mean relative error using Sun's approximation Eq. (2) comes out to be 1.62%. Our ANN based approximation and Liu's approximation Eq. (7) both performs better than Sun's approximation Eq. (2) with mean relative errors of 0.84% and 0.45% respectively.

We note that Liu's approximation is based upon the use of generating functions for the degree and excess degree distribution, whose expressions are not known for targeted link removals.

For real-world networks under random attacks, we follow a different approach. Here we do not predict the normalized minimum number of driver nodes for the entire range of the fraction of removed links. This is because of the availability of a limited dataset for training and hence, the model always performs worse than the analytical approximation. Moreover, difference estimation at l_C is also not a suitable choice as the original analytical approximation is already good for $l \le l_C$ [1]. For larger values of the fraction of removed links, the difference in n_D values between the approximation and simulation is significant. So, we choose a point $l = 0.4$ to predict the difference and subtract it from the approximation value to get a new value n_{DX}. Let the value at $l = 0.4$ be l_X. At $l = 0$, the normalized minimum number of driver nodes equals n_{DO} and at $l = 0.4$, n_D equals n_{DX}. From these two points we get,

$$n_{D,rand,ML} = n_{DO} + \frac{n_{DX} - n_{DO}}{l_X}l, \qquad (8)$$

where, $n_{D,rand,ML}$ gives the normalized minimum number of driver nodes as a function of the fraction of removed links for $l \leq l_X$. For l values greater than or equal to l_X, we calculate the normalized minimum number of driver nodes using a parabolic approximation,

$$n_{D,rand,ML} = a_{ML} l^2 + b_{ML} l + c_{ML}, \tag{9}$$

where we derive the values of a_{ML}, b_{ML} and c_{ML} from the boundary conditions. At $l = l_X$, the value and derivative of Eq. (9) equals that of Eq. (8). Hence, we get $a_{ML} l_X^2 + b_{ML} l_X + c_{ML} = n_{DX}$ and $2a_{ML} l_X + b_{ML} = \frac{n_{DX} - n_{DO}}{l_X}$. At $l = 1$ i.e. when all the links are removed, we need to control all the nodes. Hence, n_D equals one and we get, $a_{ML} + b_{ML} + c_{ML} = 1$. Using these boundary conditions we get, $a_{ML} = \frac{n_{DO} - 1 + \frac{n_{DX} - n_{DO}}{l_X}}{-l_X^2 + 2l_X - 1}$, $b_{ML} = \frac{n_{DX} - n_{DO}}{l_X} - 2a_{ML} l_X$ and $c_{ML} = 1 + a_{ML}(2l_X - 1) - \frac{n_{DX} - n_{DO}}{l_X}$. Finally, we express machine learning based normalized minimum number of driver nodes for real-world networks under random attacks as,

$$n_{D,rand,ML} = \begin{cases} n_{DO} + \frac{n_{DX} - n_{DO}}{l_X} l, & l \leq l_X \\ a_{ML} l^2 + b_{ML} l + c_{ML}, & l \geq l_X \end{cases} \tag{10}$$

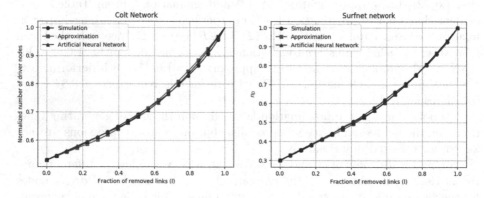

Fig. 5. Comparison of different methods to get the normalized values of minimum number of driver nodes n_D needed to control the network as a function of the fraction of removed links in real-world networks under random attacks. Simulations are based on 10,000 realizations of attacks.

Figure 5 compares our ANN-based approximation and Sun's approximation Eq. (2) with simulations for two real-world networks. We observe that ANN-based approximation fits better with the simulations. To analyze this comparison, Table 8 quantifies the performance using mean absolute and mean relative errors for 10 considered real-world networks. It can be noticed that our ANN-based approximation performs the best in the Colt network with a mean relative

Table 8. Performance indicators for real-world networks under random attacks.

Network	Mean absolute error		Mean relative error	
	Approximation	ANN	Approximation	ANN
Colt	0.0079	0.0043	0.0106	0.0058
Surfnet	0.0072	0.0052	0.0128	0.0090
EliBackbone	0.0256	0.0160	0.0454	0.0274
Garr200912	0.0121	0.0094	0.0156	0.0130
GtsPoland	0.0081	0.0046	0.0127	0.0068
Ibm	0.0072	0.0086	0.012	0.015
Arpanet19706	0.0046	0.0062	0.0073	0.0123
GtsHungary	0.0082	0.0072	0.0098	0.0088
BellCanada	0.0105	0.0071	0.0197	0.0122
Uninet	0.0207	0.0166	0.0338	0.0275

error of 0.58% and the least in the Uninet network with a mean relative error of 2.75%. Moreover, the ANN-based model does not always perform better than the analytical approximation. For example, in Ibm and Arpanet19706, the mean relative errors using ANN-based model are larger than the analytical approximation based mean relative errors. This can be explained based on the availability of a limited amount of training dataset for real-world networks. Among the 40 test real-world networks, the machine learning based approximation performs better than the analytical approximation in 28 networks. The results of the remaining networks are presented in the master thesis report [4].

6.3 Out-In Degree-Based Attack

In this section, we will derive an analytical approximation for the normalized minimum number of driver nodes n_D as a function of the fraction of removed links l for out-in degree-based attacks. Out-in degree of a link is defined as the sum of the out-degree of a source node and the in-degree of a target node. First, we compare different out-in based-attack strategies to select the most efficient one. In the first strategy, we remove links based on the increasing order of out-in degrees, second, if the out-in degrees are the same then links are removed based on the increasing order of out-degrees and finally, in the third strategy, we remove the links based on the decreasing order of out-in degrees. Based on simulations, we found that the first two strategies overlap and are the most efficient ones. So, for the remainder of this section, we will use the first strategy in which we remove links based on the increasing order of out-in degrees. It is to be noted that after removing a link, we re-calculate the out-in degrees in order to determine the next link to be removed.

Case 1: $l \leq l_C$ Similar to [1], when the fraction of removed links is less than or equal to the fraction of critical links, we assume a linear relationship between the minimum number of driver nodes and the fraction of removed links such that,

$$n_{D,out\_in} = \frac{N_{DO} + lL}{N}. \tag{11}$$

Case 2: $l \geq l_C$ When the fraction of removed links is greater than or equal to the fraction of critical links, we approximate the minimum number of driver nodes using a quadratic equation,

$$f(l) = n_D = gl^2 + hl + i, \tag{12}$$

where g, h and i can be derived from the boundary conditions. For the first boundary condition we assume , at $l = l_C$, n_D equals $\frac{N_{DO} + l_C L}{N}$. Second, at $l = 1$, n_D equals one. Third, we assume that the derivative equals zero at $l = 1$. Using these boundary conditions we get, $g = \frac{x-1}{l_C^2 - 2l_C + 1}$, $h = -2g$ and $i = 1 - g - h$ where $x = \frac{N_{DO} + l_C L}{N}$. Finally, for out-in degree-based attacks we can write,

$$n_{D,out\_in} = \begin{cases} \frac{N_{DO} + lL}{N}, & l \leq l_C \\ gl^2 + hl + i, & l \geq l_C \end{cases} \tag{13}$$

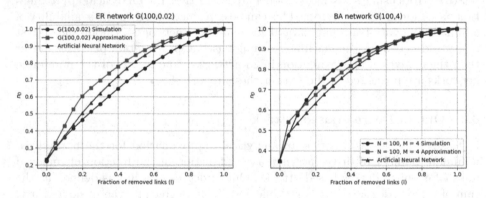

Fig. 6. Performance comparison of the machine learning based approximation Eq. (16) with the analytical approximation Eq. (13) to get the normalized values of minimum number of driver nodes n_D needed to control the networks as a function of the fraction of removed links in synthetic networks under out-in degree-based attacks.

Figure 6 shows the performance of our analytical approximation Eq. (13) for Erdős-Rényi and Barabási-Albert networks. We notice that the analytical approximation fits better with the simulations for Barabási-Albert networks. The same is also evident from Table 9 in which we show the performance of some synthetic networks. We notice that the mean relative errors are less than 3% for BA networks and greater than 10% for ER networks. We also analyze the performance of our approximation in real-world networks. Figure 7 shows

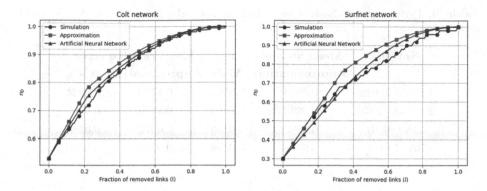

Fig. 7. Performance comparison of the machine learning based approximation Eq. (16) with the analytical approximation Eq. (13) to get the normalized values of minimum number of driver nodes n_D needed to control the networks as a function of the fraction of removed links in real-world networks under out-in degree-based attacks.

the performance of our approximation for the Colt and Surfnet networks. It can be observed that the approximation fits fairly well with the simulations. Furthermore, we analyze the performance of 10 considered real-world networks in Table 10. We notice that the mean relative errors are less than 10% in 8 out of 10 real-world networks. Moreover, the approximation performs the best in the GtsHungary network and the least in the Uninet network with mean relative errors of 1.53% and 13.61% respectively.

Next, we use ANN to further improve the performance of the analytical approximation Eq. (13). We will use ANN to predict the difference in the values of the normalized minimum number of driver nodes between the approximation value and the simulation value at l_C. We will then subtract this difference from the approximation value to get a new value n_{DX} that is closer to the simulation. At $l = 0$, the minimum number of driver nodes can be found from Eq. (13) and at $l = l_C$, the value is n_{DX}. From these two points, we get,

$$n_{D,out\_in,ML} = n_{DO} + \frac{n_{DX} - n_{DO}}{l_C} l, \tag{14}$$

where $n_{D,out\_in,ML}$ gives us the machine learning based normalized minimum number of driver nodes for $l \leq l_C$. For $l \geq l_C$, we assume a quadratic relationship for the normalized minimum number of driver nodes such that,

$$f_{ML}(l) = n_{D,out\_in,ML} = g_{ML}l^2 + h_{ML}l + i_{ML}, \tag{15}$$

To get the values of g_{ML}, h_{ML} and i_{ML}, we again use three boundary conditions. n_D equals n_{DX} at $l = l_C$. At $l = 1$, n_D equals one. The derivative $f'_{ML}(1)$ is assumed to be equal to zero at $l = 1$. Using these boundary conditions we get, $g_{ML} = \frac{n_{DX}-1}{l_C^2 - 2l_C + 1}$, $h_{ML} = -2g_{ML}$ and $i_{ML} = 1 - g_{ML} - h_{ML}$. Hence, the machine learning based approximation for the minimum number of driver nodes can be expressed as,

Table 9. Performance indicators for synthetic networks under out-in degree-based attacks.

Network	Mean absolute error		Mean relative error	
	Approximation	ANN	Approximation	ANN
ER (50, 0.048)	0.0959	0.0568	0.1786	0.0924
ER (100, 0.02)	0.0828	0.0463	0.1380	0.0680
BA (50, 4)	0.0193	0.0189	0.0278	0.0266
BA (100, 4)	0.0201	0.0308	0.0276	0.0400

$$n_{D,out\_in,ML} = \begin{cases} n_{DO} + \frac{n_{DX}-n_{DO}}{l_C}l, & l \leq l_C \\ g_{ML}l^2 + h_{ML}l + i_{ML}, & l \geq l_C \end{cases} \qquad (16)$$

In Fig. 6, we compare the performance of ANN-based approximation Eq. (16) with the analytical approximation Eq. (13) and simulations in case of synthetic networks. While we notice that the ANN-based approximation improves the performance in case of Erdős-Rényi networks, it does not always improve the performance of Barabási-Albert networks as the original analytical approximation Eq. (13) already fits well. In terms of mean absolute errors and mean relative errors, Table 9 compares the performance of both approximations. We observe that for $ER(100, 0.02)$ network, the mean relative error decreases from 13.80% to 6.80% with ANN-based approximation. We notice similar improvements for other ER networks as shown in Table 9. For BA networks, we do not always see an improvement which is also evident in $BA(100, 4)$ network in which the mean relative error increase from 2.76% to 4.0% as the original approximation already fits well with the simulations.

Table 10. Performance indicators for real-world networks under out-in degree-based attacks.

Network	Mean absolute error		Mean relative error	
	Approximation	ANN	Approximation	ANN
Colt	0.0210	0.0102	0.0267	0.0129
Surfnet	0.0469	0.0280	0.0609	0.0395
EliBackbone	0.0846	0.0373	0.1188	0.0539
Garr200912	0.0229	0.0213	0.0262	0.0242
GtsPoland	0.0256	0.0357	0.0309	0.0447
Ibm	0.0665	0.0682	0.0922	0.0951
Arpanet19706	0.0416	0.0340	0.0522	0.0519
GtsHungary	0.0140	0.0135	0.0153	0.0148
BellCanada	0.0546	0.0657	0.0742	0.0917
Uninet	0.0956	0.0586	0.1361	0.0829

Figure 7 compares the performance of ANN based approximation Eq. (16) with the analytical approximation Eq. (13) and simulations for real-world networks. The performance of all the considered 10 real-world networks is shown in Table 10. We notice that the ANN-based approximation Eq. (16) performs better than the analytical approximation Eq. (13) in 7 out of 10 considered real-world networks.

All the simulations are performed on a PC with the following specifications - 8 GB RAM and Intel Core i5 processor with 2 cores. With these specifications, for a dataset consisting of 232 networks, it costs less than 0.6 s to train the linear regression and random forest models whereas, it costs approx. 2–3 seconds to train the artificial neural network model. Once the models have been trained, after getting the average values of 10,000 simulations as inputs to the models, it costs less than 0.5 s to get the predictions.

7 Conclusion

In this work, we used various machine learning methods to quantify the minimum number of driver nodes N_D as a function of the fraction of removed links l. We studied the robustness of network controllability using machine learning based approximations on both synthetic and real-world networks under random and targeted attacks. We also derived an analytical approximation for out-in degree-based attacks. In case of targeted critical link attack, we first compared the performance of ANN, RF and LR models and conclude that the LR model performs the least due to the nonlinear relationship between the input features and the output difference. ANN model performed slightly better than the RF model. Our machine learning based approximation outperformed the analytical approximation in both synthetic and real-world networks. However, for real-world networks, our approximation performed better than the original analytical approximation in 75% of the networks. For random, attacks our approximation performed better than the analytical approximation in 70% of the real-world networks. We also compared our machine learning based approximation with Liu's approximation and Sun's approximation for ER networks under random attacks. Liu's approximation performed better than both machine learning based approximation and Sun's approximation. We also derived analytical approximation for out-in degree-based attacks. For synthetic networks, the approximation performed better in case of BA networks than ER networks. Furthermore, in 8 out of 10 considered real-world networks, the mean relative errors are less than 10%. We further improved our analytical approximation for out-in degree-based attacks using ANN and the mean relative errors reduced to less than 6% in 7 out of 10 real-world networks.

References

1. Sun, P., Kooij, R. E., He, Z., Van Mieghem, P.: Quantifying the robustness of network controllability. In 2019 4th International Conference on System Reliability and Safety (ICSRS), pp. 66–76. IEEE, November 2019

2. Liu, Y.Y., Slotine, J.J., Barabási, A.L.: Controllability of complex networks. Nature **473**(7346), 167–173 (2011)
3. Kalman, R.E.: Mathematical description of linear dynamical systems. J. SIAM Ser. A Control **1**(2), 152–192 (1963)
4. Dhiman, A.K.: Measuring the robustness of network controllability. M.Sc. Thesis, Delft University of Technology (2020)
5. Cowan, N.J., Chastain, E.J., Vilhena, D.A., Freudenberg, J.S., Bergstrom, C.T.: Nodal dynamics, not degree distributions, determine the structural controllability of complex networks. PloS ONE **7**(6), e38398 (2012)
6. Socievole, A., De Rango, F., Scoglio, C., Van Mieghem, P.: Assessing network robustness under SIS epidemics: the relationship between epidemic threshold and viral conductance. Comput. Netw. **103**, 196–206 (2016)
7. Trajanovski, S., Martín-Hernández, J., Winterbach, W., Van Mieghem, P.: Robustness envelopes of networks. J. Complex Netw. **1**(1), 44–62 (2013)
8. Wang, X., Pournaras, E., Kooij, R.E., Van Mieghem, P.: Improving robustness of complex networks via the effective graph resistance. Eur. Phys. J. B **87**(9), 221 (2014). https://doi.org/10.1140/epjb/e2014-50276-0
9. Koç, Y., Warnier, M., Van Mieghem, P., Kooij, R.E., Brazier, F.M.: The impact of the topology on cascading failures in a power grid model. Phys. A Stat. Mech. Appl. **402**, 169–179 (2014)
10. Lin, C.T.: Structural controllability. IEEE Trans. Autom. Control **19**(3), 201–208 (1974)
11. Hopcroft, J.E., Karp, R.M.: An n5/2 algorithm for maximum matchings in bipartite graphs. SIAM J. Comput. **2**(4), 225–231 (1973)
12. Nie, S., Wang, X., Zhang, H., Li, Q., Wang, B.: Robustness of controllability for networks based on edge-attack. PloS One **9**(2), e89066 (2014)
13. Pu, C.L., Pei, W.J., Michaelson, A.: Robustness analysis of network controllability. Physica A Stat. Mech. Appl. **391**(18), 4420–4425 (2012)
14. Knight, S., Nguyen, H.X., Falkner, N., Bowden, R., Roughan, M.: The internet topology zoo. IEEE J. Sel. Areas Commun. **29**(9), 1765–1775 (2011)
15. Himsolt, M.: GML: a portable graph file format, p. 35. Technical report 94030, Universitat Passau (1997)
16. Brandes, U., Eiglsperger, M., Herman, I., Himsolt, M., Marshall, M.S.: GraphML progress report structural layer proposal. In: Mutzel, P., Jünger, M., Leipert, S. (eds.) GD 2001. LNCS, vol. 2265, pp. 501–512. Springer, Heidelberg (2002). https://doi.org/10.1007/3-540-45848-4_59
17. NetworkX. Network analysis in python. https://networkx.github.io/
18. Tirpak, T.M.: Telecommunication network resource management based on social network characteristics. U.S. Patent Application No. 12/463,445 (2010)
19. Harary, F.: The determinant of the adjacency matrix of a graph. SIAM Rev. **4**(3), 202–210 (1962)
20. van der Hofstad, R.: Random graphs models for complex networks, and the brain. Complex. Sci. **1**, 199–246 (2019)
21. Erdős, P., Rényi, A.: On the evolution of random graphs. Publ. Math. Inst. Hung. Acad. Sci **5**(1), 17–60 (1960)
22. Albert, R., Barabási, A.L.: Statistical mechanics of complex networks. Rev. Mod. Phys. **74**(1), 47 (2002)
23. Barabási, A.L., Ravasz, E., Vicsek, T.: Deterministic scale-free networks. Phys. A Stat. Mech. Appl. **299**(3–4), 559–564 (2001)
24. Cohen, R., Erez, K., Ben-Avraham, D., Havlin, S.: Resilience of the internet to random breakdowns. Phys. Rev. Lett. **85**(21), 4626 (2000)

25. Cetinay, H., Devriendt, K., Van Mieghem, P.: Nodal vulnerability to targeted attacks in power grids. Appl. Netw. Sci. **3**(1), 34 (2018)
26. Holme, P., Kim, B.J., Yoon, C.N., Han, S.K.: Attack vulnerability of complex networks. Phys. Rev. E **65**(5), 056109 (2002)
27. Huang, X., Gao, J., Buldyrev, S.V., Havlin, S., Stanley, H.E.: Robustness of interdependent networks under targeted attack. Phys. Rev. E **83**(6), 065101 (2011)
28. Mengiste, S.A., Aertsen, A., Kumar, A.: Effect of edge pruning on structural controllability and observability of complex networks. Sci. Rep. **5**(1), 1–14 (2015)
29. Van Mieghem, P., et al.: A framework for computing topological network robustness. Delft University of Technology, Report 20101218 (2010)
30. Lou, Y., He, Y., Wang, L., Chen, G.: Predicting network controllability robustness: a convolutional neural network approach. IEEE Trans. Cybern. **2**, 1–12 (2020)

Configuration Faults Detection in IP Virtual Private Networks Based on Machine Learning

El-Heithem Mohammedi[1,2]([⊠]), Emmanuel Lavinal[1], and Guillaume Fleury[2]

[1] IRIT, University of Toulouse, Toulouse, France
{elheithem.mohammedi,emmanuel.lavinal}@irit.fr
[2] IMS Networks, Castres, France
{emohammedi,gfleury}@imsnetworks.com

Abstract. Network incidents are largely due to configuration errors, particularly within network service providers who manage large complex networks. Such providers offer virtual private networks to their customers to interconnect their remote sites and provide Internet access. The growing demand for virtual private networks leads service providers to search for novel scalable approaches to locate incidents arising from configuration faults. In this paper, we propose a machine learning approach that aims to locate customer connectivity issues coming from configurations errors, in a BGP/MPLS IP virtual private network architecture. We feed the learning model with valid and faulty configuration data and train it using three algorithms: decision tree, random forest and multi-layer perceptron. Since failures can occur on several routers, we consider the learning problem as a supervised multi-label classification problem, where each customer router is represented by a unique label. We carry out our experiments on three network sizes containing different types of configuration errors. Results show that multi-layer perceptron has a better accuracy in detecting faults than the other algorithms, making it a potential candidate to validate offline network configurations before online deployment.

Keywords: Configuration faults detection · Machine learning · Virtual private networks · BGP/MPLS networks

1 Introduction

Demands from companies to interconnect their remote sites and provide them with Internet access have continued to increase strongly in recent years. BGP/MPLS IP Virtual Private Networks (VPNs) [15] remain the most reliable and widely used technology for this purpose. It can also be complementary to SD-WAN solutions (Software-Defined Networking in a Wide Area Network) which can be used as an overlay technology to optimize WAN edge infrastructures.

© Springer Nature Switzerland AG 2021
E. Renault et al. (Eds.): MLN 2020, LNCS 12629, pp. 40–56, 2021.
https://doi.org/10.1007/978-3-030-70866-5_3

Within network service providers, VPNs multiplication leads to an increase in the configuration's complexity, which can cause several faults that impact the reachability of the customers' sites. Traditional methods to detect network configuration faults generally require to specify a list of formal constraints to verify the configuration's validity (e.g., checking type correctness or cross-elements dependencies). Ensuring the constraints completeness and manually updating them are well known issues in this context. In this paper, we propose another approach based on machine learning to detect and locate connectivity incidents related to configuration faults in BGP/MPLS IP VPNs. Although relying on machine learning for network management is not new [20], to the best of our knowledge, it has never been applied for configuration fault detection in the context of IP VPNs.

The approach we propose relies on a supervised learning paradigm in which a model is trained thanks to thousands of network configurations that are labeled as being either valid or not. This classification problem targets specifically the reachability state of a customer's site. After the learning process, the goal is to be able to identify if an unknown configuration provided as an input allows all the customers' sites in a VPN to connect to each other and, if it is not the case, the system should locate the sites that are not reachable. We validated our work by testing the learning model on three algorithms: decision tree, random forest and multi-layer perceptron. On a network of 100 client routers, results show that we obtain an accuracy in detecting faults between 60 and 80% depending on the learning algorithm.

This paper is organized as follows. In Sect. 2, we briefly review previous related work. In Sect. 3, we present BGP/MPLS IP VPNs, including their main configuration elements and possible errors. We formulate the learning problem in Sect. 4 and we explain our approach of data collection and feature engineering in Sect. 5. Section 6 provides experimental tests and evaluation results before concluding the paper in Sect. 7.

2 Related Work

A lot of work on faults localization in the field of computer networks has been done for more than fifteen years. Various techniques have been used such as rule-based expert systems, decentralized probabilistic management or temporal correlation [6]. Other proposals address network configuration verification [3] or data-plane verification such as Header Space Analysis [9] and Veriflow [11] that verify if safety properties hold for the current network state. These approaches present scalability issues and they require that the verified configuration has been deployed in a live network.

The increase in the size and complexity of networks makes fault detection and localization a difficult task that requires new scalable approaches, which inspired researchers to apply machine learning methods for this purpose. Some of these works target specific networks such as cellular or wireless sensor networks [10,12]. Others are more general, such as [8] in which the authors proposed a supervised learning model for incident management. Their model is composed of two sub-models: a classification model to categorize network incidents and a regression model to predict the duration of incidents, both are fed by data collected from application databases related to incident handling processes (customer relationship management, network monitoring system, etc.). However, configuration errors are not taken into account. In [19], a passive monitoring approach is used to identify and localize the link where the fault occurred. The authors proposed to capture the loss rate, the end-to-end delay and the aggregated transmission rate for each source/destination node from traffic in both normal working conditions and failure scenarios. The centralized network monitor then transmits these metrics as features to the machine learning based fault manager. Again, the data comes from network state and not configuration.

DeepBGP [2] is an example of related work that targets network configurations. The authors rely on a Graph Neural Network (GNN) model to generate BGP configurations given the feedback from a validation unit (enhanced by an Evolution Strategies optimizer) to train the neural network. Although this paper presents a very interesting approach (such as the use of a GNN that we plan to study in future work), it does not focus on configuration faults detection, but on network configuration generation.

Compared to these related work, we propose to identify and locate incidents caused by configuration faults. We rely on a supervised learning approach, looking at the network configuration as a whole (as opposed to individual devices), which allows us to detect failures on multiple devices, even if there is a single configuration error. Another specificity is the application domain: we focus on connectivity incidents within network service providers that offer L3 virtual private networks to their customers.

3   BGP/MPLS IP Virtual Private Networks

IMS Networks is an Internet Service Provider (ISP) that offers, among other services, IP Virtual Private Networks (VPNs) to its customers [15]. This service allows the operator to interconnect a set of customer sites through its backbone while guaranteeing quality of service and isolation of flows for each customer. The operator's backbone relies on a Multiprotocol Label Switching (MPLS) architecture [16] that assigns a label for each route within a VPN. A customer data packet is tagged with the label corresponding, in the customer's VPN, to the route that is the best match to the packet's destination and is further encapsulated with another MPLS label used to tunnel the packet across the backbone. The routes for a particular VPN are exchanged thanks to the Multi-Protocol Border Gateway Protocol (MP-BGP). Figure 1 illustrates a simple BGP/MPLS IP VPN architecture composed of two customers (A and B), each having two sites. Each VPN site must contain one Customer Edge (CE) router attached to one or more Provider Edge (PE) routers:

- A CE router is connected to the provider's network via an access service. It routes traffic between the customer's site and the backbone using a routing protocol such as eBGP (external BGP) or OSPF (Open Shortest Path First).

- A PE router is an edge backbone router to which CE devices are connected to. It contains a VPN Routing and Forwarding table (VRF) for each VPN. The routes within a VRF are learned from the CE it is attached to, as well as from other PEs via the MP-BGP protocol. The PE router also acts as an ingress/egress label edge router for the MPLS domain.

Routers in the provider's network that are not attached to CE devices are known as "P routers". They implement primarily MPLS forwarding and control protocols.

Configuring a complete BGP/MPLS IP VPN architecture is a complex task as it involves configuring many parameters and many types of protocols. Moreover, the size of the network can also be an issue since a network service provider can have a large amount of customers deployed on a number of sites ranging from a few to several hundreds. In the next subsections, we will explain in more depth the main configuration steps that are required to deploy and run this VPN service and the main faults that can occur in this process.

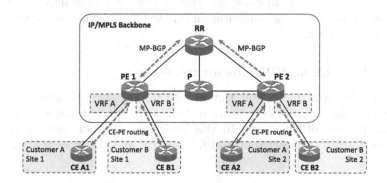

Fig. 1. BGP/MPLS IP VPN architecture overview.

3.1 BGP/MPLS IP VPNs Configuration

To implement BGP/MPLS IP VPNs, it is necessary to go through two main stages: first, the configuration of the provider's backbone and second, the configuration of the customers' VPNs.

Backbone Configuration. Once the topology and all the routers' interfaces are configured, an Interior Gateway Protocol (IGP) must be configured between all P and PE routers (such as OSPF). Next, MPLS forwarding must be activated, as well as the Label Distribution Protocol (LDP) to distribute labels between neighbors in order to establish label switched paths. Finally, MP-BGP must be configured between all PE routers so that they can afterwards advertise customer routes. To avoid scalability issues (in which every PE router peers with every other PE router in a full mesh), an alternative is to use a route reflector (RR) to centralize MP-BGP sessions (cf. Fig. 1).

Once the operator's backbone is configured, it is generally stable and rarely changes.

VPNs Configuration. When configuring a VPN instance for a specific customer, a VRF must be configured on every PE router connected to that customer's site. This is done, notably, by specifying a route distinguisher (RD) to ensure route uniqueness, along with route targets (RT) to determine which VRFs it should export or import routes to or from. The VRF must also be associated to the appropriate interface on which the CE is connected to. Regarding CE-PE routing, we can use eBGP, thereby configuring a direct peering between CE and PE routers (for a specific VRF). Otherwise, another routing protocol can be used (e.g., OSPF) and, in this case, route redistribution between the MP-BGP protocol and this other routing protocol must be configured for that VRF.

It should be noted that, contrary to the backbone's configuration, several customer sites (CE routers) can be added, updated or deleted each day, which requires human expertise to always check the changes before committing the new configuration. Besides, adding a single customer's site to the network requires almost one hundred lines of configuration on the CE and PE routers, which is a very error-prone task.

3.2   BGP/MPLS L3 VPNs Incidents

Within the Network Operation Center (NOC) at *IMS Networks*, we have observed that the majority of network incidents are due to configuration faults, hence the interest of this work. In this article, we consider several types of faults that impact the reachability of CE routers and therefore lead to VPNs disruptions.

Faults on CE-PE Routing. An error on a customer's IP subnet on a CE or misconfiguring the CE-PE routing (e.g., eBGP peering or route redistribution) are examples of faults that make a CE unreachable to all other CEs of the VPN, even if they are connected to the same PE.

Faults on VRFs. A network engineer can forget to configure a VRF instance on a PE router or can make a mistake on a specific parameter such as the RD or RT. Once the configuration has been committed with this type of fault, all the CE routers of the concerned customer connected to this PE router will be unreachable.

Faults on MP-BGP. A peering configuration fault between a PE router and the route reflector will prevent establishing the MP-BGP session and therefore the PE will neither be able to send nor receive VPN routes. All the CEs connected to this PE will therefore be unreachable. It is also necessary to configure a BGP instance for each VRF. Forgetting this step will prevent the PE router to communicate the customer's routes to the route reflector and thus all the CE routers of that customer will be unreachable.

Table 1 summarizes these configuration faults as well as their impact on the reachability of the customers' sites.

Table 1. Main configuration faults and their impact

Faults		Affected CEs
a) CE-PE routing config. faults	Routing error between CE i and the PE	CE i
	CE i subnet misconfigured	CE i
b) VRFs configuration faults	VRF j not configured on PE i	All CEs of customer j connected to PE i
	RD/RT misconfigured for VRF j on PE i	All CEs of customer j connected to PE i
c) MP-BGP configuration faults	BGP instance not configured for VRF j on PE i	All CEs of customer j connected to PE i
	MP-BGP peering error between PE i and the RR	All CEs that are connected to PE i

4   Problem Formulation

In this section, we formulate the learning problem and illustrate it on a simple example.

The general goal of our work is to build a system that can learn if a BGP/ MPLS IP VPN configuration is valid or not. More precisely, the system should identify if the target configuration allows every customer's site to be reachable. If not, it should be able to pinpoint the sites that are not reachable due to a configuration error.

Let us consider the topology presented in Fig. 1 and two scenarios in which configuration errors cause reachability issues for one or more CE routers: *i)* if the MP-BGP session between PE1 and RR is not properly configured, then CE_A1 and CE_B1 connected to PE1 will be unreachable regardless of their VRF configuration (case *c.2* in Table 1); *ii)* an incorrect RD or RT configured on VRF B on PE2 will prevent routes emanating from CE_B2 to be associated to customer's B VPN and therefore CE_B2 will be unreachable from the customer's other sites (case *b.2* in Table 1).

Learning if a CE is reachable or not is a *classification problem*; that is predicting whether it belongs to a particular category. In our context, there are only two categories (or classes): reachable or not reachable (in fact the reachability property does not depend only on the network's configuration but also on its state; however we use the term "reachable" for the sake of simplicity). Although there are two classes, this is not a binary classification problem since one configuration includes many CEs and a single fault can generate reachability issues for several CEs. In the above topology, the system should be able to classify if each CE is reachable or not, hence a vector of 4 labels: [CE_A1, CE_B1, CE_A2, CE_B2]. This is known as a multi-label classification task [18] in which an input x is mapped to an output binary vector y (with a value of 0 or 1 for each label in y). Therefore each label represents a particular CE, the value 0 indicating that it should be reachable (according to the network's configuration) and 1 indicating that the CE has a reachability issue. Back to the previous example, the fault scenarios *i)* and *ii)* should generate the binary vectors [1, 1, 0, 0] and [0, 0, 0, 1], respectively.

Figure 2 illustrates the general workflow of our approach from data collection and features selection to model validation. After analyzing and selecting the features, we generate a dataset from labeled configurations and feed it to the learning model. We repeat these steps by adjusting features and model parameters until we get decent results. It is important to note that this workflow is done offline before using the verification service. Once the model has been trained, it is deployed, a network engineer submits an L3 VPN configuration and the results should be able to predict if it is correct or not before committing the changes on a production network.

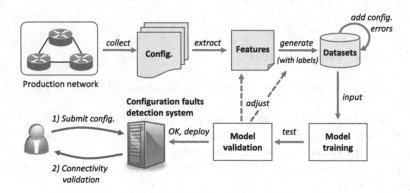

Fig. 2. General workflow from data collection to model deployment.

5   Configuration Data Collection and Analysis

In the learning paradigm, data collection, analysis and feature extraction are very important steps since the efficiency of the learning process is directly correlated to the amount and quality of data. Finding proper features is the key to fully release the potential of data [20].

5.1   Data Collection and Feature Engineering

A complete BGP/MPLS IP VPN configuration is composed of many parts related to physical topology, backbone IGP, MPLS/LDP protocols, VRF tables, MP-BGP sessions and CE-PE routing.

As explained in the previous section, in this work we focus on configuration faults that impact VPN connectivity between the customers' sites. *IMS Networks*' experience in troubleshooting connectivity issues shows that most of the incidents come from routing configuration errors on CE and PE devices, and very few come from the backbone's configuration (IGP, MPLS/LDP). In addition, the backbone's configuration rarely changes while the VPNs' configuration can be updated on a daily basis. Therefore, after having collected and analyzed a complete BGP/MPLS IP VPN configuration, we decided to focus on configuration parameters covering CE-PE routing, VRF tables and MP-BGP. We formalized these parameters as features having either a binary value indicating if the configuration parameter is present or not (e.g., MP-BGP active for VRF v on PE p), or a numerical value with a specific range (e.g., IP prefix with a range from 1 to 32 and RD/RT with a range from 1 to 100 as we assume that the maximum number of VRFs is 100). As an example, nine features in our data model are necessary to add a new VRF with a single CE, which is equivalent to approximately one hundred lines of configuration in the PE router. The selected features have a direct impact on the connectivity issues described in Sect. 3.2. This is an important aspect of feature engineering as the extracted data must be relevant with respect to the problem that is being addressed [4,20].

Configuration collection and analysis, as well as features extraction constitute the first steps of the overall workflow illustrated in Fig. 2.

5.2   Dataset Generation

In communication networks, a large amount of data is available: traffic traces, performance metrics, security alerts, logs, etc. However, network configurations is the exception since once a configuration is in production, there is no need for the service provider to maintain hundreds of other configurations. Additionally, it is very difficult to obtain a large number of configuration errors in a production network and it is unrealistic to inject faults in the network just to have training data. We thus decided to generate configuration data based on existing configurations specified by and deployed at *IMS Networks*. This allows us to have a dataset large enough to be used by the learning model and diverse enough to contain both correct and incorrect configurations.

To be able to generate realistic configurations, we defined three network architectures of different sizes which we used to create three different datasets: *a)* a small network with 5 PEs, 20 CEs and 3 customers; *b)* a medium-sized network with 8 PEs, 50 CEs and 5 customers; and *c)* a larger network with 10 PEs, 100 CEs and 10 customers. In the three networks, CEs are connected to PEs and assigned to customers randomly. According to this random allocation, the features are initialized with either, a value derived from the device or customer id (e.g., IP address, route distinguisher), or a binary value indicating if the configuration parameter is present or not (e.g., VRF v not configured on PE p). The result is one valid global configuration for each network architecture. Since we rely on a supervised learning paradigm, we have to assign labels to the datasets to establish ground truth. In our context, a label exists for each CE in the network to identify if it is reachable or not (from a configuration point of view). When generating the global configuration, all the labels are initialized to 0 since there is no configuration error.

For each network architecture, in order to train the model with sufficient data, we reproduced the configuration 150000 times. We then generated faults on these configurations (cf. Sect. 3.2). We considered 10 types of faults distributed in 3 categories: CE-PE routing faults (customer's IP subnet address or mask error, eBGP peering error), VRF faults (VRF not configured, bad RD or RT) and MP-BGP faults (BGP not configured, peering error with RR, VPN address family extension not activated, BGP instance not configured for VRF). First, we combined all the faults (CE-PE routing, VRF and MP-BGP faults) within the same dataset and second, we considered the different fault categories independently from each other. Therefore, for each network architecture, we generated several datasets with different categories of faults. For each dataset, the faults are generated by randomly selecting the type of fault and randomly selecting the device on which to apply the fault. During this process we made sure to uniformly distribute the errors over the configurations and that, at the end, each CE would be unreachable in 50% of the data points. Finally, for each fault added to a configuration, we updated the output labels accordingly (i.e., setting the labels of the unreachable CEs to 1).

6   Learning to Detect Reachability Issues

This section covers the two last steps of the workflow illustrated in Fig. 2: model training and validation. We start by explaining the algorithms we relied on to train our model and then we discuss the test results that vary according to the algorithm, network size and fault type.

6.1   ML Algorithms and Evaluation Metrics

In order to implement our learning problem, we have tested three different ML algorithms: decision trees, random forest and multi-layer perceptron. We chose these three algorithms as they are commonly used in the machine learning community when dealing with supervised classification problems.

- Decision Tree (DT) is one of the simplest approaches of supervised learning. During the training, it builds a decision tree which represents a function that takes as input a vector of features values and returns a "decision" (or "class") as output value [17].
- Random Forest (RF) is an extension of the decision tree classifier, it contains a set of individual decision trees that operate as an ensemble. The random forest output class is the most predicted class chosen by each individual decision tree [5].

- Multi-Layer Perceptron (MLP), or feedforward neural network, is a machine learning approach that is based on artificial neural networks. It contains at least three layers of neurons: an input layer, one or more hidden layers and an output layer. Neurons in layer n are connected to neurons in layer $n + 1$ with a certain weight. These weights are adjusted during the training process using back propagation [4,14].

We tested DT and RF using Scikit-learn [13], a Python module that offers implementations of various machine learning algorithms. For MLP, we used Keras [7], a high-level neural networks API built on top of TensorFlow [1].

After training the models, to evaluate their performance, we relied on conventional metrics used in classification problems: accuracy, precision, recall and F1-score. To understand these metrics in our context, we need to define True Positive (TP), True Negative (TN), False Positive (FP) and False Negative (FN) classes. TP and TN represent correctly predicted outcomes for, respectively, positive instances (a CE with an error, i.e. not reachable) and negative instances (a CE with no error, i.e. reachable). On the contrary, FP and FN describe incorrect predictions: predicting a reachability problem that actually does not exist, and predicting a reachable CE that in fact is not.

In the following definitions, we will refer to a *positive CE* as a CE containing an error, i.e. not reachable (labeled 1 in the dataset); and a *negative CE* as a CE with no error, i.e. reachable (labeled 0 in the dataset):

- *Accuracy* represents the proportion of true predictions (i.e. the number of CE whose reachability state is correctly predicted) among the total number of predictions (i.e. the total number of CEs).

$$Accuracy = \frac{TP + TN}{TP + TN + FP + FN}$$

- *Precision* is the ratio of the number of positive CEs correctly predicted over the total number of CEs predicted as positives.

$$Precision = \frac{TP}{TP + FP}$$

- *Recall* is the ratio of the number of positive CEs correctly predicted over the total number of actual positive CEs.

$$Recall = \frac{TP}{TP + FN}$$

- *F1-score* is the harmonic average of Precision and Recall.

$$F1\text{-}score = \frac{2 \times Precision \times Recall}{Precision + Recall}$$

6.2   Results and Discussion

Our experiments were performed on the datasets described in Sect. 5.2: three network sizes (20, 50 and 100 client routers) and for each one, different types of configuration faults. Each dataset contains 150000 configurations and is divided in two: one for training (containing 70% of the data points), one for testing (containing 30% of the data points).

Training Time. We first start by measuring the training time of each algorithm for each network size and for different levels of configuration faults (first, VRF tables only; second, adding MP-BGP; third, adding CE-PE routing). The results are shown in Table 2 (experiments were performed in a VM with 4 vCPU and 48 GB of RAM). We can see that the number and type of faults does not have an influence on the training time. This time is fairly constant for the three algorithms whether faults belong to one, two or three categories. Also, the results show that MLP training takes more time than DT and RF (approximately by factors three and two compared to RF for the medium and large networks respectively). However, the total training time for the larger network with the three categories of faults is less than twenty two minutes which is very reasonable for offline training. Indeed, as shown in Fig. 2, the training is done offline before deploying the system, that is before validating any new configuration update on the operational network. Therefore the approach does not have any real time constraints.

Table 2. Comparison of models training time

Faults on:	Small network			Medium network			Large network		
	DT	RF	MLP	DT	RF	MLP	DT	RF	MLP
VRFs	0 m 8 s	1 m 42 s	8 m 49 s	0 m 46 s	4 m 47 s	15 m 32 s	2 m 31 s	10 m 2 s	21 m 20 s
+ MP-BGP	0 m 10 s	1 m 53 s	8 m 45 s	0 m 42 s	4 m 52 s	15 m 57 s	1 m 56 s	9 m 3 s	21 m 37 s
+ CE-PE rt.	0 m 16 s	2 m 31 s	8 m 29	0 m 68 s	6 m 18 s	14 m 40 s	3 m 13 s	11 m 39 s	21 m 34 s

Overall Performance of Each Algorithm. In order to evaluate the general performance of each algorithm, in Fig. 3 and Fig. 4, we plotted the precision, recall, F1-score and accuracy values calculated using the test datasets for the small and large networks, respectively. These results were obtained with the datasets containing all the configuration errors described in Sect. 5.2 (i.e., 10 types of faults).

Overall, we observe that the DT performance is lower than that of RF and MLP, particularly as the network gets larger (F1-score of 60% for the DT on the large network). Besides, MLP results are more efficient than RF results on both graphs, with a bit more on the second one (between RF and MLP, the F1-score increases by 6% on the small network, and by 17% on the larger network).

Looking at precision and recall, we can see that the F1-score on the large network (Fig. 4) cannot reach 80% due to a low recall value (for both RF and

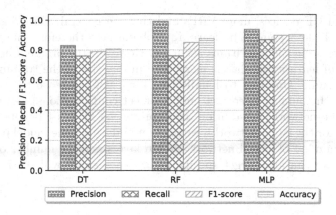

Fig. 3. Precision/Recall/F1-score/Accuracy for the small network dataset.

MLP). A low recall value indicates a high number of false negatives, that is configuration errors that are not detected (i.e., predicting a CE router as being reachable while it is actually not). This issue with the recall value will be further discussed in the subsection related to the impact of configuration error types. On the contrary, the precision is very high (almost 1 for the small network and close to 0.9 for the large network) which means that there are very little false positives, that is a CE router predicted as unreachable is almost always indeed unreachable.

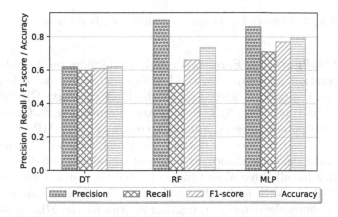

Fig. 4. Precision/Recall/F1-score/Accuracy for the large network dataset.

Impact of Network Size. Figure 5 shows the F1-score value for the small, medium and large networks with 10 types of configuration faults. It is clear that the performance of all three algorithms (DT, RF and MLP) is decreasing as the network gets larger. Increasing the size of the network means adding more PE and CE devices, more VRFs and thus, in general, more features and labels for each data point. The learning problem gets therefore more difficult, hence a decrease in the algorithms accuracy. However, the problem here is that the number of features and labels increases but the size of the training dataset remains the same. One way to handle this issue if to increase the size of the dataset (i.e., the number of correct and incorrect configurations) as we increase the size of the network. In addition, to improve the MLP's F1-score (which is close to 80% for the larger network), we plan on updating some model parameters such as the number of neural network hidden layers and the number of training epochs; and we can also target specific error types as we will see in the next subsection.

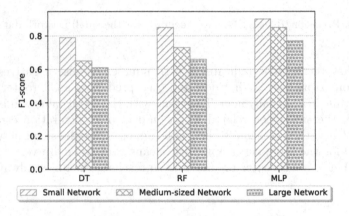

Fig. 5. F1-score for the small, medium and larger network.

Impact of Configuration Errors. To understand more precisely the impact of the configuration error types on the learning algorithms, we calculated the F1-score for each algorithm according to the fault category: CE-PE routing, VRF and MP-BGP (cf. Sect. 3.2). Results are shown in Fig. 6. One can easily observe that the performance on a dataset containing only MP-BGP errors is better than the performance on a dataset containing only VRF errors which is again better than the performance on a dataset containing only CE-PE routing errors, and this for all three algorithms. This means that models learn better about MP-BGP and VRF faults than CE-PE routing faults. This is confirmed by Fig. 7 that shows the detailed evaluation metrics for the MLP model.

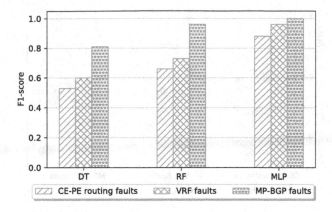

Fig. 6. F1-score for the larger network according to the fault type.

The first reason that can explain this fact is related to the impact of the configuration error on the rest of the network. As presented in Table 1, a fault on MP-BGP peering between a PE and a RR generates reachability issues on all the CEs connected to this PE (i.e., multiple labels), while a fault on a VRF does not impact all the CEs connected to the PE, but only the CEs belonging to the same customer. Finally, a fault on CE-PE routing affects only the concerned CE (i.e., one label). It is therefore easier for the learning algorithms to spot MP-BGP configuration faults and, to a lesser extent, VRF faults.

Other reasons that can explain this difference are the relationships between the features in the dataset and their data types. In CE-PE routing, the IP address and the mask of a customer's subnet are only configured once on the CE router and there is no link between this information and the rest of the configuration. Moreover, these features are numerical data, with a high cardinality (the IP address space) and no numerical relationship between the values. Therefore the learning algorithm cannot identify if these two features (IP address and mask) are correct or not. Regarding the RD/RT configuration parameters in a VRF, although these features are also numerical, they should be identical on the different sites of the same customer. This logical relationship is learned in the training process and therefore an error on those parameters can be detected by the ML algorithm. The recall value plotted in Fig. 7 confirms these assumptions: false negatives are more present in the case of CE-PE routing (i.e., unable to detect an error or the customer's IP subnet) than in the VRF configuration (i.e., better performance in detecting RD/RT errors). MP-BGP configuration elements are, on the other hand, binary features and have an impact on the reachability of multiple CE routers, hence an accuracy of almost 1 in Fig. 7.

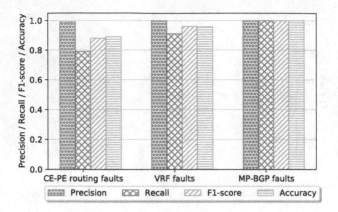

Fig. 7. Precision/Recall/F1-score/Accuracy using MLP for the larger network according to the fault type.

According to these results, we argue that MLP has a better accuracy in detecting configuration faults than DT and RF. Furthermore, we find that the MLP approach can have a much higher performance if it is not used to detect configuration errors on numerical data that have no direct or indirect relationship with other parts of the network configuration (such as customers' IP addresses). This type of configuration data should rather be verified by other means, such as rule-based systems.

7   Conclusion and Future Work

In this paper, we proposed to use supervised machine learning approaches in order to detect and locate reachability incidents between customers' edge routers interconnected by virtual private networks. We focused specifically on incidents caused by configuration errors in BGP/MPLS IP VPNs that remain complex networks for service providers. In contrast to existing rule-based approaches to verify configurations, which involve manual requirements engineering to specify and maintain the set of rules, machine learning approaches rely mainly on correct and incorrect configurations to learn and detect configuration errors. The learning system takes as input, before deployment, a new configuration and predicts connectivity issues before they occur on the production network. Experimental results have shown better performance and better scalability with the approach based on neural networks. We are therefore currently working on enhancing this model by integrating more configuration features, more fault types and larger networks.

In future work, in addition to configuration data, we plan to include state data such as network monitoring events and performance measures. With such data, we will have to integrate online training in the workflow. Furthermore, in line with the neural network approach (MLP), considering a Graph Neural Networks (GNN) model seems a natural direction for future work to represent the network topology and the routing control plane in order to detect other network incidents and identify their root cause.

References

1. Abadi, M., et al.: Tensorflow: a system for large-scale machine learning. In: 12th USENIX Conference on Operating Systems Design and Implementation. OSDI 2016 (2016)
2. Bahnasy, M., Li, F., Xiao, S., Cheng, X.: DeepBGP: a machine learning approach for BGP configuration synthesis. In: Proceedings of the Workshop on Network Meets AI & ML. NetAI 2020, Association for Computing Machinery (2020)
3. Beckett, R., Gupta, A., Mahajan, R., Walker, D.: A general approach to network configuration verification. In: ACM Special Interest Group on Data Communication. SIGCOMM 2017 (2017)
4. Boutaba, R., et al.: A comprehensive survey on machine learning for networking: evolution, applications and research opportunities. J. Internet Serv. Appl. **9**(1), 1–99 (2018). https://doi.org/10.1186/s13174-018-0087-2
5. Cutler, A., Cutler, D., Stevens, J.: Random forests. In: Zhang, C., Ma, Y. (eds.) Ensemble Machine Learning: Methods and Applications, pp. 157–175. Springer, Boston (2012)
6. Dusia, A., Sethi, A.S.: Recent advances in fault localization in computer networks. IEEE Commun. Surv. Tutorials **18**(4), 3030–3051 (2016)
7. Gulli, A., Pal, S.: Deep Learning with Keras. Packt Publishing Ltd., Birmingham (2017)
8. Isa, M.: Albarda: designing supervised learning-based incident management model. Case study: broadband network service provider. In: International Conference on ICT for Smart Society (ICISS) (2019)
9. Kazemian, P., Varghese, G., McKeown, N.: Header space analysis: static checking for networks. In: 9th USENIX Conference on Networked Systems Design and Implementation. NSDI 2012 (2012)
10. Khanafer, R.M., et al.: Automated diagnosis for UMTS networks using bayesian network approach. IEEE Trans. Veh. Technol. **57**(4), 2451–2461 (2008)
11. Khurshid, A., Zhou, W., Caesar, M., Godfrey, P.B.: Veriflow: verifying network-wide invariants in real time. In: First Workshop on Hot Topics in Software Defined Networks. HotSDN 2012 (2012)
12. Moustapha, A.I., Selmic, R.R.: Wireless sensor network modeling using modified recurrent neural networks: application to fault detection. In: IEEE International Conference on Networking, Sensing and Control, pp. 313–318 (2007)
13. Pedregosa, F., et al.: Scikit-learn: machine learning in python. J. Mach. Learn. res. **12**, 2825–2830 (2011)
14. Ramchoun, H., Amine, M., Idrissi, J., Ghanou, Y., Ettaouil, M.: Multilayer perceptron: architecture optimization and training. IJIMAI **4**(1), 26–30 (2016)
15. Rosen, E., Rekhter, Y.: BGP/MPLS IP Virtual Private Networks (VPNs). RFC 4364, IETF, February 2006
16. Rosen, E., Viswanathan, A., Callon, R.: Multiprotocol Label Switching Architecture. RFC 3031, IETF, January 2001
17. Russell, S., Norvig, P., Davis, E.: Artificial Intelligence: A Modern Approach. Prentice Hall, New Jersey (2010)

18. Sharma, S., Mehrotra, D.: Comparative analysis of multi-label classification algorithms. In: First International Conference on Secure Cyber Computing and Communication (ICSCCC) (2018)
19. Srinivasan, S.M., Truong-Huu, T., Gurusamy, M.: Machine learning-based link fault identification and localization in complex networks. IEEE Internet of Things J. **6**(4), 6556–6566 (2019)
20. Wang, M., Cui, Y., Wang, X., Xiao, S., Jiang, J.: Machine learning for networking: workflow, advances and opportunities. IEEE Netw. **32**(2), 92–99 (2018)

Improving Android Malware Detection Through Dimensionality Reduction Techniques

Vasileios Kouliaridis[1(✉)] ⓘ, Nektaria Potha[1], and Georgios Kambourakis[2] ⓘ

[1] Department of Information and Communication Systems Engineering,
University of the Aegean, Aegean, Greece
{bkouliaridis,npotha,gkamb}@aegean.gr
[2] Joint Research Centre, European Union, 21027 Ispra, Italy
georgios.kampourakis@ec.europa.eu

Abstract. Mobile malware poses undoubtedly a major threat to the continuously increasing number of mobile users worldwide. While researchers have been trying vigorously to find optimal detection solutions, mobile malware is becoming more sophisticated and its writers are getting more and more skilled in hiding malicious code. In this paper, we examine the usefulness of two known dimensionality reduction transformations namely, Principal Component Analysis (PCA) and t-distributed stochastic neighbor embedding (t-SNE) in malware detection. Starting from a large set of base prominent classifiers, we study how they can be combined to build an accurate ensemble. We propose a simple ensemble aggregated base model of similar feature type as well as a complex ensemble that can use multiple and possibly heterogeneous base models. The experimental results in contemporary Androzoo benchmark corpora verify the suitability of ensembles for this task and clearly demonstrate the effectiveness of our method.

Keywords: Mobile malware detection · Dimensionality reduction · Machine learning

1 Introduction

As mobile devices have gradually become an indispensable part of our current and future lives, mobile malware has proved an immense threat to their users both from a security and privacy viewpoint [1–3]. Nowadays, the Android operating system (OS) is the most prevalent mobile platform, with a market share that exceeds 74% [4,5], thus making it an ideal target for malware writers. Moreover, Android malware applications (apps) have become more sophisticated and their writers have learned to hide their malicious code, which per se makes timely detection an issue [6]. Up to now, state-of-the-art mobile malware detection approaches in the literature have been evaluated using older datasets, including Contagio mobile [7], MalGenome [8], and Drebin [9]. This raises an issue of whether such schemes can accurately detect current pieces of malware.

© Springer Nature Switzerland AG 2021
E. Renault et al. (Eds.): MLN 2020, LNCS 12629, pp. 57–72, 2021.
https://doi.org/10.1007/978-3-030-70866-5_4

Mobile malware detection approaches nowadays lean primarily toward static anomaly-based detection [10–14] and comprise two basic phases, namely, the training and the detection or testing one. That is, this kind of solutions employs machine learning to detect malicious behavior, i.e., deviation from a model built during the training phase [10]. Naturally, the popularity of static analysis techniques arises from the fact that such methods do not require the app to be running, hence they are usually faster and rather easy to implement.

In the current paper, we use a wide range of base verifiers covering the most prominent classification algorithms in the relevant literature. Moreover, we apply and compare two of the most common dimensionality reduction techniques, namely Principal Component Analysis (PCA) and t-distributed stochastic neighbor embedding (t-SNE) on a collection of concurrent malware from the AndroZoo dataset [15], dated from 2017 to 2020. Then, we propose two ensemble learning approaches. First, a simple ensemble which combines the outputs of base models. These outputs are extracted dealing exclusively with either the original or a transformed feature set, respectively. Second, a more complicated ensemble approach that aggregates the answers of a larger size and possibly heterogeneous set of base models constructed from both original and transformed (Original, PCA and t-SNE) feature sets. Experimental results on Andozoo benchmark dataset exhibit the effectiveness of both the aforementioned approaches.

The main contributions of this study are:

- We propose a simple ensemble approach by aggregating the output of each instance separately, for a number of malware detection base models. The combination of base models achieves the best results in comparison to each particular base model on the Androzoo corpus examined. This evidently demonstrates that an ensemble of classifiers based on a larger size and probably heterogeneous base models is the most appropriate and able to handle challenging malware detection scenarios, and thus can further improve the performance of each individual base classifiers.
- We examine the usefulness of two well-known dimensionality reduction technique namely, PCA and t-SNE when exclusively applied on malware detection base verifiers as well as ensembles, respectively. It is demonstrated that both transformations are able to considerably increase the performance of each base model as well as the proposed ensembles. However, the implementation of t-SNE is more effective than PCA transformation and assists base models and malware detection ensemble methods to further increase their effectiveness in all the examined cases.
- We compare the detailed experimental results on malware detection under the Androzoo dataset with state-of-the-art methods based on the same settings. The performance of the approaches presented in this paper is quite competitive to the best results reported so far for this malware detection corpus, demonstrating that the proposed methods can be an efficient and effective alternative toward more sophisticated malware detection systems.

The rest of this paper is organized as follows. The next section presents previous work in malware detection, while Sect. 3 describes our proposed methods.

The performed experiments are analytically presented in Sect. 4. The last section discusses the main conclusions and suggests future work directions.

2   Previous Work

So far, several corpora have been used in the literature to evaluate mobile malware detection approaches. This section offers a review of works published from 2015 to 2020, categorized in three benchmark datasets, namely Drebin, VirusShare, and AndroZoo. Note that the focus is on works that present classification results based on features stemming from static analysis.

Drebin [9] is one of the oldest datasets, used in various state-of-the-art mobile malware detection approaches [16]. On the downside, Drebin is outdated and therefore newer malware samples are needed to accurately assess detection performance. Below are some notable works which employed this dataset.

- Ali-Gombe et al. [17] presented *AspectDroid* an hybrid analysis system, which analyzes Android apps to detect unwanted or suspicious activities. The proposed system employs static bytecode instrumentation to provide efficient dataflow analysis. However, static instrumentation is unable to detect apps which use anti-unpacking and anti-repackaging obfuscation mechanisms.
- Arshad et al. [18] introduced a hybrid malware detection scheme, namely *SAMADroid*. According to the authors, SAMADroid delivers high detection accuracy by combining static and dynamic analysis, which run both in a local and remote fashion. Machine Learning (ML) classifiers were used to detect malicious behavior of unknown apps and to correctly classify them. The authors' scheme was tested on Android ver. 5.1, which is considered outdated.

VirusShare is another well-known dataset containing not only mobile malware samples, but also others from various platforms, including Windows and Linux. It is updated regularly and contains samples dated from 2012 onward.

- Xu et al. [19] proposed *HADM*. Their method converted classification features extracted during static analysis into vector-based representations. Other features fetched during dynamic analysis, namely system calls, were transformed into vector-based and graph-based representations. Deep learning techniques were used to train a neural network for each of the vector sets. Finally, the hierarchical multiple kernel learning technique were applied with the purpose of combining different kernel learning results from diverse features, and thus improve the classification accuracy. According to the authors, their model is weak against code obfuscation techniques because dynamic analysis is guided based on the results obtained from static analysis.
- Fang et al. [20] suggested a hybrid analysis method which performs dynamic analysis on the results of static analysis. During the static analysis phase, they decompiled the app's APK file to extract permissions from the manifest file as well as any occurrence of API features existing in *smali* files. Regarding

dynamic analysis, they generated user input and logged the system calls. The authors used ML to classify apps, but they provide limited information about the examined malware samples.

Much like VirusShare, AndroZoo [15] is a growing collection of Android apps collected from diverse sources, including the official Google Play store. AndroZoo is updated regularly and it currently contains over 12M samples. To our knowledge, the only work that exploits this dataset for conducting among other static analysis is given by Kouliaridis et al. [21]. Specifically, they introduced an online open-source tool called *Androtomist*, which performs hybrid analysis on Android apps. The authors focused on the importance of dynamic instrumentation, as well as the improvement in detection achieved when hybrid scrutiny is used vis-à-vis to static analysis. In their experiments, the authors compared feature importance between three datasets, namely Drebin, VirusShare, and AndroZoo. Finally, they elaborated on features which seem to be commonly exploited in malware and seldom in benign apps.

Several works in the literature use a mixed dataset, i.e., one containing samples from two or more corpora, to evaluate their approach. Below, we refer to the most important ones.

- Martinelli et al. [22] introduced *BRIDEMAID*. Their system operates in three consequent steps, namely static, meta-data, and dynamic. During static analysis, BRIDEMAID decompiles the apk and analyzes the source code for finding possible similarities in the executed actions. This is done with the help of *n-grams*. Dynamic analysis exploits both ML classifiers and security policies to control suspicious activities related to text messages, system call invocations, and administrator privilege abuses. The authors picked malware apps from the Contagio Mobile [7] dataset, which nowadays is considered obsolete.
- Surendran et al. [23] implemented a Tree Augmented Naive Bayes (TAN) model that combines the classifier output variables pertaining to static and dynamic features, namely API calls, permissions and system calls, to detect malicious behavior. Their experiments showed an accuracy of up to 97%. The authors did not provide information about the Android version they employed or how they generated user input events for the purposes of dynamic analysis.

Overall, the effect of dimensionality reduction techniques on Android malware detection, which is the focus of the current paper, has hitherto received scarce attention in the literature. We were able to only pick out the work of Vega et al. [24] who collected and analyzed malware samples from the Malgenome dataset with six dimensionality reduction techniques, namely Principal Component Analysis, Maximum Likelihood Hebbian Learning, Cooperative Maximum Likelihood Hebbian Learning, Curvilinear Component Analysis, and Isomap and Self Organizing Map. On the other hand, the authors did not evaluate the classification performance of these methods.

3 The Proposed Method

The core idea of our proposed approach is to exploit powerful low-level features, namely app's permissions and intents, and apply common techniques toward reducing dimensionality and extracting more compact and less sparse representations of samples. In a more detailed description, we use a set of popular and widely used classification algorithms as malware detection base models. More, specifically, we implemented eight classifiers handling the dataset examined either based only on the original feature set or utilizing entirely the transformed set of features accrued by applying dimensionality reduction techniques. Then, simple heterogeneous ensemble approaches are proposed by fusing the output of all base models for each instance separately. These meta-models are developed exploring two options. The first one is to use exclusively the available base models extracted from either the initial feature set of data or the modified set of features. The second option concerns to enrich the meta learner with a larger and possibly heterogeneous set of base models. Thus, we consider the answers of all the available malware detection base models extracted (based on both the Original and reduced feature set). In this way, a heterogeneous ensemble included a mixed set of base models is formed.

3.1 Dimensionality Reduction

Dimensionality reduction is one of the most useful processes that decisively contributes to analyze large volumes of data providing a simple way to transform them from the original high-dimensional and sparse feature space into a small set of new features. Several algebraic techniques have been applied in time series analysis for dimensionality reduction providing a less sparse representation of signals. This way, the reduced space is less redundant, and the resulting data are more compact and less noisy. In this study, we consider the two most widely used dimensionality reduction techniques, namely PCA and t-SNE.

– Principal Component Analysis (PCA) is one of the simpler and well-known linear transformation techniques. Specifically, it is one of the most multivariate and state-of-the-art statistical techniques in the field of dimensionality reduction achieving to reduce the dimensions of a d-dimensional dataset by projecting it onto a new (k)-dimensional subspace (where $k < d$) following some of the most important linear algebra concepts in order to increase the computational efficiency, while retaining most of the information. More specifically, PCA analysis aims to identify patterns in data detecting the correlation between variables and yielding the directions or eigenvectors (the principal components) that maximize the variance of the data. In other words, by applying PCA the observations are represented by their projections as well as the set of variables are represented by their correlations. It is also worth noticing that PCA based on a deterministic technique in a consequence of utilizing a strictly mathematical approach. To make it clearer, it is important to be mentioned that in the approximation of a small dimensional space, PCA can be accomplished by a matrix algebra technique obtaining the eigenvectors and eigenvalues from the covariance matrix or correlation matrix [25]. In this way, a less sparse matrix with significantly lower dimension in comparison with the original matrix is built.

– t-distributed stochastic neighbor embedding (t-SNE) is a nonlinear dimensionality reduction technique which leads to a powerful and flexible visualization of high-dimensional data. It uses the local relationships between points to create a low-dimensional mapping. This allows it to capture non-linear structure. It creates a probability distribution using the Gaussian distribution that defines the relationships between the points in high-dimensional space. Then, it employs the Student t-distribution to recreate the probability distribution in low-dimensional space. Unlike methods like PCA, t-SNE is non-convex, meaning it has multiple local minimal and is therefore much more difficult to optimize. This technique enables the correct visualization of data which lie on curved manifolds or which incorporate clusters of complex shape. In this way, t-SNE opens the way toward a visual inspection of nonlinear phenomena in the given data. t-SNE is a more recent DR technique that belongs to the class of non-parametric techniques [26].

4   Experiments

4.1   Description of Data

In the context of malware detection between 2010 and 2019, several corpora were built covering multiple degrees of difficulty and incorporating older or newer malware/goodware instances. These corpora are usually exploited to evaluate new malware detection approaches. In this paper, we consider the most contemporary benchmark corpora, namely AndroZoo [15]. This is a well-known and widely used real-world collection of Android apps collected from assorted sources, including the official Google Play app market [27]. Particularly, the collection of AndroZoo apps we used in the context of this work is dated from 2017 to 2020 and enclosed 1K malware apps, each of which has been cross-examined by a large number of antivirus products. It is important to be mentioned that AndroZoo can be considered as a challenging corpora since it includes new and more sophisticated malware samples in comparison to older datasets, namely Drebin [9]. We also chose a set of 1K benign apps from Google Play.

Static analysis was performed on all the collected apps using the open-source tool Androtomist [21]. Specifically, each app was decompiled to get the *Manifest.xml* file and log permissions and intents to create a feature vector. Each vector is a binary representation of each distinct feature. For example, given two apps, a1 and a2, where the first uses permissions p1, p2, p3, and intent i1 and the second uses permissions p2, p3, p4 and intents i1, i2, the analysis leads to a 6-dimensional feature vector (p1, p2, p3, p4, i1, i2), and thus the feature vectors for these two apps will be (1, 1, 1, 0, 1, 0) and (0, 1, 1, 1, 1, 1), respectively. Naturally, the scrutiny of a real-world app results to a much more lengthy vector. Precisely, the analysis of the largest set of malware and benign apps used in our experiments, i.e., 1K malware instances along with all of the 1K benign apps collected from Google Play, yielded 1,002-dimensional feature vectors.

4.2 Experimental Setup

As already pointed out, the entire dataset of malware apps used contains a collection of 1K malware apps randomly selected by AndroZoo corpus. Moreover, 1K benign apps were downloaded by the Google Play to comprise the negative category.

To set the base malware detection models eight well-known classifiers were applied, namely AdaBoost, k-nearest neighbors (k-NN), Logistic Regression (LR), Naive Bayes (NB), Multilayer Perceptron (MLP), Stochastic Gradient Descent (SGD), Support Vector Machine (SVM) and Random forests (RF). It is important to note that seven of these classification algorithms applied fall under eager learning. In this category, supervised learning algorithms attempt to construct a general model of the malware detection samples, building upon the training data. Apparently, the effectiveness of such classifiers is completely determined by the size, quality and representative of the training set. On the other hand, k-NN is a weak learner (known as lazy learner) as that makes a decision in terms of information extracted per sample separately, without needing the training part of data to construct a general model. The construction of each eager classification model was built following the 10-fold cross-validation technique. In this technique, a number of 10 different randomly segmented and equally sized sub-datasets is generated from the initial set of data. To extract each malware detection base model, we considered the set of parameter settings with the default values.

As already mentioned, apart from the original samples, two very popular and widely used dimensionality reduction technique were applied, namely PCA and t-SNE. Each base model was evaluated on either PCA or t-SNE new feature set. In this way, we examined the performance of the eight base models considered by handling the Original, PCA and t-SNE set of features, separately. To extract each malware detection base model, we relied on the set of parameter settings with the default values.

A simple meta-model is developed combining the output of all base classifiers applied following two options: in the former, we combine the answers of the 8 base models (AdaBoost, k-NN, LR, NB, MLP, SGD, SVM and RF) based exclusively on the Original, PCA and t-SNE representations, separately. In the latter, a complicated and possibly more heterogeneous ensemble model is constructed to combine the outputs of all malware detection base models. More specifically, the answers of 24 base models based on Original, PCA and t-SNE representations are totally combined to build a mixed meta-model. For each one of the above ensemble malware detection models presented, the outputs of base models is merged per instance separately by following two common aggregate functions namely, average (AVG) and majority vote (MV) technique.

The vast majority of the state-of-the-art methods in the detection of malware cases are mainly evaluated by using binary measures of correctness. Noticeably, these measures always provide a binary answer, either a positive (malware class) or a negative (benign class) one for each examined instance separately. This indicates that the information about the distribution of positive and negative instances is necessary for the sake of setting a threshold value. In this paper, we follow exactly the same evaluation procedure to achieve compatibility of comparison with previously reported results. More specifically, the classification performance measure of accuracy is considered, where TP, TN, FP, and FN represent correspondingly True Positives, True Negatives, False Positives, and False Negatives.

– Accuracy : $\frac{TP+TN}{TP+TN+FP+FN}$. The number of correctly classified patterns over the total number of patterns in the sample.

For each dataset, the set of the extracted scores based on the test instances are normalized in the interval of [0,1] per classification model per examined method. To this direction, the estimation of the threshold is set equal to 0.5. Moreover, we use the AUC of the receiver-operating characteristic curve as the main evaluation measure [28].

– Area Under Curve (AUC): The higher positive-over-negative value ranking capability of a classifier.

The AUC metric quantifies the effectiveness of each examined approach considering all possible threshold values. In general, the AUC value is extracted by examining the ranking scores rather than their exact values produced when a method is applied to a dataset. Noticeably, the estimation of the AUC measure is based on all possible thresholds.

4.3 Results

To evaluate the improvement in classification effectiveness when dimensionality reduction techniques are in force, we employ 8 well-known classifiers as well as ensemble approaches on benchmark Androzoo corpora following two options. The first one is to use the entire set of features on the dataset used (the case of $a = 1$). The second option is to select a random subspace of the initial feature set. In this case, we use a fix rate equal to 0.5 ($a = 0.5$). Table 1 reports the results of AUC and Accuracy (AC) measures of all malware detection base models (AdaBoost, k-NN, LR, NB, MLP, SGD, RF and SVM), utilizing either Original or transformed (based on PCA or t-SNE techniques) feature sets of data when $a=1$ or $a = 0.5$, respectively.

As concerns the performance of dimensional reduction techniques, it seems that the proposed malware detection base models based on PCA or t-SNE transformation are particularly effective and outperform the corresponding original ones in the most of the cases, especially when the whole feature set is considered ($a = 1$). It is discernible that t-SNE aids malware base models to achieve improved results in comparison to PCA transformation when $a = 1$.

Table 1. AUC scores of the proposed malware detection base models on the Androzoo corpora

DR method	AdaBoost		k-NN		LR		NB		MLP		RF		SGD		SVM	
	AUC	AC	AUC	AC	AUC	AC	AUC	AC	AUC	AC	AUC	AC	AUC	AC	AUC	AC
Original	0.864	0.806	0.812	0.767	0.866	0.794	0.835	0.720	0.877	0.803	0.875	0.814	0.767	0.763	0.504	0.501
Original (50%)	0.876	0.807	0.840	0.772	0.862	0.782	0.842	0.731	0.881	0.809	0.880	0.801	0.823	0.799	0.499	0.480
PCA	0.861	0.805	0.814	0.774	0.866	0.729	0.835	0.690	0.878	0.771	0.880	0.800	0.769	0.704	0.505	0.484
PCA (50%)	0.873	0.810	0.870	0.800	0.816	0.790	0.755	0.708	0.827	0.788	0.880	0.812	0.714	0.692	0.591	0.589
t-SNE	0.883	0.820	**0.885**	0.844	0.820	0.799	0.792	0.770	0.849	0.803	0.882	0.837	0.799	0.744	0.619	0.604
t-SNE (50%)	0.867	0.802	0.867	0.809	0.736	0.690	0.755	0.701	0.820	0.800	0.882	0.804	0.678	0.678	0.564	0.538

On the other hand, we also observe that the performance of both PCA and t-SNE models is negatively affected by utilizing the fixed rate of features ($a = 0.5$). It seems that these very challenging conditions significantly affect the performance of transformed models. However, the important improvement in the original models when $a = 0.5$ in almost all cases (except of LR and SVM models) verifies the previous outcome that dimensionality reduction techniques are better able to handle a larger size of feature set than the original ones. The same patterns are consistent in both the examined performance measures.

Clearly, the top-performing model seems to be the k-NN, which achieves the best results in both the examined performance metrics, especially when combined with the t-SNE technique. In particular, the use of t-SNE assists the k-NN model to become more stable surpassing all other base models. The same pattern applies to both AUC and AC performance measures.

In addition, we also consider the combination of the base models by fusing their answers for each sample separately. More specifically, two fusion functions are applied namely, average (AVG) and Majority Vote (MV) technique. Table 2 shows the effectiveness in terms of AUC and AC measures of the proposed ensemble malware detection models when either AVG or MV functions are considered on our AndroZoo dataset. Note that the ensemble approaches are not only tested on the original ($AVG_{original}$, $MV_{original}$), but also on transformed (in terms of PCA (AVG_{PCA}, MV_{PCA}) and t-NSE (AVG_{tsNE}, MV_{t-SNE})techniques) feature sets when $a = 1$ and $a = 0.5$, respectively. Moreover, the AVG_{Mixed}, MV_{Mixed} ensemble models are examined. These models are extracted by combining the entire set of base models (Original, PCA and t-SNE) when either $a = 1$ or $a = 0.5$. In this case, the output of 24 malware detection base models is combined according to AVG and MV aggregate functions.

As can be seen in Table 2, the proposed AVG_{Mixed} model is the most effective one in all cases improving the best reported results for the specific dataset. Its performance is higher when $a = 1$ in comparison to the case where a fix rate of features is considered ($a = 0.5$). This sounds reasonable since transformed base models are less enhanced in smaller feature set. Nevertheless, in case of $a = 0.5$, AVG_{Mixed} also provides very good results. This indicates that the size

of base models is an important factor that influences the performance of the
presented ensemble methods. Given that our ensembles include the entire set
of base models, it seems rational that their performance is improved when the
number of base models augments.

It is important to be mentioned that ensemble models seem to be positively
influenced when dimensionality reduction transformation are considered. Ensembles based on PCA and t-SNE base models are better than the corresponding
ones based on original models for both $a = 1$ and $a = 0.5$. Again, t-SNE models
surpass PCA models with a wide margin in the most of the cases. This shows
that the proposed t-SNE ensemble models are better able to handle malware
detection cases and are clearly better options than PCA ones. These patterns
are consistent in both the evaluation measures.

In general, the version of ensemble models based on AVG is the most effective in all cases achieving more balanced performance on both AUC and AC
measures. On the other hand, the performance of MV models seems not to
be highly competitive. At the same time, MV ensemble models is competitive
enough achieving its best performance when the output of mixed malware detection base models is considered. It appears again that combining the output of
mixed base models achieves the best performance. This verifies that ensembles of
classifiers based on multiple, possibly heterogeneous models, can further improve
the performance of individual malware detection models.

Table 2. Comparison of the AUC of both ensemble methods

DR method	Ensemble (AVG)		Ensemble (MV)	
	AUC	AC	AUC	AC
Original	0.878	0.821	0.815	0.755
Original (50%)	0.879	0.821	0.819	0.767
PCA	0.892	0.840	0.854	0.796
PCA (50%)	0.886	0.836	0.838	0.799
t-SNE	0.940	0.910	0.897	0.855
t-SNE (50%)	0.900	0.840	0.887	0.830
Mixed	**0.951**	0.917	0.912	0.877
Mixed (50%)	0.942	0.899	0.890	0.862

Figure 1 illustrates the performance (AUC) of AVG and MV ensemble models when malware detection base models are based on Original, PCA, t-SNE representations for $a = 1$ and $a = 0.5$, respectively. The AVG_{Mixed} and MV_{Mixed}
ensemble models are also reported.

Apparently, the best performing model not only for the whole feature set but
also when a fixed rate of features is randomly selected ($a = 0.5$), is the AVG_{Mixed}
model constructed by averaging the output of mixed malware detection base
models (Original, PCA and t-SNE models). Given that these ensembles include

the set of both the Original, PCA and t-SNE enhanced base models, this means that increasing the size of base models it has a positive effect on the efficiency of the proposed ensemble methods.

Again, ensembles based on original base models are outperformed by the ensembles using dimensionality reduction transformation when $a = 1$. This clearly shows that increasing the feature set helps PCA and t-SNE techniques to improve their performance. Apparently, t-SNE ensemble models are better and more stable alternative than PCA ones with a noticeable margin in all cases indicating that t-SNE models better suits in malware detection cases.

In general, averaging the output of base malware detection models seems to be the best and more stable option achieving more balanced performance both on AUC and AC measures. This strongly indicates that the vast majority of these base models provide commonly improper votes on similar malware detection cases. Both $MV_{original}$ and MV_{PCA} perform poorly. However, it should be underlined that MV_{t-SNE} and MV_{Mixed} models are very effective surpassing the best performing base model which is based on t-SNE transformation. Again, this can be explained since the meta-learner needs as accurate base models as possible and t-SNE models are more stable and reliable than PCA ones. Moreover, it clearly demonstrates the contribution of t-SNE technique on malware detection cases.

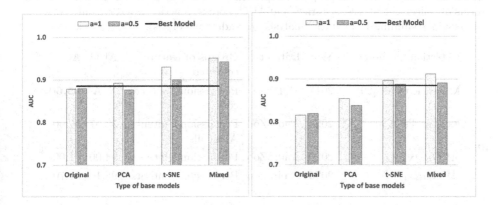

Fig. 1. The performance (AUC) of the examined ensembles when a = 1 and a = 0.5, using either AVG (left) or MV (right) fusion techniques, on Androzoo dataset for varying types of base models, respectively. The performance of the best base model is also depicted.

4.4 Comparison with the State-of-the-Art

In the course of our experiments, the performance on the Androzoo corpus is measured by the area under the receiver-operating characteristic curve as well as accuracy measures. This makes our reported results directly comparable to the ones obtained by others published works in the framework of malware detection task. The following state-of-the-art methods (ranked in chronological order) are used to estimate the competitiveness of the proposed AVG_{Mixed} model when a = 1:

- Milosevic et al. [29]: This work concentrates on the extraction of non-trivial and beneficial malicious patterns examining the usefulness of source code as well as the permissions set of features when combined with either classification or common used clustering techniques, respectively. In the experiments, the M0Droid corpus is considered and two categories of features namely, permissions and source code are utilized.
- Kouliaridis et al. [21]: In this study, a simple heterogeneous malware detection ensemble method is proposed on Adrozoo dataset. More specifically, a meta-model is constructed by averaging the output of several base models based on either static or hybrid analysis. During static analysis the authors collected features from three categories, namely permissions, intents, and API calls. The performance of this method is evaluated on several datasets, namely Drebin, VirusShare and AndroZoo.

Table 3. Comparison of the proposed approach with state-of-the-art detection works in terms of collected features, accuracy and AUC score (* For this work, we only consider the results stemming from static analysis on Androzoo corpus)

Detection method	Year	Dataset	Groups of features collected	AUC	AC
Milosevic et al. [29]	2017	M0Droid	Permissions, source code	N/A	**95.60**
Kouliaridis et al. [21]*	2020	AndroZoo	Permissions, intents, API calls	93.57	90.90
AVG_{t-SNE}	2020	AndroZoo	Permissions, intents	94.00	91.00
AVG_{Mixed}	2020	AndroZoo	Permissions, intents	95.10	91.70

Note that the published results for some of the above methods only refer to either Androzoo or M0Droid datasets. Moreover, in some works, the evaluation results are not provided on both performance measures (AUC and accuracy). Thus, we use the values of the two evaluation metrics. Table 3 demonstrates the effectiveness of the state-of-the-art methods per dataset and on AUC and AC evaluation measures, respectively.

Clearly, both the AVG_{Mixed} and AVG_{t-SNE} models examined in this study are the most effective, surpassing the reported results of Kouliaridis et al. [21]*, which also employs the AndroZoo benchmark dataset. In addition, the resulted evaluation values in [21] were extracted by averaging the answers of 8 base classification models. Under these settings, it is important to mention that these classifiers are similar to those combined for AVG_{t-SNE} model. In this way, it can be concluded that t-SNE transformation better suits in demanding malware detection cases and caters for an effective model handling only two feature categories (permissions and intents). Moreover, when the AVG_{Mixed} model is applied, the improved evaluation results in the challenging Androzoo dataset indicate that the examined model is not easily confused in challenging conditions and the combination of a larger set and mixed base models has a positive effect on the efficiency of the proposed model.

On the other hand, taking into consideration the work of Milosevic et al. [29] examined on a different dataset and feature set, it seems that the performance of both AVG_{Mixed} and AVG_{t-SNE} models is not so competitive. It is important to note, however, that the MODroid dataset comprises only 200 malware samples, which can lead to overestimated results. Moreover, MODroid is dated back in 2014, and therefore outdated vis-à-vis the AndroZoo dataset. Altogether, the improved results of the proposed AVG_{Mixed} and AVG_{t-SNE} models in the challenging AndroZoo corpora demonstrate that the presented models are not easily confused in demanding malware conditions and can capture useful malware information.

5   Conclusion

This paper introduces new insights on malware detection approaches based on ensemble learning. We utilize eight popular and extensively used base classifiers namely, eager (AdaBoost, LR, NB, MLP, SGD, SVM and RF) as well as lazy (k-NN) algorithms. This collection of verifiers provides a pluralism of malware detection scores and we attempt to take advantage of their correlations by constructing two ensembles. In this way, we explore two options. The first one is consisted by a set of homogeneous malware detection base models and learns patterns of agreement or disagreement among not only original feature set but also Principal Component Analysis (PCA) and t-distributed stochastic neighbor embedding (t-SNE) dimensionality reduction techniques. In other words, it learns dealing with the output of base models extracted utilizing exclusively the original feature set of malware detection cases or only a reduced set of features resulted by applying either the PCA or t-SNE transformation, respectively.

The second option is more knotty, handling the output of a larger and probably more heterogeneous set of base verifiers aggregating all the examined malware detection base models together (Original, PCA and t-SNE base models). Both the ensemble approaches outperform a set of strong base malware detection models in the most of the cases in terms of experiments developed on benchmark Androzoo corpora. This suggests that our ensembles are able to handle demanding malware detection scenarios and are more robust than individual models. Moreover, it seems that a relatively large and mixed size of base malware detection models is required to achieve high performance.

For each of the above ensemble variation, the output of base models is combined considering two simple fusion functions namely, average (AVG) and

Majority Voting (MV) technique. Except of the initial feature set, we examine the alternative of using a fixed rate equal to 0.5 of the initial feature set. This demonstrates that it is important to be defined which set of features will be retrieved and how many base classifiers will be considered. Our experiments show that both AVG and MV work better with mixed base models. This indicates that ensembles based on a large and heterogeneous set of base models are the best option and they provide a more stable performance. That is, in general, AVG are always the best performing models surpassing the corresponding MV ones. This strongly manifests that the vast majority of these base models provide commonly improper votes on similar malware detection cases.

We also focused on the use of two prominent dimensionality reduction techniques namely, PCA and t-SNE, dealing with either the whole initial feature set or a subspace randomly selected. It is demonstrated that the performance of both transformed malware detection models is notably reinforced than the original one when the entire number of features is handled. This verifies that these techniques are negatively affected when a smaller feature set is available. In general, t-SNE models are more effective and competitive than PCA ones indicating that t-SNE transformation helps malware detection approaches to become more stable.

The development of more sophisticated ensembles exploiting a larger set of dimensionality reduction techniques to achieve high diversity is an open research direction. Another possible future work direction is to try to further enrich the pool of our base verifiers considering not only a richer set of classification algorithms, but also several versions of the same approach with different fixed and tuned parameter settings.

References

1. Papamartzivanos, D., Damopoulos, D., Kambourakis, G.: A cloud-based architecture to crowdsource mobile app privacy leaks. In: Proceedings of the 18th Panhellenic Conference on Informatics, PCI 2014, pp. 1–6. Association for Computing Machinery, New York, NY, USA (2014)
2. Damopoulos, D., Kambourakis, G., Gritzalis, S., Park, S.O.: Exposing mobile malware from the inside (or what is your mobile app really doing?). Peer Peer Netw. Appl. 7(4), 687–697 (2014). https://doi.org/10.1007/s12083-012-0179-x. https://doi.org/10.1007/s12083-012-0179-x
3. Damopoulos, D., Kambourakis, G., Anagnostopoulos, M., Gritzalis, S., Park, J.H.: User privacy and modern mobile services: are they on the same path? Pers. Ubiquitous Comput. 17(7), 1437–1448 (2013) https://doi.org/10.1007/s00779-012-0579-1. https://doi.org/10.1007/s00779-012-0579-1
4. Mobile OS market share (2020). https://gs.statcounter.com/os-market-share/mobile/worldwide. Accessed 10 Sep 2020
5. Smartphone market share (2020). https://www.idc.com/promo/smartphone-market-share/os. Accessed 10 Sep 2020
6. Mcafee mobile threat report 2020 (2020). https://www.mcafee.com/content/dam/consumer/en-us/docs/2020-Mobile-Threat-Report.pdf. Accessed 10 Sep 2020
7. Contagio. http://contagiominidump.blogspot.com/. Accessed 10 Sep 2020

8. Zhou, Y., Jiang, X.: Dissecting android malware: characterization and evolution. In: Proceedings of the 33rd IEEE Symposium on Security and Privacy, vol. 12, no. 7 (2012)

9. Arp, D., Spreitzenbarth, M., Huebner, M., Gascon, H., Rieck, K.: Drebin: efficient and explainable detection of android malware in your pocket. In: 21th Annual Network and Distributed System Security Symposium (NDSS), vol. 12, no. (7), p. 1128 (2014)

10. Author. Details withheld to preserve blind review. anonymized

11. Yan, P., Yan, Z.: A survey on dynamic mobile malware detection. Software Qual. J. **26**, 891–919 (2018)

12. Souri, A., Hosseini, R.: A state-of-the-art survey of malware detection approaches using data mining technique. Hum.-Centric Comput. Inf. Sci. **8**, 3 (2018)

13. Odusami, M., Abayomi-Alli, O., Misra, S., Shobayo, O., Damasevicius, R., Maskeliunas, R.: Android malware detection: a survey. In: Florez, H., Diaz, C., Chavarriaga, J. (eds.) ICAI 2018. CCIS, vol. 942, pp. 255–266. Springer, Cham (2018). https://doi.org/10.1007/978-3-030-01535-0_19

14. Narudin, F.A., Feizollah, A., Anuar, N.B., Gani, A.: Evaluation of machine learning classifiers for mobile malware detection. Soft. Comput. **20**, 343–357 (2016)

15. Allix, K., Bissyandé F, T., Klein, J., Le Traon, Y.: Androzoo: collecting millions of android apps for the research community. In: Proceedings of the 13th International Conference on Mining Software Repositories, MSR 2016, pp. 468–471. ACM (2016)

16. Author. Details withheld to preserve blind review

17. Ali-Gombe, I., Saltaformaggio, B., Ramanujam, J.R., Xu, D., Richard, G.G.: Toward a more dependable hybrid analysis of android malware using aspect-oriented programming. Comput. Secur. **73**, 235–248 (2018)

18. Arshad, S., Shah, M.A., Wahid, A., Mehmood, A., Song, H., Samadroid, H.Y.: A novel 3-level hybrid malware detection model for android operating system. IEEE Access **6**, 4321–4339 (2018)

19. Xu, L., Zhang, D., Jayasena, N., Cavazos, J.: HADM: hybrid analysis for detection of malware. In: Bi, Y., Kapoor, S., Bhatia, R. (eds.) IntelliSys 2016. LNNS, vol. 16, pp. 702–724. Springer, Cham (2018). https://doi.org/10.1007/978-3-319-56991-8_51

20. Fang, Q., Yang, X., Ji, C.: A hybrid detection method for android malware. In: 2019 IEEE 3rd Information Technology, Networking, Electronic and Automation Control Conference (ITNEC), pp. 2127–2132 (2019)

21. Kouliaridis, V., Kambourakis, G., Geneiatakis, D., Potha, N.: Two anatomists are better than one-dual-level android malware detection. Symmetry **12**(7), 1128 (2020)

22. Martinelli, F., Mercaldo, F., Saracino, A.: BrideMaid: an hybrid tool for accurate detection of android malware. In: Proceedings of the 2017 ACM on Asia Conference on Computer and Communications Security (2017)

23. Surendran, R., Thomas, T., Emmanuel, S.: SamaDroid: a tan based hybrid model for android malware detection. J. Inf. Secur. Appl. **54**, 102483 (2020)

24. Vega Vega, R., Quintián, H., Calvo-Rolle, J., Álvaro, H., Corchado, E.: Gaining deep knowledge of Android malware families through dimensionality reduction techniques. Logic J. IGPL **27**(2), 160–176 (2018)

25. Deerwester, S., Dumais, S.T., Furnas, G.W., Landauer, T.K., Harshman, R.: Indexing by latent semantic analysis. J. Am. Soc. Inf. Sci. **41**(6), 391–407 (1990)

26. Bunte, K., Biehl, M., Hammer, B.: A general framework for dimensionality-reducing data visualization mapping. Neural Comput. **24**(3), 771–804 (2012)

27. Google play. https://play.google.com/. Accessed 10 Sep 2020
28. Fawcett, T.: An introduction to ROC analysis. Pattern Recogn. Lett. **27**(8), 861–874 (2006)
29. Milosevic, N., Dehghantanha, A., Choo, K.K.R.: Machine learning aided android malware classification. Comput. Electr. Eng. **61**, 266–274 (2017)

A Regret Minimization Approach to Frameless Irregular Repetition Slotted Aloha: IRSA-RM

Iman Hmedoush[(✉)], Cédric Adjih, and Paul Mühlethaler

Inria, Le Chesnay-Rocquencourt, France
{iman.hmedoush,cedric.adjih,paul.muhlethaler}@inria.fr

Abstract. Wireless communications play an important part in the systems of the Internet of Things (IoT). Recently, there has been a trend towards long-range communications systems for IoT, including cellular networks. For many use cases, such as massive machine-type communications (mMTC), performance can be gained by going out of the classical model of connection establishment and adopting the random access methods. Associated with physical layer techniques such as Successive Interference Cancellation (SIC), or Non-Orthogonal Multiple Access (NOMA), the performance of random access can be dramatically improved, giving the novel random access protocol designs. This article studies one of these modern random access protocols: Irregular Repetition Slotted Aloha (IRSA). Because optimizing its parameters is not an easily solved problem, in this article, we use a reinforcement learning approach for that purpose. We adopt one specific variant of reinforcement learning, Regret Minimization, to learn the protocol parameters. We explain why it is selected, how to apply it to our problem with centralized learning, and finally, we provide both simulation results and insights into the learning process. The obtained results show the excellent performance of IRSA when it is optimized with Regret Minimization.

Keywords: IRSA · Regret minimization · Random access

1 Introduction

1.1 Communications in the Internet of Things

In the past years, one has witnessed an increase in the technological demands on embedded systems and sensors, that led to the emergence of the Internet of Things (IoT). The Internet of Things provides a comprehensive set of solutions that enables the seamless interconnection of a smart community of devices and sensors. This article focuses on one of the most important challenges for IoT networks: communications. This has been a highly visited topic for research, development, and standardization in the past decade because the differences of IoT applications are translated into a large variety of requirements and constraints.

© Springer Nature Switzerland AG 2021
E. Renault et al. (Eds.): MLN 2020, LNCS 12629, pp. 73–92, 2021.
https://doi.org/10.1007/978-3-030-70866-5_5

Some of these constraints are the ability to support millions of connected devices, the necessity for low power consumption, and the need for high throughput, low latency, and high reliability.

In IoT, there has been a recent trend in the use of long-range low power networks: with cellular networks (in 4G with NB-IoT, and LTE-M, in 5G with URLLC, and NR-light in future 5G Rel. 17), or with networks in the unlicensed band (LoRaWAN, SigFox). As a consequence, there have been new directions in the design of IoT protocols to satisfy the critical requirements of IoT applications, including modern variants of random access methods. One such method is the focus of this article.

1.2 Modern Random Access Protocols for IoT Communications

A recent family of random access protocols has emerged as a promising solution, for modern random access: it is Irregular Repetition Slotted Aloha (IRSA) [4], and its generalization with coding, Coded Slotted Aloha (CSA) [5]. It has become the focus of the IoT protocol designers' attention, since it has been shown that it could asymptotically reach the optimal throughput of one retrieved packet per slot, in the classical random access collision model (where the maximum throughput of slotted ALOHA is $\frac{1}{e}$). NOMA variants like PDMA/IRSA [6] exist.

The principle is that the users send multiple copies of each of their data packets to the receiver that uses Successive Interference Cancellation (SIC) to resolve the collisions. The transmission is done in slots. Each copy (also known as a replica) contains the same payload and the same preamble with additional information about the positions of its copies in other slots. Once one packet is received without collision, SIC exploits this information to reconstruct the physical signal that corresponds to the decoded packet and subtracts this physical signal at the positions of its other copies.

1.3 Irregular Repetition Slotted Aloha (IRSA)

IRSA is a recent member of the CSA family which is an optimization of Contention Resolution Diversity Slotted Aloha (CRDSA) [11]. In CRDSA, the users would repeat their packets twice. In IRSA, the users are allowed to choose the number of repetitions (*repetition degree*) according to a probability distribution. In classical IRSA, we consider a MAC frame of M slots and N users who send their packets towards a central node. The channel load is defined as $g = \frac{N}{M}$ (i.e. average number of users per slot). The receiver uses SIC to decode the collided packets. Each user chooses its repetition degree based on the user degree distribution $\Lambda = (\Lambda_0, \Lambda_1, \dots \Lambda_D)$, where Λ_i is the probability to use the degree i and D is the maximum degree.

At the end of each frame, the receiver starts performing iterative decoding using SIC. At each decoding iteration, the receiver starts by searching for the non-colliding packets (referred to as the singletons). After finding, decoding, and recovering all the singletons in the frame, the receiver removes their physical copies from their positions in the frame. This suppresses some collisions and

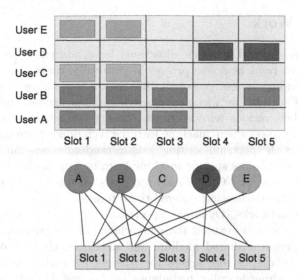

Fig. 1. IRSA representation: transmissions of users in slots (top), coding theory representation to model the decoding process (Tanner graph, bottom). Notice that transmissions are in the same frequency channel, hence when two users are transmitting on the same slot, there is a collision.

in turn, can make new singletons appear in the next decoding iteration. The iterative decoding continues until the receiver can not find new singletons or the whole frame is decoded. Figure 1 represents a simple example used to illustrates the SIC decoding. The figure shows a frame where 5 users compete to send their packets on 5 slots. The receiver starts by searching for the singletons. As can be seen from the figure, slot 4 has only one packet from user D, so it can be decoded on slot 4. Then, with SIC, it can be subtracted from slot 5. This allows the packet from user B to be a new singleton on slot 5 so that it is decoded from slot 5 and subtracted with SIC in slots 1, 2, and 3. Now the packet of user A becomes a new singleton on slot 3 so it is decoded on slot 3 and subtracted from slots 1 and 2 as well. We can see that the packets of users C and E cannot be retrieved since none of their replicas had ever become a singleton and they form a stopping set. At this point, the decoding process stops.

The remaining of this paper is organized as follows: Sect. 2 introduces the related work of applying machine learning algorithms to random access protocols. Then, Sect. 3 explains the system model and our problem statement. In Sect. 4, we introduce our learning approach: IRSA-RM, and detail the algorithm structure. Finally, we present our numerical results in Sect. 5 and conclude in Sect. 6.

2   Related Work

In the literature, several research directions have addressed topics that are related to modern random access protocols. Naturally, there also exists a large literature on random access protocols themselves that dates back to several decades. In this section, we focus on modern random access, and on research works that applied various reinforcement learning techniques. We identify the following related topics, where machine learning techniques have been used: cognitive networks with spectrum sensing, classic random access protocols in IoT networks, and finally, more specific machine learning approaches to protocols of the IRSA family itself, or NOMA-based protocols. In the following, we describe some of the related articles that covered these topics.

In cognitive networks, Dynamic Spectrum Access is a wireless network paradigm where the users exploit their knowledge of the environment in order to successfully access a shared medium and maximize their throughput. The problem of dynamic spectrum access for wireless networks has been recently explored with machine learning techniques in [12] and [13]. It is shown that the problems of joint user association and spectrum access are typically combinatorial and non-convex, and require near-complete and accurate information to obtain the optimal strategy. According to [14], developing efficient learning approaches to optimize medium access has been the center of attention of many research works. In particular, Deep Q-Learning (DQL) provides promising solutions for the Dynamic Spectrum Access problem specifically, for IoT networks. In [15], a novel distributed dynamic spectrum access algorithm based on deep multi-user reinforcement learning (DRL) has been proposed. The users transmit over shared channels using a random access protocol. Time is slotted but no SIC is used in the receiver to resolve collisions. In this proposed approach, each user maps its current state (the history of selected actions and past network state observations) to a certain action (shared orthogonal channel) based on a trained deep-Q network. Their objective is to maximize a utility function. The proposed algorithm enables the user to learn good policies in an online distributed manner. The authors of [16] address the problem of collisions and idle time of random access protocols by designing a fully distributed IoT protocol to improve the device access to the shared medium. The proposed online learning scheme is based on designing optimized dictionaries of transmission patterns to avoid collisions between users. The dictionary contains a subset of the possible binary vectors of a length equal to the total number of slots in the frame. The goal is to select an optimized set of transmission patterns from the dictionary, where the dictionary is common to all users. The proposed scheme provides some URLLC guarantees for IoT applications that require the same time latency, energy efficiency, and low coordination overhead.

The dynamic multi-channel access was also considered in [9], where the user selects a channel, at each time slot, from multiple correlated channels. Each user can observe the state of the chosen channel only at a given time slot, which means that the current state of the system is not fully observable, hence the problem is modeled as a Partially Observable Markov Decision Process (POMDP).

The goal of the study is to design an adaptive DQN framework that can adapt to time variations and maximize the long-term expected reward for each user.

Another work direction for designing efficient IoT protocols is to enhance the existing MAC protocols so that they fit the new requirements of IoT networks. Optimizing the performance of MAC protocols has been addressed in many research studies during the last fourty years. Some of these studies have introduced machine learning techniques to variants of the ALOHA protocol family. A novel Q-learning based on Informed Receiving Protocol has been introduced in [3]. ALOHA-QIR provides some intelligence to the nodes to access the slots that have a lower probability of collision. The nodes keep hopping to different slots to learn the optimum ones. In this ALOHA variant, the nodes keep listening during the hopping while the receiver is informed by the preferred slots of each node by sending "ping packets", so the receiver can turn off when needed. The classical Q-learning algorithm with a simple reward design (± 1) is used to learn the optimum slots to select. The proposed approach help to achieve over two times the maximum throughput of Slotted ALOHA. In [17], the IRSA MAC protocol is optimized using online learning. By considering the base station as the decision-maker, the performance of IRSA is optimized by maximizing a utility function that reflects the number of decoded packets. The problem of optimal resource allocation (slots allocation) is formalized as a Multi-Armed-Bandit (MAB) problem. The authors use the Bayesian UCB algorithm to solve the MAB problem and compare it with other commonly used methods. The degree distribution is also optimized by fixing the degrees and optimizing the probabilities to select them (e.g., of the form $\Lambda_2, \Lambda_3, \Lambda_8$).

3 System Model and Assumptions

3.1 System Description

We consider IRSA as an access protocol for users (devices) sharing a communication channel to a receiver. We assume that the receiver is a single base station. As for other protocols of the CSA family, the access time of the channel is divided into slots with equal duration. The duration of the slot is equal to the time needed to transmit a packet (including propagation delays, etc.). In our system, however, we assume that a *frameless IRSA* is used (as in [7]), as opposed to the framed, classical version of the IRSA protocols. In framed IRSA, there is a frame of a predefined length, where each user randomly selects slots, and at the end of which the decoding is performed. In frameless IRSA, there is a very large set of slots, potentially infinite. In our model, for practical reasons, this set of slots has always a fixed size of M slots and it is called a contention round. We divide the contention round into virtual frames where each virtual frame size is about 20% of the contention round. When the user decides to send a packet, the user is associated with a virtual frame. The active user sends the replicas of its packet during the virtual frame period only. The goal of introducing the virtual frame is to facilitate the decoding process and rewards computations which will be precisely explained in the next section.

At each time slot, the number of active users is determined by a Poisson arrival rate μ (e.g. the number of active users on one slot is a random variable N_a with distribution $\Pr(N_a = k) = \frac{\mu^k e^{-\mu}}{\mu!}$). In our case, to be consistent with the literature, the arrival rate μ is also denoted network load G. It is the average number of active users on one slot. Each active user selects a repetition degree to use from a set of multiple allowed degrees which are identical for all users. At the base station, SIC is used to resolve the collisions. Figure 2 shows the frameless IRSA structure with all active users and their associated virtual frames where they are allowed to send their packets. Virtual frames can overlap and the transmissions from different active users can cause collisions as seen in the figure. Unlike the classical IRSA decoding, (explained in Sect. 1.3), the base station performs online decoding by decoding each received slot instead of waiting for the whole frame to end and start the decoding process.

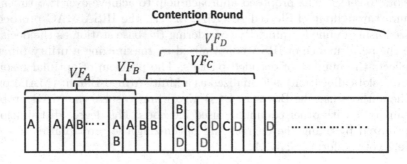

Fig. 2. IRSA frameless structure

3.2   Problem Statement

Many studies have explored and analyzed the performance of IRSA in an IoT network. The main purpose of these studies is generally to find and study a better variant of this protocol. Given any IRSA system, the goal is often to find an optimized user degree distribution that maximizes a certain metric (throughput, achievable load, etc.). This problem is usually formulated as an optimization problem which can be described as follows:

$$
\begin{aligned}
\underset{(\Lambda_i)}{\text{maximize}} \quad & C(\Lambda_0, \Lambda_1, .., \Lambda_D) \\
\text{subject to} \quad & 0 \le \Lambda_i \le 1 \ \forall i \\
& \sum_{i=0}^{i=D} \Lambda_i = 1
\end{aligned}
\tag{1}
$$

where C is the system criteria that needs to be optimized and D is the maximum degree.

Depending on the system model, more constraints can be added to the optimization problem. Notice that the system is initially a stochastic optimization problem, as the performance depends on the random variables of the users' arrivals and their degree selections. Pioneering work in [4] adapted the Density Evolution (DE) method to analyze the asymptotic performance of different (framed) variants of the IRSA protocol. It is based on an analogy with LDPC codes for which DE was initially introduced, as a tool for analyzing the asymptotic network capability to approach the error-correcting codes [18]. DE allows evaluating how many packets will be decoded, through iterations of functions, by modeling the decoding process steps: this is valid asymptotically when the frame size grows to infinity. Then one can compute the function C of the problem in Eq. (1) (through function iterations).

For several IRSA variants, the problem may still be difficult to solve for instance in the case of the non-convexity or non-linearity of the constraints. But the optimization problem formulation may be useful. For instance, in our prior work [19], we revisited the variant K-IRSA proposed elsewhere: K-IRSA is a variant of classical IRSA with multiple packet reception capability at the receiver. Using a variant of the previous optimization problem Eq. (1), DE was used to write a new constraint on the edge degree distribution and the system criterion was to maximize the achievable system load. The OP was solved by converting the new constraint into a finite set of linear constraints and using the bisection method on linear problem formulations.

4   IRSA-RM: IRSA Based on Regret Minimization

4.1   Problem Formalization

We adopt the frameless IRSA structure (explained in Sect. 3.1). For simplicity of presentation, we formulate the problem with two classes of users. The two classes have different access priorities. The users of the same class share the same degree distribution (Λ_i). The base station uses SIC to perform the slot by slot online decoding. Our objective is to find the best degree distribution for a known Poisson arrival rate $\mu = G$, that maximizes the weighted throughput of both classes. Formally, our problem could be written as an optimization problem using Eq. (1):

$$
\begin{aligned}
\underset{(\Lambda_i)}{\text{maximize}} \quad & \alpha_0 T_{C_0} + \alpha_1 T_{C_1} \\
\text{subject to} \quad & 0 \leq \Lambda_i \leq 1 \ \forall i \in [0, 1, 2, .., D] \\
& \sum_{i=0}^{i=D} \Lambda_i = 1
\end{aligned}
\tag{2}
$$

Where: T_C is the throughput of the class C and α_0, α_1 are constant weights indicating the importance of the throughput each class.

One can think of solving such a problem by using the DE tool since frameless IRSA transmission and decoding using SIC can still be represented by a bipartite graph [7]. Since the BS performs the decoding for each slot, a part of the bipartite graph will be available at any time thus, the density evolution equations will not necessarily represent the decoding state in the middle of the contention round. Because classical DE [4] is valid only asymptotically and the finite length analysis can be computationally expensive, we adopt another direction in this article. We propose a new learning framework for optimizing the transmission strategy of frameless IRSA. We consider a method of offline learning. We assume multi-agent settings and we apply the method of Regret Minimization, where each user wants to minimize its regret by taking better next decisions.

4.2   Reinforcement Learning Approaches and Regret Minimization

In this article, we use *Reinforcement Learning* (RL) to find good solutions of the problem Eq. (2). A classic reference on reinforcement learning in general is [8]. As in many network problems, the decisions taken by one node, device, or one user can be modeled as a Markov Decision Process (MDP) [8, section 3]: each participant of in the network is an *agent*, that makes decisions, denoted as *actions*, based on some current *state* from the environment. *Rewards* for each taken action are computed and are used to adjust the future choice of actions. Classical algorithms such as Q-Learning [8, section 6.5], Multi-Armed Bandits [8, section 2], and others, are well-known.

Applying those to random access introduces several challenges: the first one is that there are several agents instead of just one (*Multi-Agent* Reinforcement Learning, MARL, see [8, section 15.10]), the second one is, by definition of random access, each agent only knows part of the network state, if only because it does not know the actions of other agents (*Partial Observable* Markov Decision Process, POMDP, see [8, section 17.3]).

Learning in a multi-agent setting is indeed a complex task: the impact of the decision taken by one agent may depend on the decisions taken by other agents in the system. Thus, first, classical RL approaches for a single agent can create difficulties, such as non-stationarity and oscillations when applied to multi-agent systems. Second, controlling multiple agents poses additional challenges compared to single-agent systems such as the definition of the collective goal of the agents, the heterogeneity of the agents, the ability to operate with a large number of agents, and partial observability [20].

Frameless IRSA is such a multi-agent system, subject to partial observability. Numerous learning approaches have been proposed in the literature to handle POMDP, including Deep Reinforcement learning (DRL) [20]. Many of the proposed algorithms in the literature lose their proof of convergence in a MARL setting, and there does not necessarily exist a general theory characterizing the cases under which every MARL algorithm is successful [21]. Their convergence or non-convergence dynamics is a topic of study by itself, with also strong links with game theory [1, 21]. Indeed, while applying Q-Learning to frameless IRSA, we experienced non-convergence, which lead us to select an algorithm whose multi-agent dynamics have been well studied: Regret Minimization (RM) [2].

Regret Minimization is an algorithm where each agent maintains a set of *weights* for actions. Once normalized, the weights indicate the probability that the agent selects each action. At a given time, after the action selection by one agent according to weights, each such action i changes the environment state and has a corresponding reward which is provided by the environment. At the same given time, an optimal action could have been played by the agent instead, which would have resulted in an optimal reward r. The difference between the optimal reward r_{opt} and the actual reward r_i gives the loss of the agent at that given time: $\ell_i = r_{opt} - r_i$, which is a measure of *regret* for selecting action i.

The Polynomial Weights algorithm is one of the Regret Minimization algorithms that assigns weights for each action and uses the "loss" concept to update the weights after each playing round [2]. Formally, it is as follows [2, page 13]:

$$\text{Initially: } w_i^{(1)} = 1 \text{ and } p_i^{(1)} = \frac{1}{|X|} \text{ for } i \in X$$

At time $t-1$: an action $i \in X$ is selected according to $(p_j^{(t-1)})_{j \in X}$

the reward of action i is computed: $r_i^{(t-1)}$

the potential reward of the best action is: $r_{opt}^{(t-1)}$ (3)

the loss is computed as: $\ell_i^{(t-1)} = r_{opt}^{(t-1)} - r_i^{(t-1)}$

Update for time t: $w_i^{(t)} = w_i^{(t-1)}(1 - \eta \ell_i^{(t-1)})$

$$p_i^{(t)} = \frac{w_i^{(t)}}{\sum_{i \in X} w_i^{(t)}}$$

with:

X: the set of possible actions.

$w_i^{(1)}$: the initial associated weight to the action i.

$w_i^{(t)}$: the associated weight to the action i at time t.

$p_i^{(1)}$: the initial probability to use an action i out of $|X|$ actions.

$p_i^{(t)}$: the probability to use an action i at time t.

η: the learning parameter (akin to a learning rate).

The weights update is based on two main parameters: the learning parameter to control the speed of the weight changes, and the loss which specifies the impact of the played action by computing how far was the played action from the optimality. The weights of the actions are used to compute the probability to use each of the actions in the next playing round.

Notice that richer variants of Regret Minimization have been proposed, such as Counterfactual Regret Minimization (CFR) [10]; with IRSA, they would be well suited for agents with richer interactions, for instance, agents taking decisions on the transmission of each replica (instead of selecting a degree once).

4.3 Applying of Regret Minimization to Frameless IRSA

Back to our initial problem, we assume that the network consists of users competing in the same slotted wireless channel to transmit packets towards one base station using the frameless IRSA protocol. Users are grouped in classes of different priorities. The users of one class also share the same degree distribution. As mentioned, each user has partial observability about the network, because it does not know on which slots the other agents are transmitting (nor about collisions). However, they have additional information: an important assumption is that the base station is maintaining a discretized estimate $\bar{G}$ of the load $G = \mu$ (Poisson agent arrival rate) in the system and broadcasting it to each agent.

Our objective is to maximize the total throughput of users, for each given network load G; where the throughput of each class is actually weighted by a different factor (so that some classes carry more weight, as a priority mechanism). We assume that each active agent when it decides to send a packet has to send the packet and its replicas within a virtual frame which is associated to the agent. It is interesting to look at the base station perspective: it observes singletons on some slots, collisions on some other slots, and performs SIC for each packet already decoded.

We adapt the Polynomial Weight RM algorithm detailed in Eq. (3) to solve our problem. To emphasize the learning aspect, here, the term "agent" will be used as an equivalent to "user". In order to map the problem features to RM, we have the following assumptions and system model:

- A centralized *offline* learning approach based on Regret Minimization is considered. A large number of simulations or *episodes* (as in Q-Learning) are run. Each episode corresponds to a long contention round. It is intended that after learning has finished, the weights could be used in an actual network, or in our case, are actually used as distributions Λ in further simulations without learning.
- The base station is assumed to broadcast a discretized estimate $\bar{G}$ (with finite number of possible values) of the actual load G: currently $\bar{G}$ is the measured average number of users per slot, since the beginning of the contention round.
- The action of each agent is: selecting the number of repetitions (the degree).
- As per Polynomial Weights RM, each of the agents maintains weight tables (denoted w) which are used to compute the probability to select each action, akin to a probability distribution Λ. We extend it: one different table of weights is used depending on some state. The agents consider the load estimate given by the base station as the environment state, and it is discretized to constitute a finite set of possible states. For each different discretized load estimate, the agent uses and updates a different set of weights $(w_i(\bar{G}))_{i \in X}$.
- Additionally, the agents of the same class are sharing the same weight tables w table in the learning process. Updates of the weights after each selected action are thus shared within agents of one class[1]. The goal of using the same w table for all the agents of the same class is to drive the agents inside one class to act cooperatively, and to work coordinately towards the collective goal. This may depart from usual assumptions in RM and evolutionary dynamics.

---

[1] Note that then the learning also behaves as if one class would be one agent by itself. The algorithm, and our implementation, also works with non-shared tables.

- At the moment an agent selects a degree, the results of this action are unknown until some time has elapsed (see Sect. 4.4). Thus an approach with delayed updates similar to *n-step Sarsa* [8, section 7.2] is used.
- The main challenge for applying the Polynomial Weight algorithm is to compute the loss. The loss computation is directly related to the rewards calculation. As our goal is to optimize the joint throughput, we opt to directly link the rewards to the number of decoded agents in each class and set "reward = number of decoded agents". Defining an IRSA reward is otherwise difficult.
- The number of decoded and non-decoded users is available at the global simulator level during our centralized learning process.

We further detail delayed updates and reward computation in the next sections.

4.4 Delayed Updates

Each agent sends within its virtual frame size on one side, and the base station decodes slot by slot on the other side. Therefore, the base station needs to wait at least, for the end of the virtual frame to decide if one agent can be decoded or not. Hence, to accurately compute a reward, one delay needs to be introduced: it is illustrated in Fig. 3. As shown in the figure, the agent A starts to be active at time τ and sends its packets using the action i, e.g. sending i replicas, ($i = 4$ in this case), within the associated virtual frame spanning the time from τ to $\tau + VF_A - 1$. Only at time $\tau + VF_A$, one is certain that all replicas of A have been sent. But decoding can be further delayed: during this virtual frame, other agents could become active and transmit in overlapping virtual frames (possibly shifted in time, see Fig. 2) and could induce collisions that need to be resolved in order to recover one of the replicas of A. But in turn, those other agents might collide with agents whose virtual frames are occurring even later, etc.[2] For practical

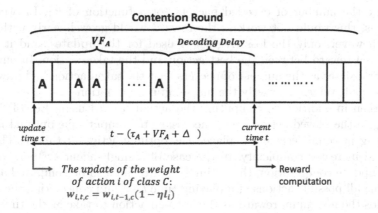

Fig. 3. Reward computation and update delays

---

[2] It is indeed possible to construct a frameless IRSA scenario where one user can be only decoded after an arbitrarily large delay.

purposes, an additional *decoding delay* denoted Δ has to be introduced after which the base station would consider the slots definitely non-decodable. As a result, in our simulator, we compute the rewards of an action selected at time τ, only at time $t = \tau + VF_A + \Delta$, and perform the RM weight update at that time. This is similar to n-step Sarsa [8, section 7.2], with $n = VF_A + \Delta$.

4.5 Reward and Loss Computation

We assume that there are two classes C_0 and C_1, and that the class C_0 has always a higher priority than the class C_1. This priority difference is introduced in the reward computation of each class. As the base station decodes the slots up to time t, (see Fig. 3), it computes the number of decoded agents of each class up to the time t. The associated reward of an agent A who played an action i at the time τ is computed at the time t as follows:

$$
\begin{aligned}
r_i(A) &= P_{C_0,t} & \text{if} & & A \in C_0 \\
r_i(A) &= \alpha P_{C_1,t} + (1-\alpha)P_{C_0,t} & \text{if} & & A \in C_1
\end{aligned}
\tag{4}
$$

where:
$r_i(A)$: is the associated reward of action i from the agent A in the class C.
$P_{C,t}$: is the number of decoded packets of agents of class C up to time t.
α: is the parameter that weights the priority of classes.

On Eq. (4), notice that as α is smaller, the priority of class C_0 is higher. Notice also that the reward of each agent is computed based on the collective amount of decoded packets of all agents (of the same class), and hence they act cooperatively. This is in opposition to the selfish behavior of the agents if the reward was based exclusively on the individual performance of each agent.

Now, more importantly, in IRSA, reward computation is difficult, because the decoding process is iterative: it is difficult to assert if an individual action is responsible for undecoded packets. A straightforward reward is used here: essentially the number of decoded packets (or a function of it). In other RL algorithms, this would not work as the reward would grow linearly with time. In RM, however, only the loss (regret) is used for the updates, and it is the difference of reward between the best action and the taken action. In our case, the loss translates as the number of packets that the taken action had prevented to be decoded. Which is exactly the meaningful information.

But then in addition to computing the actual reward using Eq. (4), corresponding to the played action i, it is necessary to compute the optimal reward that the agent could receive if it played the optimal action at time τ. This has a cost and increases complexity. In the case of a single-agent system, with no delay in update computation, the optimal reward can also be computed at time t by trying all possible actions at a playing time $t-1$ and considering the action that yields the maximum reward as the optimal action to take at the time $t-1$. If the action space is large, this process could already be costly.

However, it is more complicated in a multi-agent system where 1) reward computation is delayed (here: by necessity), 2) where other agents are also interacting in the environment in the interval between the action of one node and

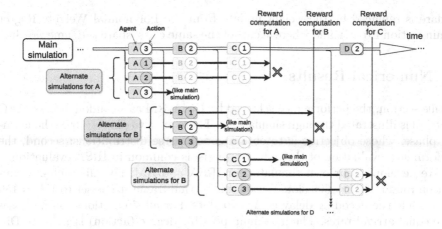

Fig. 4. Alternate simulations for agents A, B, C, and D

its associated reward computation. To handle this, in practice, we maintain one main simulation where each action selected by one agent is actually performed. But we also maintain *alternate simulations* (equivalent to "alternate realities" in mundane terms), that differ from the main simulation only by one action of one agent. Each action of an agent indeed results in creating one new associated alternate simulation for each of its other possible actions (initialized as a copy of the main simulation). At the time of the reward computation for the agent, the reward is computed in each of its alternate simulations: since in its alternate simulations the only difference is the action of the agent (not those of other agents), the difference of reward between different actions can be immediately ascribed to the actions themselves. Figure 4 illustrates alternate simulations, in a scenario where 3 actions $1, 2, 3$ are possible, and where agent A selects action 3, agent B selects action 2, agent C selects action 1, and agent D selects action 2 in the main simulation. The alternate simulations correspond to simulations where one agent selects each of the 2 alternate actions.

Consider an agent A of class C that selected an action $i \in X$ at time τ. Its optimal reward is computed as follows:

$$r_{\text{opt}}(A) = \max_{j \in X} r_j(A) \text{ with } X = \{0, 1, \ldots, D\} \tag{5}$$

And the loss of playing an action i, by an agent A at time τ is computed using the following equation:

$$\ell_i(A) = \frac{r_{\text{opt}}(A) - r_i(A)}{N} \tag{6}$$

where:
r_i, is the associated reward of action i, which is computed using Eq. (4) and the knowledge of the class C of node A.
N: is a normalizing factor, taken to be the total number of users in the system.

We summarize the design of our offline regret minimization-based learning algorithm: it is an adaptation of n-step Sarsa [8, section 7.2], where the Q table update is replaced by the weight update from the Polynomial Weights Regret Minimization Eq. (3), and where agents of the same class, share the same weights.

5   Numerical Results

In this section, the performance achieved by IRSA-RM as a random access MAC protocol is illustrated through simulations. There are two phases: first, the learning phase, whose objective is to obtain good degree distributions; second, the performance evaluation of these distributions as common in IRSA evaluation.

We developed our own simulator for IRSA and RM. For all results, a contention round of $M = 500$ slots is used, the virtual frame size is set to $VF = 150$ slots, while the decoding delay is $\Delta = 50$ slots. For all simulations, both classes have equal arrival rates. The maximum possible degree (action) is $D = 10$. Different cases of class priority are studied; the results are obtained for two classes always, and two different values of the priority parameter: $\alpha = 0.1$ and $\alpha = 0.3$: in both cases, the class C_0 has a higher priority than the class C_1.

We start with the learning phase, whose objective is to find good degree distributions with respect to Eq. (2), interpreted through Eq. (4). In this phase, agents are restricted to one fixed subset X of actions from the set of all possible actions $X \subset \{0, 1, 2, \ldots, D\}$. Several such subsets are selected. For each action subset, several learning processes are run: the total user Poisson arrival rate $G = \mu$ is fixed during each of them, and one learning process is run for each G taken from 0.1 to 1.2 with step 0.1. At the end of each learning process, for each class C, the RM algorithm yields some weights $(w_{C,i}(\bar{G}))_{i \in X}$ from which probabilities to select actions are derived $(p_{C,i}(\bar{G}))_{i \in X}$ which are directly interpreted as lambda distributions (e.g. $\Lambda_i^{\mathrm{RM}}(X, C, \bar{G}) \triangleq p_{C,i}(\bar{G})$). Each learning process is run for $E = 5000$ episodes, and the learning rate η is set to 0.04.

We then compute the performance when applying the obtained distributions. Our main metric is the throughput, e.g. how many decoded packets are recovered per slot. We evaluate the throughput for different loads: but for these simulations, the load does not represent a Poisson arrival rate, but an exact load $g = \frac{N}{M}$, e.g. there are $g \times M$ users exactly, as common in IRSA performance evaluation. We represent throughput versus load in figures, as is done in [4, fig. 5] for instance.

For each selected action set, the throughputs of each of the classes C_0 and C_1 are computed from an average of 300 simulations for a given load g. Note also that for a given load g, one uses the distribution $(\Lambda_i^{\mathrm{RM}}(X, C, \bar{G}))_{i \in X}$ obtained in the learning phase by first selecting the $\bar{G} \in \{0.1, 0.2, 0.3 \ldots 1.2\}$ closer to g.

The scaled throughputs are represented in Fig. 5. The scaling factor is 2, to account for the fact that the actual load of one class is $\frac{1}{2}g$, and to make it comparable to classical IRSA (without classes). Thus, the graph for a "perfect" random access protocol would be a line $y = x$ for $x \in [0, 1]$. The priority parameter α is set to 0.1: this means that the agents of class C_1 would trade 10 lost packets of class C_1 for 1 successfully decoded packet of class C_0.

We selected various action sets with different features: some with a continuous set of degrees $(0, 1, 2, 3, 4)$, some with high degrees, some with a mix of both high

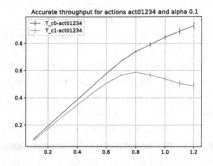

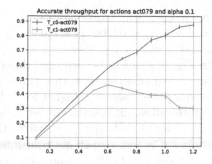

(a) Achieved throughput of both classes, actions [0,1,2,3,4]

(b) Achieved throughput of both classes, actions [0,7,9]

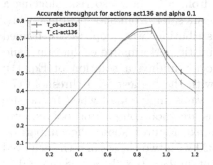

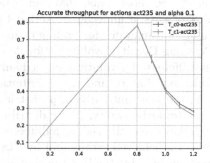

(c) Achieved throughput of both classes, actions [1,3,6]

(d) Achieved throughput of both classes, actions [2,3,5]

Fig. 5. Achieved throughput for frameless IRSA with two classes after using a Regret Minimization based offline learning algorithm

and low degrees. As the class C_1 tends to a strategy that weights 10 times more than the throughput of the class C_0, we expect it to choose the actions that limit the collisions with the packets of the class C_0, if possible, until around the throughput of class C_0 is somewhere 5× to 10× higher.

The Fig. 5 reports the results for 4 different action sets. We are interested in assessing the quality of the priority mechanism introduced by having different distributions for the classes: it can be measured from the gap between the achievable throughput of the two classes. From the results (confirmed by others not presented here), we find that the first defining feature is the inclusion or not of the action 0 in the action set: Fig. 5a and Fig. 5b represent results from two different subsets of actions that include action 0. We can see that the usual sharp decrease of throughput with IRSA around $g = 1$, does not occur for class C_0: only the throughput of class C_1 decreases at higher loads. The priority mechanism is thus working very well, as class C_1 leaves room for class C_0, indeed as its distribution is: $\Lambda^{RM}(\{0,1,2,3,4\},1,1.2) = (\Lambda_0 = 0.594, \Lambda_1 = 0.088, \Lambda_2 = 0.086, \Lambda_3 = 0.101, \Lambda_4 = 0.130)$, with $\Lambda_0 = 0.594$, around 60% of its transmissions are suppressed at load $g = 1.2$.

In Fig. 5c and Fig. 5d, we used different actions sets, this time, without the action 0. We observe that this time, the class C_1 experiments the classical IRSA sharp decrease at a higher load. Introducing action $= 1$ in the set, seems to slightly allow differentiation between classes: for action set $X = \{1,3,6\}$, $\Lambda_1(C_0) = 0.412$ and $\Lambda_1(C_1) = 0.590$, therefore a noticeable amount of packet transmission is just one single transmission (e.g. no repetition). Transmissions with such degree=1 colliding on the same slot cannot be retrieved by SIC, hence automatically result in lost packets (and lost slots). Therefore, at higher loads, the protocol has to find a balance between this phenomenon (wasting slots), and higher degree repetitions, that can benefit from SIC, but also risk blocking a number of slots, if undecoded. Action 1 appears safer, from shown values of Λ_1.

In [4], a framework for finding degree distributions was proposed for framed IRSA, using Density Evolution (for deterministic performance evaluation), and using Differential Evolution (as a heuristic for finding a solution of Eq.(1)): this method aims to find the distribution with the highest load threshold G^*, that is, the load until which, packet loss is vanishingly small when frame size increase towards infinity. These distributions are good comparison points, even though they are optimized for a different context. Figure 6 shows the comparison between the achieved throughput by IRSA-RM with two classes and the achieved throughput by using the IRSA degree distribution $\Lambda_2 = 0.5, \Lambda_3 = 0.28, \Lambda_8 = 0.22$ from [4] (named there "$\Lambda_3(x)$") which we refer to as "external distribution". The Fig. 6a shows a higher achieved throughput for the class C_0 using the learned set of actions $\{0,1,2,4,6\}$ with IRSA-RM compared to an external distribution. This is due to the priority mechanism: using the action 0 a sizeable amount of time, the class C_1 which leaves free slots for the class C_0. This same effect appears in Fig. 6b, thanks to degree $= 1$. In contrast, the achieved throughput for both classes in Fig. 6c and Fig. 6d is comparable to throughput of the external distribution (always close, and for load $g > 0.8$, better for $\frac{1}{4}$ of the points, otherwise worse). Both IRSA-RM and and the external distribution achieve the same maximum load 0.8. This comparison proves that our learning algorithm operate very well in its mission to find good distributions.

Next, we study the impact of the priority parameter α on the achieved throughput in both classes. In Fig. 7a, we compared the achieved throughput of both classes in case of $\alpha = 0.1$ and $\alpha = 0.3$ when using the action set $\{0,7,9\}$. As previously in Fig. 5a, with $\alpha = 0.1$, and action 0 in the action set, the priority mechanisms work well. The gap between the achievable throughput of both classes decreases (in blue) when α is increased to 0.3, as expected. Indeed $\Lambda_0(C_1, g = 1.2)$ decreases from 0.759 to 0.591 (hence action 0 is less used). When action 0 is not available, as as in Fig. 7b with action set $\{1,3,5\}$, the best option for the class C_1 is to favor the other class is select action 1 (as $\Lambda_1(C_1) = 0.594$ for $g = 1.2, \alpha = 0.1$). As explained previously, the impact is still limited as shown by the small gap, and small difference when $\alpha = 0.3$ (and $\Lambda_1(C_1) = 0.538$ for

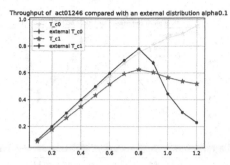

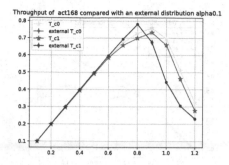

(a) Throughput comp. between actions $\{0, 1, 2, 4, 6\}$ and an external distribution

(b) Throughput comparison between actions $\{1, 6, 8\}$ and an external distribution

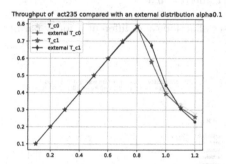

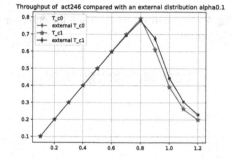

(c) Throughput comparison between actions $\{2, 3, 5\}$ and an external distribution

(d) Throughput comparison between actions $\{2, 4, 6\}$ and an external distribution

Fig. 6. Throughput comparison between different set of actions and an external distribution ($\Lambda_2 = 0.5, \Lambda_3 = 0.28, \Lambda_8 = 0.22$)

$g = 1.2, \alpha = 0.3$). Indeed, the class C_1 has no other choice than sending at least one replica, which will always occupy some slot(s).

Finally, Fig. 8 reports the convergence of the RM learning process. The learning parameter was set to $\eta = 0.04$. Remember that the learning algorithm updates the weights of the actions $(w_i)_{i \in X}$ for each selected action after the proper update delay, and that these weights are used to compute the probabilities $(p_i)_{i \in X}$ of selecting each action according to Eq. (3). Again these are equivalent to a degree distribution Λ. In Fig. 8a, for a network load $G = 0.8$ and $\alpha = 0.1$, we show the evolution of the probabilities during learning for the action set $\{0, 1, 3, 6\}$ at the end of each episode. The probabilities of selecting the smaller degrees 0 and 1 are dropping while the probabilities to use the larger degrees 3 and 6 are rising. The changes stop around episode 3300 where the probabilities start to plateau (it is also true for class C_1). Disregarding action 0 (and to some extent, action 1) is the result of class C_0 attempting to maximize its throughput. On the contrary, probabilities of action 0 and 1 have the inverse behavior for class C_1; notice that because G is not so high, action 0 is still not

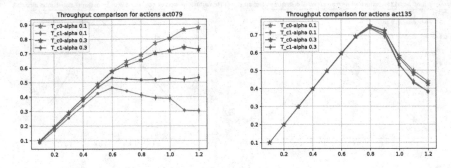

(a) Throughput of actions $\{0, 7, 9\}$ for $\alpha =$ 0.1 and $\alpha = 0.3$

(b) Throughput of actions $\{1, 3, 5\}$ for $\alpha =$ 0.1 and $\alpha = 0.3$

Fig. 7. Throughput comparison for different set of actions and different priority parameter values

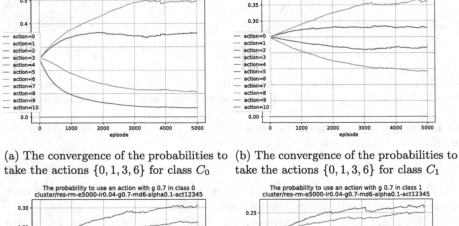

(a) The convergence of the probabilities to take the actions $\{0, 1, 3, 6\}$ for class C_0

(b) The convergence of the probabilities to take the actions $\{0, 1, 3, 6\}$ for class C_1

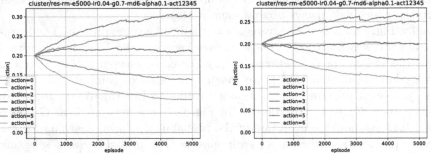

(c) The convergence of the probabilities to take the actions $\{1, 2, 3, 4, 5\}$ for class C_0

(d) The convergence of the probabilities to take the actions $\{1, 2, 3, 4, 5\}$ for class C_1

Fig. 8. The convergence of the probabilities to take the actions for both classes and for different sets of actions

the most selected. In Fig. 8c and Fig. 8d, we show the convergence of the probabilities for another set of actions without the action 0 and for a network load $G = 0.7$ and $\alpha = 0.1$. The probabilities show a form of convergence around the episodes 3100–3200 for both classes. Notice that the learning rate could be a function of the episodes as in 1/(episode index), but for practical purposes, our fixed learning parameter appears sufficient for our learning phase.

6 Conclusion

In this article, we studied one of the modern random access protocols: Irregular Repetition Slotted Aloha (IRSA) in its frameless version. We adapted a reinforcement learning approach based on Regret Minimization to optimize the transmission strategy of this protocol, and thus, proposed the protocol "IRSA-RM". RM is well suited to IRSA, as in both cases, one uses a set of probabilities of selecting a given number of repetitions Λ. The learning is performed offline: it learns the main protocol parameters (the user degree distribution) for a set of predefined network loads. After the learning phase, the parameters can be later used in a network: assuming that the estimate of the load is broadcast by the base station, each device will select the set of parameters that were learned with the closest load. We detailed precisely the mapping between our problem, optimizing IRSA, and the centralized learning approach with RM, including delayed updates, reward computation, and alternate simulations, the introduction of priority classes, etc. Simulation results show a great performance of IRSA when it is optimized with Regret Minimization, and how IRSA-RM behaves for different types of actions (degrees) sets. Future work includes considering richer actions, more sophisticated RM techniques such as CFR, and applying Deep Reinforcement Learning techniques.

References

1. Bloembergen, D., Tuyls, K., Hennes, D., Kaisers, M.: Evolutionary dynamics of multi-agent learning: a survey. JAIR **53**, 659–697 (2015)
2. Blum, A., Mansour, Y.: Learning, regret minimization, and equilibria. In: Nisan, N., Roughgarden, T., Tardos, E., Vazirani, V.V. (eds.) Algorithmic Game Theory, pp. 79–102. Cambridge University Press (2007)
3. Chu, Y., Mitchell, P.D., Grace, D.: ALOHA and Q-Learning based medium access control for Wireless Sensor Networks. In: Proceedings of ISWCS 2012, pp. 511–515, August 2012. ISSN 2154–0225
4. Liva, G.: Graph-based analysis and optimization of contention resolution diversity slotted ALOHA. IEEE Trans. Commun. **59**(2), 477–487 (2011)
5. Paolini, E., Liva, G., Chiani, M.: Coded slotted ALOHA: a graph-based method for uncoordinated multiple access. IEEE Trans. Inform. Theory **61**(12), 6815–6832 (2015)
6. Srivatsa, C.R., Murthy, C.R.: Throughput analysis of PDMA/IRSA under practical channel estimation. In: 2019 IEEE 20th International Workshop on Signal Processing Advances in Wireless Communications (SPAWC), pp. 1–5, July 2019. ISSN 1948–3252

7. Stefanovic, C., Popovski, P., Vukobratovic, D.: Frameless ALOHA protocol for wireless networks. IEEE Commun. Lett. **16**(12), 2087–2090 (2012)
8. Sutton, R.S., Barto, A.G.: Reinforcement Learning: An Introduction. Adaptive Computation and Machine Learning Series, 2nd edn. The MIT Press, Cambridge (2018)
9. Wang, S., Liu, H., Gomes, P.H., Krishnamachari, B.: Deep reinforcement learning for dynamic multichannel access in wireless networks. IEEE Trans. Cogn. Commun. Netw. **4**(2), 257–265 (2018)
10. Zinkevich, M., Johanson, M., Bowling, M., Piccione, C.: Regret Minimization in Games with Incomplete Information, page 8
11. Casini, E., De Gaudenzi, R., Del Rio Herrero, O.: Contention resolution diversity slotted ALOHA (CRDSA): an enhanced random access scheme for satellite access packet networks. IEEE Trans. Wirel. Commun. **6**(4), 1408–1419 (2007)
12. Fooladivanda, D., Al Daoud, A., Rosenberg, C.: Joint resource allocation and user association for heterogeneous wireless cellular networks. IEEE Trans. Wirel. Commun. **12**, 384–390 (2011)
13. Ge, X., Li, X., Jin, H., Cheng, J., Leung, V.C.M.: Joint user association and user scheduling for load balancing in heterogeneous networks. IEEE Trans. Wirel. Commun. **17**, 3211–3225 (2018)
14. Luong, N.C., et al.: Applications of deep reinforcement learning in communications and networking: a survey. IEEE Commun. Surv. Tutor. **21**(4), 3133–3174 (2019)
15. Naparstek, O., Cohen, K.: Deep multi-user reinforcement learning for distributed dynamic spectrum access. IEEE Trans. Wirel. Commun. **18**, 310–323 (2019)
16. Destounis, A., Tsilimantos, D., Debbah, M., Paschos, G.S.: Learn2MAC: online learning multiple access for URLLC applications. In: IEEE INFOCOM 2019 - IEEE Conference on Computer Communications Workshops (INFOCOM WKSHPS), Paris, France, pp. 1–6 (2019)
17. Toni, L., Frossard, P.: IRSA Transmission Optimization via Online Learning (2018)
18. Wang, L., Xiao J., Guanrong, C.: Density evolution method and threshold decision for irregular LDPC codes. In: International Conference on Communications, Circuits and Systems (IEEE Cat. No.04EX914), Chengdu, vol. 1, pp. 25–28 (2004). https://doi.org/10.1109/ICCCAS.2004.1345932
19. Hmedoush, I., Adjih, C., Mühlethaler, P., Kumar, V.: On the performance of irregular repetition slotted aloha with multiple packet reception. In: International Wireless Communications and Mobile Computing (IWCMC), Limassol, Cyprus 2020, pp. 557–564 (2020). https://doi.org/10.1109/IWCMC48107.2020.9148173
20. Nguyen, T.T., Nguyen, N.D., Nahavandi, S.: Deep reinforcement learning for multiagent systems: a review of challenges, solutions, and applications. IEEE Trans. Cybern. **50**(9), 3826–3839 (2020)
21. Klos, T., van Ahee, G.J., Tuyls, K.: Evolutionary dynamics of regret minimization. In: Balcázar, J.L., Bonchi, F., Gionis, A., Sebag, M. (eds.) ECML PKDD 2010. LNCS (LNAI), vol. 6322, pp. 82–96. Springer, Heidelberg (2010). https://doi.org/10.1007/978-3-642-15883-4_6

Mobility Based Genetic Algorithm for Heterogeneous Wireless Networks

Kamel Barka[1] , Lyamine Guezouli[1]([envelope]) , Samir Gourdache[1] ,
and Sara Ameghchouche[2]

[1] LaSTIC Laboratory, University of Batna 2, Batna, Algeria
{kamel.barka,lyamine.guezouli,s.gourdache}@univ-batna2.dz
[2] RIIR Laboratory, University of Oran 1, Oran, Algeria
ameghchouche-sara@edu.univ-oran1.dz

Abstract. In heterogeneous wireless networks, collaboration between mobile and static elements allows optimal exchange (in terms of latency, reliability and energy savings) ensured by the mobile elements of the data collected with precision by the static elements. In this article, we focus on this collaboration. We base our study on genetic algorithms to select the best next destination from the event area according to different criteria, such as the amount of data collected and where the passage of the UAVs (Unmanned Aerial Vehicles) is delayed. Choosing the best destination ensures better latency and a high level of received data. The events and characteristics of the event area will be sensed by the static elements (this is to ensure optimal precision since the static element will be placed in the desired location to be studied), while the mobile elements will be charged to collect the data sensed by the static elements, and to ensure their routing to the collecting station. The results confirm the effectiveness of our collaborative approach compared to a solution based on random mobility of mobile elements.

Keywords: Heterogeneous wireless networks · Mobility · Nodes collaboration · UAV · Genetic algorithm

1 Introduction

UAVs technology applied to agriculture has significant added value, since its generalization could contribute to better value for resources. According to FAO (Food and Agriculture Organization), food forecasts due to world population growth rates are of concern. This organization estimates that by 2050, it will be necessary to increase current agricultural production by 70% to feed the entire planet, so the use of remote sensing, as a technology to increase production, can make a decisive contribution to this challenge for society at the global level.

At the same time, this growth in agricultural production must be accompanied by a rationalization of the consumption which is often used indiscriminately and massively on many crops. Remote sensing can help to make the consumption of fertilizers and herbicides more efficient. There is also the issue of pesticides, which is also another

É. Renault et al. (Eds.): MLN 2020, LNCS 12629, pp. 93–106, 2021.
https://doi.org/10.1007/978-3-030-70866-5_6

point of concern, since they have a very relevant impact on the planet, and anything that is working in the line of optimizing the use of pesticides to treat pests is also a substantial contribution to the environmental challenge in the coming years.

Another fundamental challenge for the coming decades will be the proper management of the world's water resources. UAVs are an important tool for a more equitable distribution of water resources. UAVs can now be used as a source of useful and up-to-date information on water resources and infrastructure. They can be useful in quantifying and preventing droughts, floods and water-related events. They will allow us to monitor crops, to find a way to conserve nature. Ultimately, drones close the gap between satellite and field agriculture.

For agricultural management, before that, the farmer walks around his agricultural field. With the sun, rain, cold… It is impossible to calculate how many kilometers he will travel on his legs in so many days outdoors. In the past, field supervision meant walking in these fields and spending many hours supervising them. As soon as there was a problem, either a sheet was taken and analyzed or, if it was easy to detect, treatments were quickly started.

Nowadays, in a few minutes, the drone compiles all the information that the farmer would have taken several days to collect while walking in the fields. One of the challenges is how the drone must mobilize. The farmer only has to specify the parcel he wants to supervise, and the drone, through an appropriate mobility model, moves on the parcel. The flight time varies according to the hectares and the drone battery. In a short time, the drone collects all the information that would have taken several working days to the farmer.

In this article, we deploy static wireless sensors in the network to take the various necessary measurements for the farmer, then we inject drones that circulate on this parcel to collect the data sensed by the sensors. To define the drones' mobility, we define a mobility model based on genetic algorithms in order to optimize the drones' movement and to supervise the agricultural parcel in a reduced time.

The article is organized as following: The second and third parts are devoted to present the works related to random or controlled mobility models and the work related to the agricultural field. The fourth part presents the algorithm of our proposed solution based on controlled mobility. In the fifth part of the article, we explained the process of our proposal with diagrams. The results of the simulation evaluation are presented in sixth part. A conclusion and perspectives are discussed in the last part of this article.

2 Related Works

2.1 Based Random Mobility Models

In these mobility models, the nodes move toward a destination randomly selected, and the nodes reach it using a speed and a direction, which are also random [1]. Below are some examples of random models used by researchers.

Random Walk. The Random Walk [2] was created to describe the irregular and extremely unpredictable movement that can have different entities in nature. In this model a node moves from its current position to a new position, randomly choosing a direction and a speed at which to travel. The speeds and directions of the nodes are chosen in a predefined range, respectively [velocityMin, velocityMax] and $[0, 2\pi]$. Each movement occurs during a time interval t, or a constant traveled distance d, at the end of which each direction and speed is recalculated.

If a node reaches an edge of the simulation during its movement, it rebounds at an angle determined by the incoming direction. Throughout the simulation, the nodes move in this way. In this model, the nodes always move around the starting point with a probability of 1, which ensures that no one moves away. There are two main problems with this model: first, there is no information on where it has been and what its speed was, because the current speed and direction are independent of the previous ones. Also, if you choose small parameters (a small fraction of the time, or a few steps before changing direction), the movement will occur only in a portion of the simulation area.

Random Waypoint. The Random Waypoint mobility model [2] adds pauses between changing direction and speed. A node randomly chooses a starting point and stops at this point for a certain period of time. Then the node selects a random destination within the simulation area and a uniformly distributed speed in the range [Min speed, Max speed] and travels to the new destination at the selected speed. After arrival, it stops for a fraction of the time, then resumes the process. The movement of the nodes in the Random Waypoint, is like the movement in the Random Walk, if the time pause was zero and the speed intervals were the same.

A variant of Random Waypoint is defined in [3]. The purpose is to improve routing. Indeed, the authors based their solution on the exploitation of the pause time marked by a mobile entity that are close to the base station. During the pause time, the mobile entity exchanges its data with the BS (BS) and also acts as a relay between its neighbors and the BS. This is in order to expand the BS coverage.

Authors in [4] and [5] propose new protocols for wireless sensor networks (WSNs). The mobile entity (drone or actor) moves according to RWP (Random Waypoint) mobility model. Each mobile entity during its pause time creates a temporary cluster, collects and processes sensor data and performs actions on the environment based on the information gathered from sensor nodes in its cluster. Once a mobile entity detects a base station (BS) it forwards the collected data to it. The aim is to minimize both the non-covered area and the number of actor nodes deployed in the network, and to achieve low latency and high delivery ratio, while taking advantage of the resources available on the mobile entity and their mobility.

2.2 Based Controlled Mobility Models

In the Random Waypoint the nodes can move freely in the simulation area. However, in real life applications, the movement of the node is often subject to the environment in which it is located. Thus, the nodes can move in a pseudo-random way on predefined paths in the simulation area.

In controlled mobility, generally a predefined trajectory is calculated in advance and loaded into each mobile unit. This path must be followed by the mobile units without making a random movement. At the end of this planned path, the mobile unit can change direction and repeat the same process.

SDPC (Self-Deployable Point Coverage) [6] is a topological mobility model for networks containing UAVs. It is a model designed to provide connectivity between UAVs where the goal is to optimize coverage of a large number of mobile nodes located on the ground. Indeed, to ensure such a scenario, it is necessary to consider an optimal positioning of UAVs to cover the largest possible area. An application example of this model is the deployment of UAVs in a disaster area to create alternative infrastructures.

DPR (Distributed Pheromone Repel) [6] is a model that uses pheromones and localized research to ensure that drones visit areas not recently visited. The principle is like that of natural pheromones, a drone leaves virtual pheromones on the area it visited that evaporated over time. UAVs share a pheromone card to ensure the synchronization of their movements. The probability that an UAV will choose one region over another is calculated by considering the pheromone maps exchanged between UAVs. This model can be used in various applications using drones for search and rescue.

2.3 For Agricultural Fields

In the field of precision agriculture, to obtain aerial images of a field, it is necessary to refer as well to images taken either by satellites or by aircraft. Ordinary aircraft are limited by the resolution of their images and the frequency with which they fly over a field. The authors in [7] studied the use of drones to obtain precise results for agriculture. Indeed, this precision resides, for example, in targeting areas where fertilizers and pesticides are applied in areas that do not need them. The authors compared the results precision with those taken by an ordinary aircraft.

In the same field of precision agriculture, drones or UAVs equipped with sensors, multispectral cameras and specific fertilizers are used to reduce human activities and face parasites or sudden climate changes issues by offering new solutions for continuous monitoring. The authors in [6] propose a recruitment protocol based on a bio-inspired approach that seems to have good performance in terms of parasites killed and to be scalable in terms of bytes sent over the FANET network. To highlight the performance of the proposed protocol, the authors vary different parameters in order to appropriately adjust its implementation. There are many sensors that perform measurements for a variety of plants, soils and climate variability. And High-altitude Access Point (HAP) use the results of these sensors to improve production. In addition, Internet of Things Underground is used for real-time detection and monitoring, specially designed for planting and field operations.

3   Mobility Model Based Genetic Algorithm for Heterogeneous Wireless Networks (MGH)

In this section, we present our proposal to ensure node mobility in heterogeneous environment, note that this proposal is a continuation of work already published in [4].

3.1 Event Area Modeling

In this step, the BS represents the event area through a set of reference points $Z = \{P_1, P_2,..., P_n\}$.

To identify the coordinates of these points, the BS (with the coordinates (0,0,0)) divides the area into several identical and contiguous cubes (see Fig. 1), so that the length of each edge is calculated as follows:

$$k = \sqrt{2(R^2 - r^2)} \qquad (1)$$

With k: the length on the edge of each cube,

   R : the smallest communication range of the drone

   r : the communication range of the static sensor.

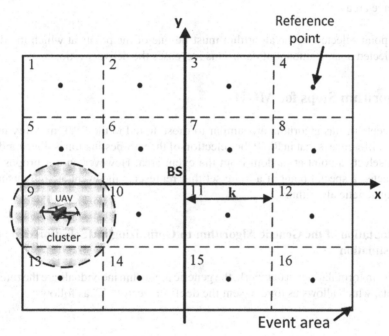

Fig. 1. Top view of reference points representing the event area

Since the reference points represent the gravity centers of these cubes, the event area can be represented by this set:

$$Z = \left\{ P_{ij}\left(i\frac{k}{2}, j\frac{k}{2}, z\right), i\,et\,j \in \mathbb{Z}\,et\,z : constant \right\} \qquad (2)$$

3.2 Cluster Creation

For this step, we followed the same mechanism of cluster creation presented in [4]. This cluster creation mechanism will be done at the pause time of the drone.

3.3 New Destination Selection

After the destruction of the current cluster, the drone chooses a new destination (another reference point). In order to ensure the best destination choice, we consider that the drone uses a genetic algorithm to select the point that verifies the following requirements:

- Minimize the distance between the current reference point and the new chosen, this reduces energy consumption and also latency
- For a period of time (T), this genetic algorithm must select a point that allows the drone to fly over the BS, allowing it to send the collected data to the BS. This increases the delivery ratio and reduces latency.
- The point selected by the algorithm must be one of the oldest points chosen by the other drones, we must avoid selecting the points recently chosen. This maximizes the coverage area

The point selected by the algorithm must be one of the points at which the drones have collected a large amount of data. This increases the delivery ratio.

4 Algorithm Steps for MGH

The concepts of this algorithm are similar to those found in the RWP mobility model. The only difference is that in RWP the selection of the new destination is done randomly where it selects a point at random from the event area. However, in our proposal, we have selected a special point of a finite set that represents the event area, by using the following genetic algorithm:

4.1 Adaptation of the Genetic Algorithm to Optimizing the Selection of a New Destination

1) Problem formulation: In our work, the genetic algorithm individuals are the reference points, which allows us to represent the decision vector "X" as follows:

$$\vec{X} = \begin{pmatrix} x \\ y \\ z \end{pmatrix} \tag{3}$$

We have defined the optimization problem of selecting a new destination as a problem of three (3) objectives that are:

- The last time of passage (t_{last}): is the last time when a drone created a cluster on this point
- The quantity of data collected (d_c): is the number of data that a drone collected during its last passage through this point

- Distance (d_E): is the Euclidean distance between this point and another reference point

with the first two to maximize and the last to minimize. We can then represent the objective function vector f(X) as follows:

$$f(\vec{x}) = \begin{cases} maxf_i(x) & i = 1, 2 \\ minf_i(x) & i = 3 \end{cases} \tag{4}$$

With:

$$f(\vec{x}) = \begin{pmatrix} maxf_1 = max(t_{last}) \\ maxf_2 = max(d_c) \\ minf_3 = min(d_E) \end{pmatrix} \tag{5}$$

$$orf(\vec{x}) = \begin{pmatrix} maxf_1 = max(t_{last}) \\ maxf_2 = max(d_c) \\ minf_3 = max(-d_E) \end{pmatrix} \tag{6}$$

2) Coding the chromosomes: In our case, the decision variables are binary variables representing the reference point numbers, we can represent them by a size vector (l) which is calculated by the following equation:

$$l = floor(log_2(|Z| - 1)) + 1 \tag{7}$$

With |Z| the cardinality of the set Z (i.e., the number of reference points) and floor designates the function integer part. So, in this case the set Z contains 32 reference points then $l = 5$, i.e. the chromosome containing at most 5 bits. Thus, point number 2 is represented by chromosome 00010.

3) Creation of the initial population: In general, the initial population is created during the deployment of drones, where we use a mechanism that ensures a good balance between diversity and good quality solutions. The creation of the initial population is as follows:

The size of the initial population is fixed. It is composed of eight (8) reference points. To ensure diversity, we randomly select two points from each region (east, west, north and south). To facilitate the task, we have divided the area into 4 squares, then we randomly choose 8 points so that each two points belong to a square.

4) Points evaluation: In this step, the reference points of the initial population are evaluated by using the following fitness function:

$$\begin{cases} f_i^j = \dfrac{\Delta t^j * dc^j}{(x_j - x_i)^2 + (y_j - y_i)^2} \\ \Delta t = time - t_{last}^j \end{cases} \tag{8}$$

With Δt is the time that has elapsed since the last time the point P_j was visited, and dc is the last amount of data collected at the point P_j,

5) Point selection: In this step, we choose the strongest reference points (i.e., best fitness scores) to produce the best performing children. To do this, we sort the population's points in descending order, then we select the first four points.
6) Crossover operator: In this step, we applied the crossover operation on each parent pair of four points chosen in the selection step. This crossing uses two parents to form two Offspring. Both parents are chosen according to the Euclidean distance between them (Eq. 9) where the distance between the parent couple is as short as possible (i.e.; each two close points represents a couple of parents).

$$De\left(P_i, P_j\right) = \sqrt{\sum_{t=1}^{3} \left(c_t' - c_t\right)^2} \tag{9}$$

With c_t and c'$_t$ represent the coordinates of the reference points P_i and P_j respectively.

Indeed, these new reference points (individuals) are formed by crossing between two parents. A random position k is chosen between $[1; l-1]$ where l is the length (Eq. 7). The crossing is done by exchanging the bits from position $k + 1$ to l, i.e., these bits have passed from one individual to another to form two new individuals. This crossing then takes place at this location with a probability P_c.

7) Mutation Operator: mutation is the occasional random (i.e., low probability) modification of the bit value (i.e., inversion of a bit). We draw for a random bit in the interval [0, l], a random number between 0 and 1 and if this number is less than pm then the mutation takes place.
8) Replacement operator: Then we apply the replacement operator who decides to replace 50% of the population P. Since the individuals in the population are sorted in descending order, it is therefore sufficient to replace the points of the last half of the population by the new points (Offspring), and the size of the population remains the same. In this way we keep the best reference points (elitism).
9) Termination: We must repeat the procedure from the evaluation stage until the pause time has elapsed (i.e., the stopping criteria has been met) in which case the drone chooses the population's first reference point as the new destination.

4.2 Passage Over BS

When a drone moves from one reference point to another (Fig. 2), it may be that it does not fly over the BS for a long period of time, which results in data loss (i.e., reduce the delivery ratio), so this data does not reach the BS in a suitable time (i.e., latency becomes important). For this purpose, our MGH algorithm has a mechanism that selects a reference point (P) allowing the drone to fly over the BS at least once during a predefined period of time T. This mechanism is explained as follows:

It is assumed that the drone is at reference point $P_1(x_1,y_1,z_1)$. In order for this drone to fly over the BS $(0,0,z)$, it is therefore sufficient to refer to the objective function (f), allowing to select a point from the set of points $P(x,y,z)$ which verifies the following constraints:

$$\begin{cases} P(x, y, z) \in Z \\ x = M * x_1 \\ y = M * y_1 \\ z = z_1 \\ M \in \mathbb{R}^- \end{cases} \tag{10}$$

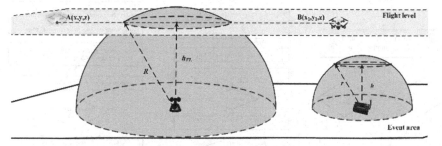

Fig. 2. Drone on point A selects point B to fly over the BS

In the worst case, the possible solution to this problem is the reference point BS$(0,0,z)$.

4.3 Updating Reference Point Parameters

We consider two types of updates, one done by the drone and another by the BS. Once the pause time of the drone expires in a reference point, this drone updates the parameters of this point (i.e.; save the new t_{last} and dc value), and as soon as this drone detects the BS, the drone sends (to the BS) both the collected data and the new parameters from where it has recently stopped. In turn, the BS returns the latest updates from other drones.

4.4 Network Operation

According to the work in [4], after deployment of the sensor nodes in the event area, each sensor node constructs a neighborhood table of the 1-hop neighbors with the default rank value. When there are many UAVs on the network, each UAV flies in its FL and starts the first phase of clustering. Each sensor node in the covered areas can hears many UAVs. To deal with situation it saves the rank assigned by each UAV heard (cluster-head) during its pause time i.e., each sensor node will construct a dynamic list of the UAVs that it heard (a sensor node can be a member in several clusters), this list contains the IDs of the UAVs heard, their rank values and their expiration dates. Hence, when a sensor node has a data to send, it looks in its list of UAVs and picks the UAV destination with which it has the smallest rank. Once a rank is expired, the node automatically removes the ID of the corresponding UAV from its list.

5   MGH Execution Process

The algorithm steps are shown in the Figs. 3, 4 and 5.

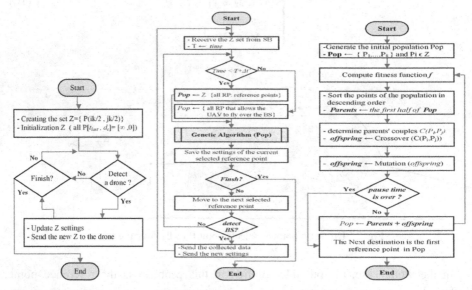

Fig. 3. Execution process on the BS

Fig. 4. Execution process on the UAV.

Fig. 5. Execution process of the Genetic Algorithm.

6   Simulation Evaluation

This section shows a series of simulation results to evaluate the performance of MGH. All simulations were carried out with the TOSSIM simulator [8].

The simulation consists of 144 stationary sensor nodes in the first configuration (resp. 961 in the second configuration) deployed uniformly in a 300×300 m event area in the first configuration (resp. 1000×1000 in the second configuration). A base station located at the center of the event area, and set of UAVs nodes are deployed in this event area. The transmission range of sensor, base station and UAV node is 30, 70 and 70 m, respectively. Each sensor node sensing data at each duty cycle (2 min) and carry the data to an UAV node that is created a cluster on a hop-by-hop basis. Each UAV flies at a constant velocity (10 m/s) according to the protocol SSP (Self-organization Smart Protocol) [4] with pause time equal 2 min. Table 1 summarizes the main parameters of the network.

Table 1. Simulation parameters

Parameter	Value
Topology	*Grid (300 × 300) //Config. 1* *Grid (1000 × 1000) // Config. 2*
Static sensor nodes	*144 // Configuration 1* *961 // Configuration 2*
UAV nodes	*1 and more*
Sensor transmission range	*30* m
UAV transmission range	*70* m
BS transmission range	*70* m
UAV velocity	*10* m/s
Pause time (T_p)	*2* min
Switchover time at the BS (T)	*6* min
Probability of crossing Pc	*1*
Probability of mutation Pm	*0.05*

6.1 Delivery Ratio Evaluation

In this Section, the focus is on the delivery ratio when we increase the number of UAV nodes deployed in the network.

Figure 6 indicates the UAV's network density influence on the ratio of the successfully delivered packet. For the first configuration (300 × 300), as represented, MGH solution outperforms SSP right from the beginning and possesses the highest ratio of the successfully delivered packet due to less amount of packet drops in the network. With MGH, a reduced number of UAVs makes it possible to reach 100% of the receptions. MGH enables a good selection of destination which increases the packet delivery ratio.

For the second configuration (event area 1000 × 1000), we notice an important loss of packets with a reduced number of UAVs, but just with 10 UAVs MGH process the highest ratio of the successfully delivered packet. Obviously, the effectiveness of MGH is due to the stop criterion (linked to the pause time) which does not allow it to exceed the cluster creation time. Also, the probability that the same destination will be selected by several UAVs is very low by this algorithm.

6.2 Latency Evaluation

In this Section, the focus is on the latency when we increase the number of UAV nodes deployed in the network. We define the latency as the average time required for a message from a source node (sensor) to reach the final destination (the base station) via the intermediate node (UAV node).

As represented in Fig. 7, MGH solution has a marginally moderate latency from SSP. We also note that, obviously, the injection of several UAVs into the environment makes it possible to optimize reception latency. This is due to the very low probability to detect a

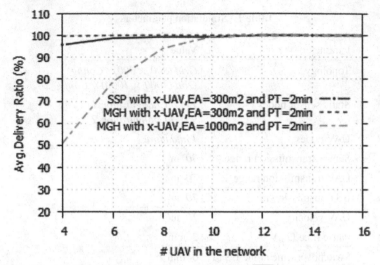

Fig. 6. Average delivery ratio for MGH and SSP

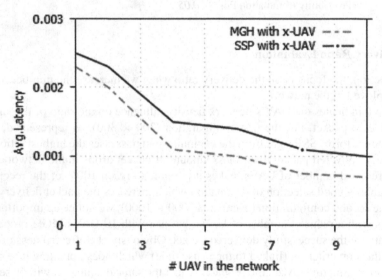

Fig. 7. Latency for MGH and SSP

base station (by UAV) when there is a small number of UAVs, and it is therefore unlikely that a UAV will transmit its collected data to a base station. However, what concerns us is that the latency is rather optimal in the MGH solution compared to the random SSP solution, this is explained by the fact that the MGH algorithm forces the UAV to join the BS (after a certain time of collecting data), and also with MGH algorithm UAV must choose a destination where the passage of the UAVs is delayed.

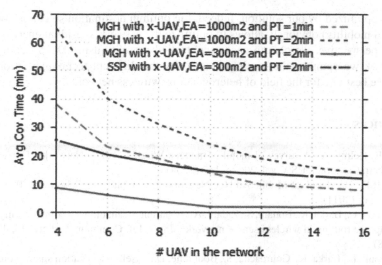

Fig. 8. The average coverage time for MGH and SSP

6.3 Average Coverage Time of the Event Area

Coverage time is defined as the time required for the UAVs to discover all the static sensors deployed in the event area.

For the first configuration (300 × 300), based on the incremental number of UAVs, Fig. 8 represents the time required for the BS to discover all the static sensors. For the random approach (SSP), the discovery will take a considerable amount of time, it is due to the random trajectory followed by the UAVs. For MGH solution, we notice a successful discovery of all static sensors after a truly minimal delay.

Nevertheless, if our study covers a large event area (second configuration 1000 × 1000), the discovery of all static sensors is better time-wise according to the MGH solution. Indeed, while the pause time was 2 min, this time is sufficient for MGH algorithm to select the next destination. However, when we decreased the pause time to one minute, we noticed an optimization in MGH solution, this is because the time to select the next destination by the MGH algorithm will not exceed the pause time specified at one minute, which has a positive effect to the event area average coverage time.

7 Conclusion

For our work, we have specified several goals to be achieved by our MGH solution, which are to ensure energy efficiency, low latency, high success rate, and suitably interactions between static sensors and UAVs. Indeed, the results obtained by simulations reflect the success in proving the effectiveness of MGH solution compared to SSP, which is based on a strictly random mobility. Low latency, and high success rate are proven by simulation results. We have been able to maintain a sufficient number of UAVs to ensure good performance. However, for the energy efficiency metric, we can deduce that this metric is optimized by our solution. Indeed, the optimal destination selection (through controlled mobility) obviously allows to optimize the network's residual energy.

As a perspective, we plan to ensure controlled mobility using Deep Learning. In future work we compare the results obtained by the different techniques in order to deduce the best one for the field of heterogeneous wireless networks.

References

1. Bai, F., Helmy, A.: A survey of mobility models, Wireless Adhoc Networks. University of Southern California, USA, vol. 206, p. 147 (2004)
2. Roy, R.R.: Handbook of Mobile Ad Hoc Networks for Mobility Models, vol. 170. Springer, New York (2011)
3. Guezouli, L., Barka, K., Bouam, S., Zidani, A.: A variant of random way point mobility model to improve routing in wireless sensor networks. Int. J. Inf. Commun. Technol. **13**, 407–423 (2018)
4. Guezouli, L., Barka, K., Gourdache, S., Boubiche, D.E.: Self-organization smart protocol for mobile wireless sensor networks. In: 2019 15th International Wireless Communications & Mobile Computing Conference (IWCMC), pp. 1002–1006 (2019)
5. Barka, K., Guezouli, L., Gourdache, S., Boubiche, D.E.: Proposal of a new self-organizing protocol for data collection regarding mobile wireless sensor and actor networks. In: 2019 15th International Wireless Communications & Mobile Computing Conference (IWCMC), pp. 985–990 (2019)
6. Sanchez-Garcia, J., Garcia-Campos, J., Toral, S., Reina, D., Barrero, F.: A self organising aerial ad hoc network mobility model for disaster scenarios. In: 2015 International Conference on Developments of E-Systems Engineering (DeSE), pp. 35–40 (2015)
7. Kuiper, E., Nadjm-Tehrani, S.: Mobility models for UAV group reconnaissance applications. In: 2006 International Conference on Wireless and Mobile Communications (ICWMC 2006), p. 33 (2006)
8. Stehr, N.J.: Drones: the newest technology for precision agriculture. Nat. Sci. Educ. **44**, 89–91 (2015)
9. Tropea, M., Santamaria, A.F., Potrino, G., De Rango, F.: Bio-inspired recruiting protocol for FANET in precision agriculture domains: pheromone parameters tuning. In: 2019 Wireless Days (WD), pp. 1–6 (2019)
10. Levis, P., Lee, N., Welsh, M., Culler, D.: TOSSIM: accurate and scalable simulation of entire TinyOS applications. In: Proceedings of the 1st International Conference on Embedded Networked Sensor Systems, pp. 126–137 (2003)

Geographical Information Based Clustering Algorithm for Internet of Vehicles

Rim Gasmi[1(✉)], Makhlouf Aliouat[1], and Hamida Seba[2]

[1] Faculty of Sciences, LRSD Laboratory, Ferhat Abbas University Setif 1,
Setif, Algeria
gasmi34000@gmail.com, aliouat_m@yahoo.fr
[2] LIRIS, CNRS Lyon 1, University of Lyon, Lyon, France
hamida.seba@univ-lyon1.fr

Abstract. Nowadays, Internet of Vehicles (IoV) are considered as the most important promoter domain in the Intelligent Transportation System (ITS). Vehicles in IoV are characterized by a high nodes' mobility, high nodes' number and high data storage. However, IoV suffer from many challenges in order to achieve robust communication between vehicles such as frequent link disconnection, delay, and network overhead. In the traditional Vehicular Ad hoc NETworks (VANETs), these problems have often been solved by using clustering algorithms. Clustering in IoV can overcome and minimize the communication problems that face vehicles by reducing the network overhead and ensure some Quality of Service (QoS) to make network connectivity more stable. In this work, we propose a new Geo-graphical Information based Clustering Algorithm "GICA" destined to IoV environment. The proposal aims to maintain the cluster structure while respecting the quality of service requirements as the network evolves. We evaluated our proposed approach using the NS3 simulator and the realistic mobility model SUMO.

Keywords: IoV · IoT · Clustering · Smart vehicles · Robust communication

1 Introduction

In the last decade, the world has seen a great development in the vehicle industry, which in turn increases the number of vehicles vastly. This development has surely made the lives of peoples easier and has even accelerated the economic growth around the world. In the other hand, it has also caused some challenges, such as traffic congestion, traffic accident, energy consumption and environmental pollution [1]. These issues led researchers to create a new concept called Vehicular Ad hoc Network (VANET) that allows vehicles to communicate with each other using Vehicle to Vehicle communication mode (V2V), with road Infrastructures using Vehicle to Infrastructure communication mode (V2I) [2].

© Springer Nature Switzerland AG 2021
É. Renault et al. (Eds.): MLN 2020, LNCS 12629, pp. 107–121, 2021.
https://doi.org/10.1007/978-3-030-70866-5_7

The emergence of the Internet of Things (IoT) opens the door to the conventional VANETs towards a new paradigm called the Internet of Vehicles (IoV). IoV is considered as a new Intelligent Transportation System paradigm [3]. IoV extends VANETs communication types, networks technologies and applications. This development creates new interactions at the road level among vehicles, humans and infrastructures. However, IoV still suffer from several drawbacks such as the high vehicles dynamicity that increases the frequent link disconnections. The high number of vehicles connected to IoV in smart city, leads to a high growth in network overhead. Clustering algorithms are the most used method to reduce network overhead and maximize network stability [4]. The concept of these algorithms is based on dividing network into several groups and each group is headed by a chosen group member. This new network scheme decreases the number of messages exchanged, where the network topology is already known and node does not need to broadcast route request to all network nodes in order to find a new route. In this paper, we propose a new Geographical Information based Clustering Algorithm "GICA" destinated to IoV environment. The proposed algorithm aims to find the stable nodes in the network in order to be selected as Cluster Head.

2   Related Work

Several works have been proposed to address scalability issues in vehicular networks. Self-Organized Clustering Architecture for Vehicular Ad Hoc Networks (SOCV) [5], is a one-hop clustering protocol that ensures safe network communications in VANETs. The authors create a dynamic virtual backbone in the network that includes Cluster-Heads and cluster-gateways nodes. These nodes guarantee efficient message propagation in the network.

Destination Based Routing (DBR) algorithm for context-based clusters in VANETs [6] combines two mechanisms. The First one is based on direction, relative velocity, interest list and final destination. The second one is a destination-based routing protocol used to enhance inter-cluster communication. Using context-based clustering that includes interest list of vehicles, the overall end-to-end communication is significantly improved.

QoS Based Clustering for Vehicular Networks (QoSCluster) [7] is a multi-metric clustering algorithm. The authors propose two kinds of metrics to calculate the weight value: QoS Metrics that include average link expiration time, average link bandwidth and average time to completion, and Stability Metrics that include the level of connectivity, relative mobility and the average relative distance. The authors compared the proposed scheme to DMCNF, the obtained results show the effectiveness and robustness of QoSCluster.

Mobility adaptive density connected clustering algorithm (MADCCA) [8] uses traffic density to elect cluster heads. The Key parameters used in MADCCA are vehicle's velocity and node density.

An efficient hierarchical clustering protocol (EHCP) for multi-hop communication is introduced in [9]. The goal of this proposal is to ensure effective resource use and to enhance the network lifetime. The authors assume that the vehicles are connected to the Internet through road-side unit gateways. By using Internet connection, each vehicle can collect information about its neighboring nodes then it executes the clustering algorithm to elect the appropriate Cluster Heads. To elect Cluster Heads, the EHCP uses two different parameters: link expiration time and the vehicle's relative degree.

Distributed Clustering Algorithm Based on Dominating Set (DS) for Internet of Vehicles, called DCA-DS is presented in [10]. The authors use the node span parameter to select the Cluster Head. The node span parameter is defined as the number of neighboring nodes that belong to no cluster, including the node itself. DCA-DS uses simple heuristic method as well as a greedy strategy. The node that has the highest span value is added to the DS, which later takes the role of Cluster Head and the rest of its neighbors become Cluster Members. The process is iteratively repeated until all the nodes are clustered.

A Segmented Trajectory Clustering-Based Destination Prediction mechanism is proposed in [11]. The authors propose destination prediction-based trajectory segmentation algorithm to segment each original trajectory to different sub-trajectories. The sub-trajectories are clustered based on the average nearest point pair distance to reveal the common characteristics or similar tracks. Then the authors use a deep neural network based on the history trajectories in order to predict destinations.

3   Motivation

The majority of clustering algorithms existing in literature are generally based on one-hop clustering. This means that the communication process is only between a cluster member and its Cluster Head (1-hop distance at most). The use of such method minimizes the Cluster Head coverage area, thereafter, several clusters are established. The high number of formed clusters in the network leads to a cluster overlapping problem and decreases the network performance. Moreover, vehicular networks are characterized by a high topology change caused by a high vehicles' mobility and a limited driving directions. Most proposed schemes in literature do not take in consideration the combination between vehicle mobility and destination information to form clusters. Moreover, many proposed clustering approaches use high number of control messages in order to exchange vehicles parameters, which causes network overloading. This leads to several collisions, especially if the proposed system uses multi-metric mechanism that needs a high number of messages. In this work, we propose a geographical information-based clustering algorithm, called "GICA", destined to an IoV environment. GICA is based on three metrics: destination, mobility and position. It uses a multi-hop clustering mechanism to connect cluster members with their Cluster Head. The proposed algorithm combines the two main characteristics of vehicular networks direction and mobility in order to select the best nodes in the network to be Cluster Heads. Furthermore, we take advantage of beacon messages that are already used to update the parameters used to select Cluster Heads.

4 Network Model

In this section, we present the Geographical Information based Clustering Algorithm "GICA" destined to IoV environments. GICA aims to find the stable nodes in the network to elect them as Cluster Heads. It is based on the following assumptions:

- We suppose that the digital map is divided into numerous regions, and each vehicle knows the organization of the city map.
- Each vehicle in the network is equipped with a GPS device and a digital map that allow it to locate its position in the map.
- Each vehicle can estimate its speed.
- Each vehicle has its own final destination matrix that is used to save and calculate the vehicles that share the same final region as it.
- All nodes that are in the same region will pass by the same regions in order to arrive to their final destination regions.

In order to find the best vehicles in the network to be elected as Cluster Heads, we propose to divide the city map into different regions. The use of such method allows to find the most durable nodes in term of time, where the node that has the higher number of neighbors, that share the same next region as it, will be selected as Cluster Head. This selected node will not lead to frequent link disconnection because they will keep a long communication period with its neighbors. Figure 1 shows an example of clusters-based city map divided into 6 regions.

Our network architecture is organized in clusters. Each vehicle can take one of the following states:

- **Not-Decided (ND):** is the initial status taken by each node or when the cluster head gives up its role.
- **Cluster Head (CH):** it is the coordinator of the group (cluster). Its role is to communicate information between nodes (intra-cluster communications), or between different clusters (inter-cluster communications).
- **Cluster Member (CM):** it is a core member node that benefits different services from the CHs.
- **Gateway:** is a relay node. It is the common access point for two or more cluster heads.

5 Cluster Head Election Metrics

To elect the clusterheads, we use the following metrics:

Destination Metric (D): is defined by the number of vehicles that share the same final region as a specific vehicle. To compute D, we use a matrix, called the final destination matrix , to save the nodes that share the same final region. All the nodes in the network are identified by an ID from 1 to n with n is the number of nodes in the network. The final destination matrix A size is n * n such

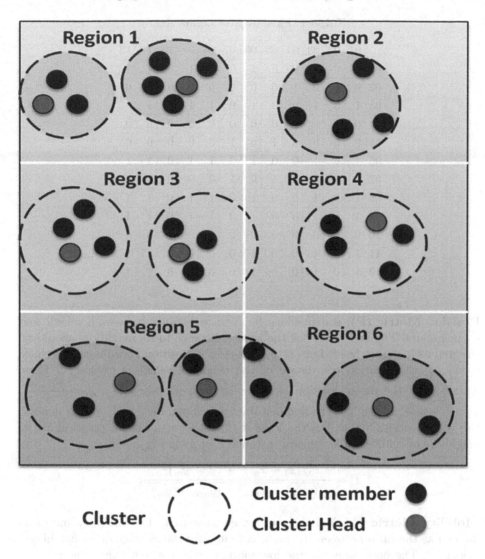

Fig. 1. clusters-based city map divided into 6 regions

that A[i][j] = 1 if vehicle i shares the same final region as vehicle j and A[i][j] = 0 otherwise. The sum of columns from 1 to n in line i indicates the number of vehicles that share the same final region as vehicle i. So, the value of D is computed as follows:

$$D = \sum_{j=1}^{n} A[i][j], \qquad (1)$$

Table 1 shows an example of final destination matrix with n equal to 12.

Table 1. Final destination matrix.

	01	02	03	04	05	06	07	08	09	10	11	12
01	1	0	1	0	0	1	0	1	1	0	1	1
02	0	1	0	1	0	0	0	0	0	0	0	0
03	1	0	1	0	0	1	0	1	1	0	1	1
04	0	1	0	1	0	0	0	0	0	0	0	0
05	0	0	0	0	1	0	1	0	0	0	0	0
06	1	0	1	0	0	1	0	1	1	0	1	1
07	0	0	0	0	1	0	1	0	0	0	0	0
08	1	0	1	0	0	1	0	1	1	0	1	1
09	1	0	1	0	0	1	0	1	1	0	1	1
10	0	0	0	0	0	0	0	0	0	1	0	0
11	1	0	1	0	0	1	0	0	0	0	1	1
12	1	0	1	0	0	1	0	0	0	0	1	1

Position Metric (P): is defined as the average distance between a vehicle and its neighbors. The node that has the lowest distance to its neighbors should be selected as a cluster head. Let (x_i, y_i) denote the position coordinate of vehicle i, (x_j, y_j) the position coordinate of the neighbor vehicle j of node i. Here, we calculate the mean position of the neighbors of node i as $x_m = \frac{\sum_{j=1}^{n} x_j}{N}$, $y_m = \frac{\sum_{j=1}^{n} y_j}{N}$, where N is the number of its neighbors. Let (x_{max}, y_{max}) denotes the vehicle position that has the longest distance to the mean position of the neighbors of vehicle i. The general distance P of node i is given by:

$$P = \frac{\sqrt{(x_i - x_m)^2 + (y_i - y_m)^2}}{\sqrt{(x_{max} - x_m)^2 + (y_{max} - y_m)^2}}, \tag{2}$$

Mobility Metric (M): In our proposed approach, The mobility metric is defined as the ratio between the node velocity and the average of its neighbors' velocities. The node that has the low mobility will not leave the cluster early, which indicates that the node is suitable to be selected as CH. Let us assume that $P_1(x_1, y_1)$ is the position of node i at time t_1 and $P_2(x_2, y_2)$ is the position of node i at time t_2. di is the distance traveled by node i over time Δt ($\Delta t = t_2 - t_1$):

$$d_i = \sqrt{(x_1 - x_2)^2 + (y_1 - y_2)^2}, \tag{3}$$

Thus, the velocity of node i over Δt, is computed as:

$$v_i = d_i/\Delta t, \tag{4}$$

The mobility metric is defined as follows:

$$M_i = \frac{v_i}{\sum_{j=1}^{n} v_j/n}, \tag{5}$$

Where n is the number of neighbors.

To select the CHs a combination of the previous defined parameters is proposed:

$$W = f_1 D + f_2 \frac{1}{M} + f_3 \frac{1}{P}. \tag{6}$$

Where f_1, f_2 and f_3 are weight factors and $f_1 + f_2 + f_3 = 1$

6   Geographical Information Based Clustering Algorithm "GICA"

6.1   Initial Network Phase

This phase defines the first period where no clusters have been formed in the network. Each vehicle starts with the initial status (ND), for each Interval Time it broadcasts a Hello message to its neighbors in the aim of exchanging with them its updated information to be used to elect the Cluster Heads such as mobility, position, region, etc. The Hello message exchange process is also used to inform the neighbors that the corresponding vehicle is still present and did not leave the transmission range. After receiving the Hello message, each vehicle modifies and updates its neighborhood table. The corresponding vehicle adds the new discovered vehicles that did not already exist in its neighborhood table and updates the information of those that are already saved in its neighborhood table. The received Hello message is also used to recalculate mobility and position parameters used to calculate the weight value. Moreover, each vehicle updates its final region matrix by adding the new vehicles that share the same final region as the corresponding vehicle or by delating the vehicles that leave the communication range or change their final region. The Hello message includes the vehicle ID, current status, current position, current mobility, final region, weight value and CH ID (Fig. 2).

Algorithm 1: Initial network phase

Node-status ND;
Wait for a Hello message;
Receive(Hello message);
Get position(x,y);
Get final region();
Calculate destination(D)metric based on Eq.1;
Calculate mobility(M)metric based on Eq.2;
Calculate position(p)metric based on Eq.(3&4&5);
Calculate node weight based on Eq.6;
Exchange Hello message;

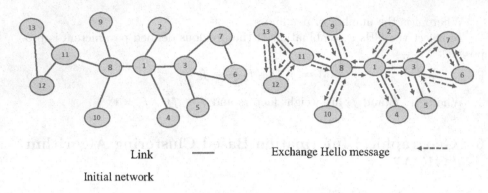

Link ——— Exchange Hello message ◄-----

Initial network

Fig. 2. Initial network phase

6.2 Clustering Phase

Our clustering election process is based on k-hop cluster structure, where the CH is selected based on the calculated weight value at k-hop distance. After receiving all the needed information to calculate the weight value (destination, mobility, position). Each vehicle calculates its weight value and compares it with those of its neighbors. The vehicle that owns the highest weight selects itself as CH and sends a message CHMsg containing its ID and its weight value to all its neighbors. When a ND vehicle receives a CHMsg message, it changes its status to CM in order to assume its role as a cluster member and updates the ID of its CH. After that, each vehicle sends an AcceptMsg to its CH in order to add it to its membership table. In the case where a vehicle receives numerous CHMsg messages from its neighbors, it chooses the CH that has the highest weight, then it records all the IDs of the CHs that invited it to join their cluster according to the order of their weight. On receiving an AcceptMsg the CH adds the vehicle in its membership list (Fig. 3).

Algorithm 2: Clustering

Wait for a Hello message;
Receive (Hello message);
Calculate its weight;
Get neighbor vehicle weight();
while *Clustering phase is active* **do**
    Compare neighbor vehicle weight with its weight ;
    if *its weight is the highest among its neighbors weights* **then**
        | Broadcast CHMsg message;
    else
        | Do nothing ;
    end
end

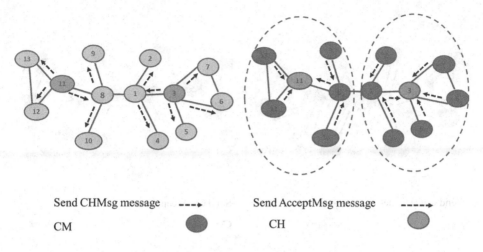

Send CHMsg message   ---→   Send AcceptMsg message   ---→

CM   ⬤   CH   ⬤

Fig. 3. Clustering phase

6.3 Gateways Selection Process

Each Cluster Head is responsible to choose its gateway nodes, after adding all the nodes and declare them as CM, the CHs broadcast an Invit message to their neighbors, the nodes that receive more than one CHMsg from different CHs are considered as gateway candidates. The gateway candidates respond with GatAccept message. To select the best gateway nodes the CH chooses the candidates that have the highest weight value (Fig. 4).

Algorithm 3: Gateway selection process

Wait for CHMsg message;
Receive CHMsg;
if *the receiver node receives more than one CHMsg from different CHs* **then**
  | the receiver node is gateway candidate;
  | Wait for GatMsg message;
  | Receive GatMsg;
  | Send GatAccept message to CH;
end

6.4 Maintenance Phase

Three state transitions are possible for a vehicle in the clustering process:

- An ND node wants to join an existing cluster to become CM: in this case, the vehicle broadcasts Hello messages periodically. If the ND node hears a Hello message from a CH neighbor, it sends a JOINMsg message to the CH neighbor. This allows the ND node to join the cluster. If the ND node receives multiple responses from different CHs, the one with the highest weight will be chosen.
- A CM leaves the cluster: If the CH does not receive a Hello message from a specific CM, then this CM is considered to have quitted the current cluster and will be deleted from the list of members.

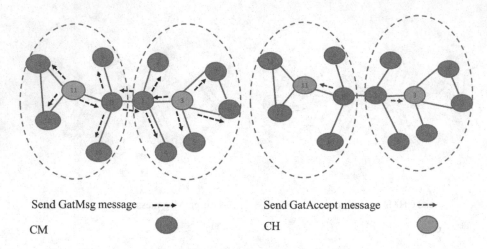

Send GatMsg message   ----▶   Send GatAccept message   ---▶

CM   ⬤   CH   ⬤

Fig. 4. Gateway selection process

– A CH leaves the current cluster: if during a specified period of time, the
CMs do not receive Hello messages from their CH, this CH is supposed to
have left the current cluster. The CM joins other clusters saved during the
cluster formation phase by sending JOINMsg. Otherwise, if no CH is saved,
it broadcasts JOINMsg to discover new CHs (Fig. 5).

Algorithm 4: Gateway selection process

---

vi, vj : vehicles ;
if *Vi want to join a new cluster* **then**
    Vi broadcast JOINMsg message;
    if *Vi receives one AcceptMsg* **then**
       | Vi ⟵ CM ;
    else
       if *Vi receives more than one AcceptMsg* **then**
         | Vi chooses the CH that has the highest weight;
       end
    end
end
if *a CH did not receive Hello message from vj for a period of time* **then**
    | CH deletes vj from its list of CMs;
end
if *a CM did not receive Hello message from its CH* **then**
    Send JOINMsg message to the saved CHs ; **if** *no CH is saved* **then**
       | Broadcast JOINMsg message;
    end
end

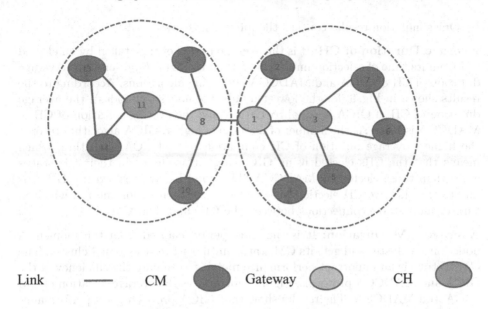

Fig. 5. Clusters formed

7 Simulation and Results

To evaluate the performance and efficiency of our scheme, we use Network Simulator 3 (NS-3) as a network simulator [12] and Simulation of Urban MObility (SUMO) [13] to generate mobility traces of vehicles and road traffic scenarios to imitate a real road network. For Media Access Control (MAC) layer, we used IEEE802.11p. We consider the channel bandwidth as 10 MHz. The proposed scheme is considered to be used in an urban area where the speed is randomly generated and its maximum value is fixed as 80 km/h. The rest of simulation parameters are given in Table 2.

Table 2. Simulation parameters.

Parameter	Value
Simulation time	300 s
Simulation area	1500×1500 m
Number of nodes	50–250 nodes
Number of regions	9 regions
Transmission range	250 m
Bandwidth	10 MHz
MAC/PHY protocol	IEEE802.11p
Vehicles' speed	30–80 km/h
Hello message	1 s

Our simulation results concern the following issues:

Average Duration of CH: It is the average period of time taken by an elected CH from its time of selection until it loses its role. Figure 6 presents the Average duration of CH of GICA and MADCCA clustering algorithms. According to the results shown in Fig. 6, for the lowest number of nodes (50 nodes), the average duration of CH of GICA is equal to 100 s, while the average duration of CH of MADCCA is 120 s. As the number of nodes gets higher, GICA algorithm records the highest average duration of CH compared to MADCCA algorithm, which means that the CHs elected using GICA algorithm maintain their CH status more than those elected by MADCCA. The good behavior shown by GICA is due to the effective CH election process that uses final region metric, which in turn ensures stable connections between the CHs and their CMs.

Average CM Duration: It is the time period elapsed from the moment a node joins a cluster and gets its CM status until it leaves the joined cluster. The CM lifetime is an important performance metric to evaluate the efficiency of the GICA and MADCCA protocols. Figure 7 introduces the average duration of CM GICA and MADCCA. The results show that GICA proves a good performance under different numbers of nodes compared to MADCCA. This can be explained by the weight value used by GICA that combines three metrics that ensure network stability (mobility, direction, position).

Overhead: We consider the messages needed for cluster formation and selection of the CH as a communication overhead. It is the amount of information (in bits) circulating in the network per unit time. As shown in Fig. 8, GICA records the

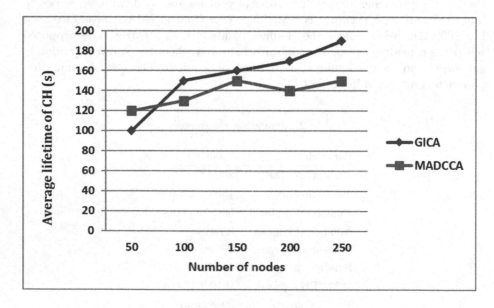

Fig. 6. The average duration of cluster head

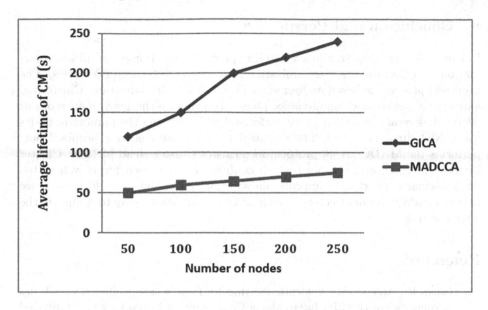

Fig. 7. The average duration of cluster member

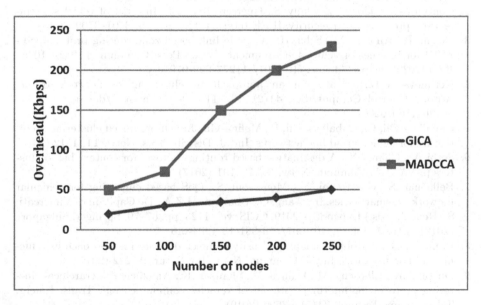

Fig. 8. Overhead of GICA and MADCCA protocols

lowest overhead comparing to MADCCA. This good performance is due to the stability of the nodes elected as CHs by our proposal that minimizes the link disconnection.

8 Conclusion and Perspective

In this paper we propose a new clustering protocol based on geographical information to maintain the communication stability in Internet of Vehicles. The proposed protocol is based on four elementary phases: initialization, Clustering, gateways selection and maintenance phase. To evaluate the performance of our protocol, several simulations were performed while varying the number of nodes using NS3 simulation tool. The obtained results show that our proposal outperforms the MADCCA [8] protocol in terms of Cluster Head lifetime, Cluster Member lifetime, and overhead which confirms that our scheme is well suited for Vehicular networks. To improve our solution in the short and long term, we expect to add more parameters in term of communication delay to minimize the response time.

References

1. Gasmi, R., Aliouat, M.: Vehicular Ad Hoc NETworks versus internet of vehicles - a comparative view. In: International Conference on Networking and Advanced Systems (ICNAS), Annaba, Algeria (2019)
2. Contreras-Castillo, J., Zeadally, S., Guerrero-Ibaez, J.: Internet of vehicles: architecture, protocols, and security. IEEE Internet Things J. **5**, 3701–3709 (2018)
3. Gasmi, R., Aliouat, M., Seba, H.: A stable link based zone routing protocol (SL-ZRP) for internet of vehicles environment. Wirel. Pers. Commun. **112**(2), 1045–1060 (2020). https://doi.org/10.1007/s11277-020-07090-y
4. Kerimova, L.E., et al.: On an approach to clustering of network traffic. Autom. Control Comput. Sci. **41**(2), 107–113 (2007). https://doi.org/10.3103/S0146411607020071
5. Caballero-Gil, C., Caballero-Gil, P., Molina-Gil, J.: Self-organized clustering architecture for vehicular ad hoc networks. Int. J. Distrib. Sens. Netw. **11**, 1–12 (2015)
6. Sethi, V., Chand, N.: A destination based routing protocol for context based clusters in VANET. Commun. Netw. **9**, 179–191 (2017)
7. Bellaouar, S., Guerroumi, M., Moussaoui, S.: QoS based clustering for vehicular networks in smart cities. In: Wang, G., Bhuiyan, M.Z.A., De Capitani di Vimercati, S., Ren, Y. (eds.) DependSys 2019. CCIS, vol. 1123, pp. 67–79. Springer, Singapore (2019). https://doi.org/10.1007/978-981-15-1304-6_6
8. Ram, A., et al.: Mobility adaptive density connected clustering approach in vehicular ad hoc networks. Int. J. Commun. Netw. Inf. Secur. **9**, 222 (2017)
9. Dutta, A.K., Elhoseny, M., Dahiya, V., Shankar, K.: An efficient hierarchical clustering protocol for multihop internet of vehicles communication. Trans. Emerg. Telecommun. Technol. **31**(5), e3690 (2019)
10. Senouci, O., Aliouat, Z., Harous, S.: DCA-DS: a distributed clustering algorithm based on dominating set for internet of vehicles. Wirel. Pers. Commun. **115**, 401–413 (2020). https://doi.org/10.1007/s11277-020-07578-7
11. Wang, C., Li, J., He, Y., Xiao, K., Hu, C.: Segmented trajectory clustering-based destination prediction in IoVs. IEEE Access **8**, 98999–99009 (2020)

12. Riley, G.F., Henderson, T.R.: The *ns-3* network simulator. In: Wehrle, K., Güneş, M., Gross, J. (eds.) Modeling and Tools for Network Simulation, pp. 15–34. Springer, Heidelberg (2010). https://doi.org/10.1007/978-3-642-12331-3_2
13. Behrisch, M., et al.: SUMO-simulation of urban mobility: an overview. In: The Third International Conference on Advances in System Simulation, pp. 63–68 (2011)

Active Probing for Improved Machine-Learned Recognition of Network Traffic

Hamidreza Anvari[✉] and Paul Lu

Department of Computing Science, University of Alberta, Edmonton, AB, Canada
{hanvari,paullu}@ualberta.ca

Abstract. Information about the network protocols used by the background traffic can be important to the foreground traffic. Whether that knowledge is exploited via optimization through protocol selection (OPS) or through other forms of parameter tuning, a machine-learned classifier is one tool to identifying background traffic protocols. Unfortunately, global knowledge can be difficult to obtain in a dynamic distributed system like a shared, wide-area network (WAN).

Previous techniques for protocol identification have focused on passive or end-point signals for classification. For example, end-to-end round trip time (RTT) can, especially when gathered as a time series, reveal a lot about what is happening on the network. Other related signals, such as bandwidth, and the number of retransmissions can also be used for protocol classification. However, as noted, these signals are typically gathered by passive means, which may limit their usefulness.

We introduce and provide a proof-of-concept of *active probing*, which is the systematic and deliberate perturbation of traffic on a network for the purpose of gathering information. The time-series data generated by active probing improves our machine-learned classifiers because different network protocols react differently to the probing. Whereas passive probing might be limiting the time series observations to a period of steady state (e.g., saturated network), active probing forces the system out of that steady state. We show that active probing improves on prior work (with passive probing of RTT) by between 7% to 16% in additional accuracy (depending on the window size), and reaching 90% averages in precision, recall, and F1-scores.

Keywords: Machine-learned classifier · Active probing · Protocol selection · Data transfer · Wide-area networks · Fairness · Shared network

1 Introduction

Information about the network protocols used by the background traffic can be important to the foreground traffic. Whether that knowledge is exploited via optimization through protocol selection (OPS) [3] or through other forms of

© Springer Nature Switzerland AG 2021
É. Renault et al. (Eds.): MLN 2020, LNCS 12629, pp. 122–140, 2021.
https://doi.org/10.1007/978-3-030-70866-5_8

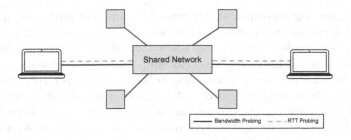

Fig. 1. Probing background traffic on shared network

parameter tuning, a machine-learned classifier is one tool to identifying background traffic protocols. Unfortunately, global knowledge can be difficult to obtain in a dynamic distributed system like a shared, wide-area network (WAN) (Fig. 1). There are many possible senders, receivers, endpoints, and potential bottlenecks.

For example, if it is known that a background data stream is already using TCP with the BBR (Version 1)[1] congestion control algorithm (CCA) (TCP-BBR), then the new foreground data stream should also use TCP-BBR (instead of, say, TCP-CUBIC) because BBR is known to be unfair to CUBIC data streams [3,12]. In contrast, two TCP-BBR data streams tend to be fair to each other. And, conversely, if a background data stream is already using TCP-CUBIC, then the new foreground data stream should also use TCP-CUBIC, or else the foreground transfer might be unfair to the background transfer. Recognizing the protocols in use by background streams before selecting the protocols for the new foreground stream can be important for maintaining high performance and fairness. An analogy is to always look both ways before crossing the street.

Previous techniques for protocol identification have focused on passive or endpoint signals for classification. For example, end-to-end round trip time (RTT) can, especially when gathered as a time series (e.g., as a side-effect of normal packet transmissions), reveal a lot about what is happening on the network. Other related signals, such as bandwidth, and the number of retransmissions (Fig. 1) can also be used for protocol classification. However, as noted, these signals are typically gathered by passive means, which may limit their usefulness.

A shared network is challenging because of overlapping patterns and complex interactions between different data streams (Fig. 4). When a network link is a bottleneck, time series data of RTTs can approximate a saturated, flat line (Fig. 4, especially B1-B1 case). Other mixtures of protocols can create large oscillations in the RTTs with patterns that are difficult to classify.

However, the innate properties of different protocols cause them to react differently when perturbed. Therefore, we introduce *active probing* as a tech-

---

[1] While BBR version 2 has been under development, at the time of writing this paper, BBR version 1 is still the only stable version publicly available; hence the one used in this study for all the experiments and evaluations.

nique to separate out the different mixture of streams and protocols on a shared network, and provide more diverse time-series features for classification. Our proof-of-concept implementation and evaluation of active probing shows that the systematic and deliberate perturbation of traffic on a network can gather additional information. In contrast, *passive probing* involves the gathering of information with an explicit goal of not interfering or perturbing the existing traffic. Unfortunately, passive probing might coincide with a network in steady state, which limits the interesting features that are observable. As we discuss below, active probing is designed to push a network out of steady state in order to observe the reaction and recovery behaviour. And that additional information can increase classification accuracy by 7% to 16% (Fig. 6), going up to 90% averages in precision, recall, and F1-scores (Fig. 9b).

By an imperfect analogy with physics and mass spectrometry, a group of objects in motion can appear to be a single cluster (e.g., less separable; appear as a flat line). But, if one could apply the same force to all of the objects (e.g., systematic perturbation), the fact that objects (e.g., protocols) have different mass (and different charges) would mean they would experience different acceleration and be identifiable. Without applying the force (e.g., active probing), it may be hard to separate out the different protocols in use. With the force, the patterns change and can be used for classification.

The outline of the remainder of the paper is as follows. In Sect. 2 we introduce the background and related work, reviewing the impact of background traffic on shared networks, and TCP Congestion Control Algorithm (CCA) schemes. Section 3, frames our methodology, including its scope, design decisions and the adopted machine-learning workflow. In Sect. 4 the experimental setup and testbed configurations are specified. In Sect. 5 we provide and discuss the evaluation results, including machine-learning performance for original model, as well as two hypothetical use-case scenarios. Section 6 concludes the paper and outlines the vision for the future study.

2   Background and Related Work

2.1   Impact of Background Traffic

One important property of data networks is the distinction between *dedicated* and *shared* models in terms of bandwidth utilization.

In a dedicated network model, a private data path is established to connect several points of presence, usually for large-scale and data intensive research projects or industrial use. For example, Google's B4 private world-wide network [15], and Microsoft's private WAN resources [18] fall in the category of dedicated networks, where dedicated bandwidth and network infrastructure are implemented. Alternatively, the bandwidth reservation techniques exist which reserve and guarantee the bandwidth for a specific use-case, usually as an overlay on a shared network infrastructure. For example, the On-demand Secure Circuits and Advance Reservation System (OSCARS) is one technique for bandwidth reservation in this category [10].

In contrast, in bandwidth-sharing networks, which are still the common practice for a large number of academic and industrial users, the available network resources, e.g. bandwidth, are shared by multiple users and applications in a competitive environment (Fig. 1). In this situation, network resources are being shared between multiple users at the same time, resulting in a dynamic workload on the network. One studied consequence of this dynamic behaviour is the emergence of periodic burstiness of the traffic over the network [13,16]. This burstiness may result in various levels of contention for network resources, which could lead to an increased rate of packet loss, and therefore decreased bandwidth utilization, both in aggregation and per individual users.

When it comes to bandwidth management and traffic engineering, the direction of research and topics of interest change based on the network utilization mode. In dedicated networks, the general line of research concerns efficient and effective resource scheduling and routing, or bandwidth reservation techniques, usually in a centralized fashion [10,15,18]. In contrast, for shared networks, fair collaboration with other network users, in an autonomous way, becomes a desired quality. Hence, the research focus shifts towards probing available bandwidth and rate adjustment, for fair and efficient utilization of the bandwidth, mainly in a decentralized, end-to-end, fashion [1].

For probing the available resources on a shared network, there are a number of studies, investigating the possibility of estimating the available bandwidth in high-speed networks [21,25]. In addition to bandwidth estimation, estimating the network workload and the type of background traffic on the network could also affect the performance of data transfer tasks. In one study, the effect of background traffic on distributed systems has been investigated [23].

In our previous work we have investigated and shown the counter-intuitive performance of some well-known tools and protocols depending on the type of background traffic in the network [2,4]. We have also investigated the impact of different TCP Congestion Control Algorithms (CCAs) on each other, with possibly drastic impact on the other traffic while running on a shared network [3,5].

In summary, the existence and nature of background traffic streams on the network would impact various performance metrics (bandwidth, latency, etc.) of a data transfer task. Hence, obtaining knowledge about the background traffic would allow for efficient adjustment of the network configurations, e.g. choosing appropriate data-transfer tools and protocols.

2.2 TCP Scheme: Congestion Control Algorithm (CCA)

Among the network protocols, TCP is the most popular reliable protocol in use in different networking contexts. In addition to reliability, TCP protocol consists of *Congestion Control Algorithm (CCA)*, a.k.a TCP Scheme. CCA is an algorithm responsible for dynamically adjusting the sending rate and other data transfer parameters. One inherent challenge in designing a CCA, is to offer efficient bandwidth utilization [1]. CUBIC [11] and BBR [7] are two popular state-of-the-art CCAs. CUBIC is the default CCA on most Linux machines, and BBR is a recent algorithm from Google, gaining popularity in recent years.

In bandwidth-sharing networks, the protocols should provide a good trade-off between efficiency and being fair to other traffic (i.e., background traffic). Fairness, at a high level of abstraction, is defined as a protocol's tendency towards utilizing an equal share of bandwidth with other concurrent traffic streams. Jain fairness [14] is a well-known metric, measuring how fair the consumers (i.e., traffic streams) are in sharing system resources (i.e., network bandwidth). While most CCAs include fairness as part of their design goals, fairness problems have been reported for both CUBIC and BBR CCAs [17,19].

As a new trend in designing CCA algorithms, machine learning (ML) techniques have been applied for designing or optimizing protocols. Remy used simulation and ML to create a new TCP CCA, via a decentralized partially-observable Markov decision process (dec-POMDP) [24]. Performance-oriented Congestion Control (PCC) is another recent study where an online learning approach is incorporated into the structure of the TCP CCA [8]. Vivace [9] and Proteus [20] are two newer CCAs based on PCC, trying to improve on its utility functions.

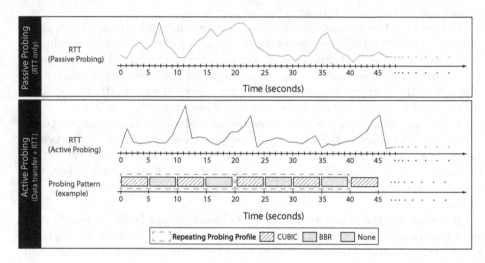

Fig. 2. Active probing: RTT measurement (ping) combined with probing profile pattern based on bandwidth interference (iperf)

3   End-to-End Traffic Recognition

In our previous study, we devised a method to classify the mixture of background traffic in a controlled environment [3]. In that approach, we periodically probe end-to-end delay (RTT) and use the resulting RTT time-series for training a classifier to identify the mixture of TCP-based traffic on the network. We leverage

	FG: BBR	FG: CUBIC
BG: BBR	Good Throughput Good Fairness	Poor Throughput Poor Fairness
BG: CUBIC	Good Throughput Poor Fairness	Good Throughput Good Fairness

Fig. 3. Interaction of CUBIC and BBR on a shared network

the developed insight to optimize protocol selection for transferring data on the network, so-called Optimization through Protocol Selection (OPS). In a follow-up research, we validated the extensibility of this approach to a variety of network configurations of bandwidth and latency [5].

In this study, as illustrated in Fig. 2, we extend the *passive probing* (RTT only) technique, by introducing an *active probing* element to the model. We hypothesize that, by stimulating the network traffic with short bursts of traffic, the resulting RTT signatures would better identify the mixture of background traffic. As discussed earlier, the systematic perturbation of the background traffic provides more information with which we can recognize and classify the protocols in use. The intuition behind this idea is that, different network protocols, by design, have distinct reaction to a competing traffic stream; hence reinforcing the unique RTT patterns to classify the background mixtures.

3.1 Scope: CUBIC and BBR TCP Schemes

While a diverse collection of reliable and unreliable tools and protocols are in use on the Internet, TCP-based protocols are predominant type of traffic on data networks. Hence, in this study we only consider TCP-based background traffic. In order to keep our work consistently comparable with our previous work, we limit our study to the mixtures of TCP CUBIC and TCP BBR as the background traffic to be recognized through classification. According to our previous study [3,5], the interaction between CUBIC and BBR schemes could be characterized as in Fig. 3.

In particular, we will consider six distinct classes of background traffic to train a classifier for them. These six classes represent various mixtures of up to two streams of CUBIC and BBR, as summarized in Fig. 5c.

3.2 Active Probing Profiles

As earlier discussed, we conduct active probing in the form of short bursts of traffic. To better trigger background streams to reveal their unique patterns, as opposed to a single traffic burst, we use a sequence of short traffic bursts with different configurations. We refer to such a sequence of short bursts as *Active Probing Profile*, illustrated as dashed boxes in Fig. 2.

We have scripted our experimental environment with the probing profile as a module. This would enable us to extend our experiments to more probing profiles in the future; for example, varying number of parallel streams, TCP and UDP streams, and more. For this study in particular, we have defined the active probing profile as a sequence of four short bursts of TCP streams of different CCAs:

1. CUBIC (5 s)
2. BBR (5 s)
3. CUBIC (5 s)
4. None (5 s)

In the process of generating data for training, as well as when deploying the classifiers, the probing profile is conducted repeatedly while RTT is being probed.

3.3   Machine Learning Process

In this section, we briefly review our machine learning process including data collection and preparation, training classifiers, and evaluation metrics. Visit our previous study for the full description on the details of our ML process, including the description on Dynamic Time Warping (DTW) distance measure used as the distance measure with K-NN classifiers [3].

Data Collection and Preparation. The first step in the process of training a classifier, is to prepare a sufficiently large training dataset, including representative samples for all the classes to be classified. In this study, we have conducted a series of scripted experiments on our controlled testbed (Sect. 4.1), probing RTT for 30 min per each background traffic class. The RTT is being probed with a 1-s sampling rate, resulting in a time-series of length ~1800 per each class. To smooth out the possible noise in measuring RTT, each probing step consists of sending 10 ping requests to the other end-host, with 1 ms delay in between, and the average value is recorded as the RTT value for that second. We repeated this experiment for the six background traffic mixtures, discussed in Sect. 3.1.

For gathering data, we use our controlled testbed, specified in Sect. 4.1, conducting the background traffic between $(S1, D1)$ and $(S2, D2)$ pair of nodes. $(S3, D3)$ pair of nodes are used to conduct active probing profile and simultaneously probing end-to-end RTT. For all data-transfer tasks, we used the *iperf* tool[2] for generating TCP traffic of the desired CCA.

A 90-s window of populated time-series are presented for all six classes for both configurations of active and passive probing in Fig. 4.

For preparing the resulting time-series for training process, we partition the long RTT time-series into smaller chunks of fixed window size parameter, denoted by w.

---

[2] http://software.es.net/iperf/.

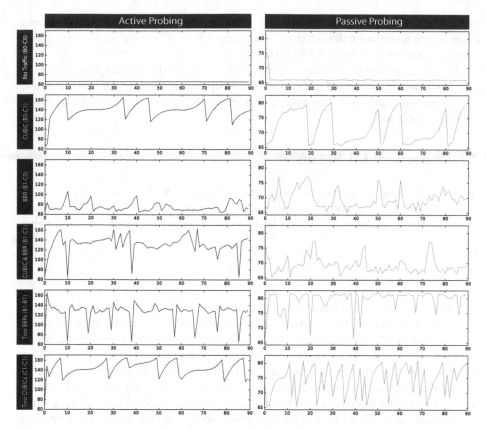

Fig. 4. Illustrative examples: active vs. passive probing impact on RTT signal: 90 s time-series signatures per traffic mixture. Active probing changes time-series patterns. Note the different ranges for Y-axis across columns.

Classification Models. We train K-NN classifiers for three configurations of K parameter, 1-NN, 3-NN, and 5-NN.

Since our input vectors to the classifier are time-series, we use Dynamic Time Warping (DTW) distance measure to appropriately measure similarity between pairs of time-series.

We feed the resulting time-series of fixed length w as the training data to our K-NN classifier models. The resulting classifier takes an RTT time-series of length w as an input vector, and predicts the class label for that time-series as the output.

Selecting the appropriate time-length parameter w is a trade-off between the amount of information provided to the classifier and the RTT probing latency. On the one hand, the larger the w becomes, the more information will be provided to the classifier, providing a better opportunity to recognize the traffic pattern. On the other hand, a larger w means it would take a more time for the probing phase of a real system to gather RTT data to use as input to a classifier. To make an appropriate decision about the parameter value w, we did a parameter-sweep experiment where we calculated the classification accuracy for all the classifiers, varying parameter w from 5 s to 60 s, with a 5 s step.

For performance evaluation, we use 5-fold cross validation, keeping 20% samples from each class as hold-out during the training, using them for testing phase. The reported results are the average accuracy over the five folds on the cross-validation scheme.

4   Experimental Setup

In this section we review the experimental setup used for test conduct and evaluation of our proposed model for end-to-end traffic recognition. Our setup consists of our network testbed and the method used for gathering data for training classifiers.

4.1   Network Testbed

For the testbed, we employ the emulated network created in our previous study [4], augmenting that network to accommodate profile-based active-probing scripts. The emulated network is implemented as an overlay on top of a physical network. All the nodes in our testbed are physically located in a dedicated computer cluster. The hardware configuration of the nodes are provided in Fig. 5b. The nodes are equipped with dual network interface cards of 1 Gb/s and 10 Gb/s native rates[3]. All the end-nodes are equipped with multiple TCP CCAs, including CUBIC and BBR, to support active-probing profiles based on switching CCAs. There is no interference from other traffic because the cluster is isolated from other networks.

4.2   Infrastructure, Software and Frameworks

In our implementation, we have used the following software and frameworks:

Infrastructure (OS). For our networking cluster, all the nodes were running Linux distribution CentOS 6.4 using kernel version `4.12.9-1.el6.elrepo.x86-64`.

Congestion Control Algorithms (CCA). We have used two CCA schemes in this study: CUBIC and BBR. For CUBIC, we use the standard version shipped with the aforementioned kernel version in our networking nodes. As mentioned earlier, BBR Version 1 is the only stable version publicly available, therefore Version 1 is the one used in this study for all the experiments and evaluations.

Network Customization. For implementing the emulated network environment we have used Dummynet network emulator [6]. Our automation scripts for

---

[3] In this paper we only present results conducted on the 1 Gb/s network.

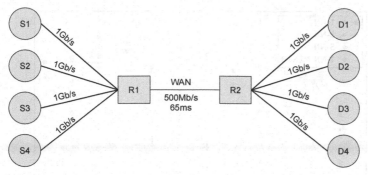

(a) Dumbbell Network Topology

Node(s)	CPU (Model/Cores/Freq.)	RAM
$S1$, $D1$	AMD Opt. 6134 / 8 / 2.30	32 GB
$S2$, $D2$	Intel Ci3-6100U / 4 / 2.30	32 GB
$S3$, $D3$	Intel Ci3-6100U / 4 / 2.30	32 GB
$S4$, $D4$	AMD E2-1800 / 2 / 1.70	8 GB
$R1$, $R2$	AMD A8-5545M / 4 / 1.7	8 GB

Traffic Class	BBR Stream#	CUBIC Stream#
B0-C0	0	0
B0-C1	0	1
B1-C0	1	0
B1-C1	1	1
B1-B1	2	0
C1-C1	0	2

(b) Nodes Configuration   (c) Background Traffic Classes

Fig. 5. Testbed architecture and experiment configuration

data-gathering and evaluation are implemented in either Python or Bash shell scripts.

Machine Learning Environment. All the ML programming and evaluation metrics are implemented in Python language. For implementing K-NN classifiers we have used scikit-learn library version 0.19.1 [22].

5   Results and Discussion

In this section, we provide and discuss the performance evaluation results of the developed classifier for characterizing the mixture of background traffic. We first discuss the general classification performance for all the six classes of background traffic (Fig. 5c). We then define and evaluate two decision scenarios as potential use-cases to deploy the trained classification models.

All the accuracy results are reported as an average over 5-fold cross validation. Across all the figures, best viewed in colour, corresponding colours are used to highlight corresponding variation of K-NN classifier model.

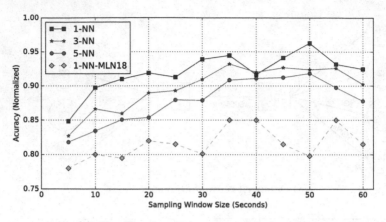

(a) Classification Accuracy per window size

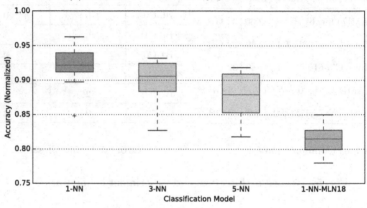

(b) Box Plot for Classifiers Accuracy

Fig. 6. Accuracy of classification models for window size w of input time-series with active probing (5-fold cross-validation). 1-NN-MLN18 represents passive probing performance as the baseline [3].

5.1   Classification Performance

The accuracy results for different classifiers per time length w of input time-series are provided in Fig. 6. Along with the results for classifiers trained using active-probing, The results for the *1-NN* classifier with passive probing [3] are also provided for reference and comparison. All active-probing classifiers outperform passive-probing classifiers by between 7% (for $w = 5$) to 16% (for $w = 50$) (depending on window size) (Fig. 6a), This performance improvement verifies our hypothesis that systematic perturbation using active probing will reinforce unique patterns in RTT time-series for different background traffic mixtures.

Across different configurations for active-probing classifiers, the 1-NN tends to offer a better overall accuracy (Fig. 6a). Also according to the box plot, 1-NN presents the most consistency in accuracy for varying window sizes, with a narrow variation in observed performance (Fig. 6b).

To further analyze the performance consistency and variation, the accuracy results along with accuracy variation across 5 folds are provided in Fig. 7 for all 4 classifiers (1-NN, 3-NN, 5-NN with active probing, and 1-NN with passive probing). For better interpretability, instead of standard error bars, the box plot is provided across 5 folds for each window size.

In most cases, 1-NN (Fig. 7a) outperforms 3-NN (Fig. 7b) and 5-NN (Fig. 7c) with active probing. However, since in 1-NN we solely rely on the first nearest neighbour, the variation across folds is relatively higher compared to 3-NN and 5-NN. As such, one might choose to sacrifice by 1%–2% on accuracy on average and deploy 3-NN for better consistency and robustness in decision making.

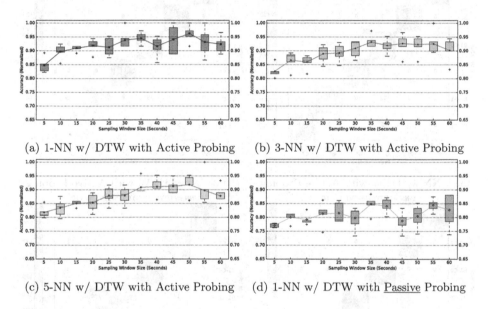

(a) 1-NN w/ DTW with Active Probing   (b) 3-NN w/ DTW with Active Probing

(c) 5-NN w/ DTW with Active Probing   (d) 1-NN w/ DTW with <u>Passive</u> Probing

Fig. 7. Accuracy variation (BoxPlot) of classification models for varying time length w of input time-series per k in K-NN with DTW - with and without active probing (5-fold cross-validation)

To further investigate the classifiers' performance and their challenges in predicting traffic classes, the confusion matrices for $w = 10$ are provided in Fig. 8, corresponding to the four classifiers in previous figures. As expected, the classifiers with active probing consistently manifest more diagonal confusion matrices, representing more accurate predictions across 6 classes. 1-NN with active probing (Fig. 8a) and 1-NN with passive probing (Fig. 8d) correspondingly represent the highest and lowest performance across all classifiers.

To take a close look at how active probing helps improve the quality of prediction, the isolated version of confusion matrices for 1-NN with and without active probing along with other classification reports are provided in Fig. 9 and Fig. 10 respectively. As one concrete observation: active probing has resolved the confusion in making a distinction between B1–C1 and B1–B1 classes (annotated in Fig. 10a). The distinction between these two classes is important, since (as we will see in following sections) it affects our ability to decide if we have homogeneous or heterogeneous protocols in use by background traffic. With active probing, the confusion in determining CCAs is mostly eliminated, and the only, less critical, confusion remains the distinction between the number of streams of the same protocol running on the network (annotated in Fig. 9a).

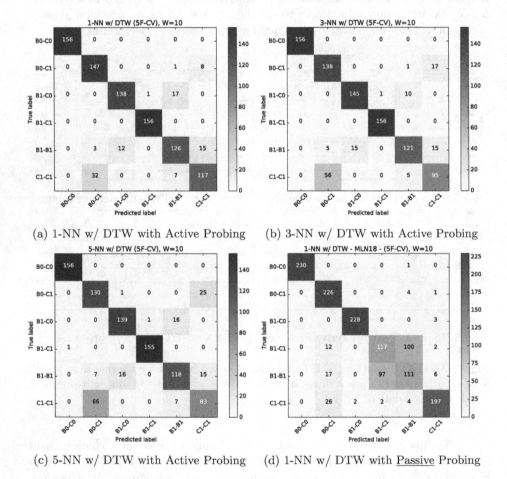

(a) 1-NN w/ DTW with Active Probing   (b) 3-NN w/ DTW with Active Probing

(c) 5-NN w/ DTW with Active Probing   (d) 1-NN w/ DTW with <u>Passive</u> Probing

Fig. 8. Confusion matrix for classification models for $w = 10$ of input time-series per k in K-NN with DTW - with and without active probing (5-fold cross-validation)

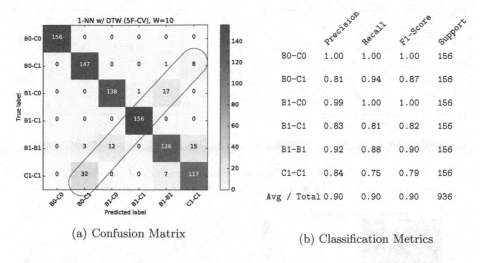

(a) Confusion Matrix

(b) Classification Metrics

Fig. 9. K-NN (DTW) performance ($w = 10$, $k = 1$) with <u>active</u> probing

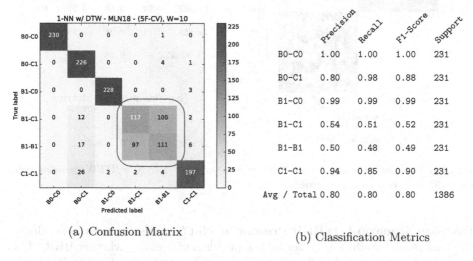

(a) Confusion Matrix

(b) Classification Metrics

Fig. 10. K-NN (DTW) performance ($w = 10$, $k = 1$) with <u>passive</u> probing

5.2 Decision Scenarios

The ultimate goal in designing a background traffic recognizer is to deploy that model in production use-cases, improving decision making performance under different networking scenarios.

In this section, we define two decision making scenarios, as it might arise. We then study how accurately the discussed classifiers, along with active probing, would perform in decision making.

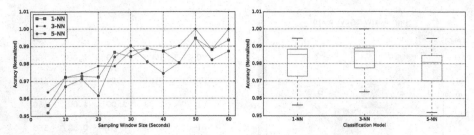

(a) Classification Accuracy per window size   (b) Box Plot for Classifiers Accuracy

Fig. 11. Decision scenario 1: is BBR present or not? Accuracy of classification models for varying time length w with active probing (5-fold cross-validation)

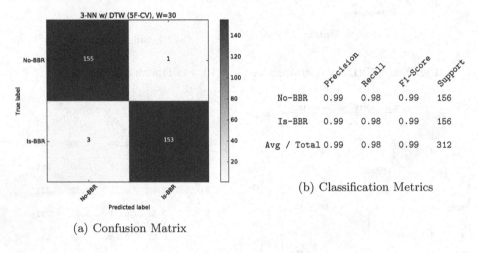

	Precision	Recall	F1-Score	Support
No-BBR	0.99	0.98	0.99	156
Is-BBR	0.99	0.98	0.99	156
Avg / Total	0.99	0.98	0.99	312

(b) Classification Metrics

(a) Confusion Matrix

Fig. 12. Decision scenario 1: is BBR present or not? K-NN w/DTW performance ($w = 30$, $k = 3$) with active probing

Decision Scenario 1: Is BBR Present or Not? In first scenario, we evaluate the background traffic recognizer for the problem of deciding whether BBR CCA exists in the background traffic mixture or no. The motivation for this decision is that, for the currently available BBR implementation, there is a known fairness issue while competing with loss-based CCAs, including CUBIC, summarized in Fig. 3 [3, 19]. As a result, identifying whether BBR traffic exists on the network would help both end-users (to decide on which protocol to use) as well as network administrators (for example, to define regulations for bandwidth allocation).

The accuracy performance and sample classification metrics for this decision scenario using active-probing are provided in Fig. 11 and Fig. 12. The accuracy of this decision scenario is between 95% and 99%, representing near perfect performance in identifying if BBR CCA is present in the network or not. While this is a simplistic scenario, the promising, near-perfect, results yield a great opportunity in scaling and deploying this technique.

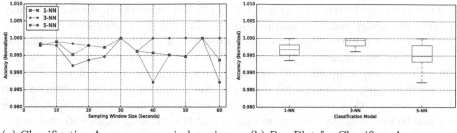

(a) Classification Accuracy per window size   (b) Box Plot for Classifiers Accuracy

Fig. 13. Decision scenario 2: homogeneous or heterogeneous CCAs? Accuracy of classification models for varying time length w with active probing (5-fold cross-validation)

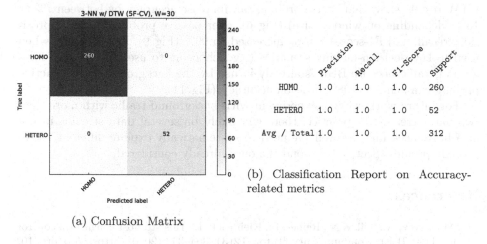

	Precision	Recall	F1-Score	Support
HOMO	1.0	1.0	1.0	260
HETERO	1.0	1.0	1.0	52
Avg / Total	1.0	1.0	1.0	312

(b) Classification Report on Accuracy-related metrics

(a) Confusion Matrix

Fig. 14. Decision scenario 2: homogeneous or heterogeneous CCAs? K-NN w/DTW performance ($w = 30$, $k = 3$) with active probing

Decision Scenario 2: Homogeneous or Heterogeneous CCAs? In second scenario, we aim to answer a more general question: whether the mixture of background traffic consists of homogeneous or heterogeneous protocols. Similar to the last scenario, this scenario would have potential use-cases among users, administrators, or infrastructure providers.

Similar to the scenario 1, applying our K-NN based model with active probing yields promising performance in this scenario as well. The results are provided in Fig. 13 and Fig. 14.

6 Concluding Remarks

The challenge for many ML problems is getting quality features by which to perform tasks such as classification. And, for our problem of network traffic recognition and protocol classification, that challenge is increased by the overlapping nature of data streams on shared networks. Passive probing techniques have been explored with reasonable success [3], but networks in steady state may not present enough patterns and features for classifiers to work well.

Therefore, we have introduced and presented a proof-of-concept evaluation of active probing. Based on the observation that the innate properties of different CCAs and network protocols will cause them to react differently when systematically perturbed, an ML-based traffic recognition system observes the RTT time-series that results. A pattern of active probing profiles (Sect. 3.2) is intended to purposefully force protocols out of their steady state and then react in a pattern that is machine learnable. In contrast to passive probing, which might only be observing the system in steady state, active probing forces an opportunity to observe the dynamic changes in the state of system.

After a detailed evaluation of different K-NN parameters, our 1-NN with DTW results show that active probing can increase accuracy by between 7% to 16% (depending on window size) (Fig. 6), over passive probing. Average precision, recall, and F1-scores can be increased to 90% (Fig. 9). Furthermore, when used in key decision-making scenarios (e.g., choosing to use BBR for the foreground traffic because BBR is already in use by the background traffic), active probing can achieve >98% decision accuracy (Fig. 11).

For future work, we will experiment with background traffic with more than two data streams. In practice, there will likely be several data streams in use at a bottleneck link. As well, we need to systematically explore different active probing profiles (Sect. 3.2) beyond the ones already considered.

References

1. Afanasyev, A., Tilley, N., Reiher, P., Kleinrock, L.: Host-to-host congestion control for TCP. IEEE Commun. Surv. Tutor. **12**(3), 304–342 (2010). https://doi.org/10.1109/SURV.2010.042710.00114
2. Anvari, H., Lu, P.: Large transfers for data analytics on shared wide-area networks. In: Proceedings of the ACM International Conference on Computing Frontiers, CF 2016, pp. 418–423. ACM, New York (2016). https://doi.org/10.1145/2903150.2911718
3. Anvari, H., Huard, J., Lu, P.: Machine-learned classifiers for protocol selection on a shared network. In: Renault, É., Mühlethaler, P., Boumerdassi, S. (eds.) MLN 2018. LNCS, vol. 11407, pp. 98–116. Springer, Cham (2019). https://doi.org/10.1007/978-3-030-19945-6_7
4. Anvari, H., Lu, P.: The impact of large-data transfers in shared wide-area networks: an empirical study. Procedia Comput. Sci. **108**, 1702–1711 (2017). International Conference on Computational Science, ICCS 2017, 12–14 June 2017, Zurich, Switzerland. https://doi.org/10.1016/j.procs.2017.05.211. http://www.sciencedirect.com/science/article/pii/S1877050917308049
5. Anvari, H., Lu, P.: Learning mixed traffic signatures in shared networks. In: Krzhizhanovskaya, W., et al. (eds.) ICCS 2020. LNCS, vol. 12137, pp. 524–537. Springer, Cham (2020). https://doi.org/10.1007/978-3-030-50371-0_39
6. Carbone, M., Rizzo, L.: Dummynet revisited. SIGCOMM Comput. Commun. Rev. **40**(2), 12–20 (2010). https://doi.org/10.1145/1764873.1764876

7. Cardwell, N., Cheng, Y., Gunn, C.S., Yeganeh, S.H., Jacobson, V.: BBR: congestion-based congestion control. Queue **14**(5), 50:20–50:53 (2016). https://doi.org/10.1145/3012426.3022184
8. Dong, M., Li, Q., Zarchy, D., Godfrey, P.B., Schapira, M.: PCC: re-architecting congestion control for consistent high performance. In: 12th USENIX Symposium on Networked Systems Design and Implementation (NSDI 2015), pp. 395–408. USENIX Association, Oakland (2015). https://www.usenix.org/conference/nsdi15/technical-sessions/presentation/dong
9. Dong, M., et al.: PCC Vivace: online-learning congestion control. In: 15th USENIX Symposium on Networked Systems Design and Implementation (NSDI 2018), pp. 343–356. USENIX Association, Renton (2018). https://www.usenix.org/conference/nsdi18/presentation/dong
10. Guok, C., Robertson, D., Thompson, M., Lee, J., Tierney, B., Johnston, W.: Intra and interdomain circuit provisioning using the OSCARS reservation system. In: 2006 3rd International Conference on Broadband Communications, Networks and Systems, BROADNETS 2006, pp. 1–8, October 2006. https://doi.org/10.1109/BROADNETS.2006.4374316
11. Ha, S., Rhee, I., Xu, L.: CUBIC: a new TCP-friendly high-speed TCP variant. SIGOPS Oper. Syst. Rev. **42**(5), 64–74 (2008). https://doi.org/10.1145/1400097.1400105
12. Hock, M., Bless, R., Zitterbart, M.: Experimental evaluation of BBR congestion control. In: 2017 IEEE 25th International Conference on Network Protocols (ICNP), pp. 1–10 (2017). https://doi.org/10.1109/ICNP.2017.8117540
13. Hong, C., et al.: Achieving high utilization with software-driven WAN. In: Proceedings of the ACM SIGCOMM 2013 Conference on SIGCOMM, SIGCOMM 2013, pp. 15–26. ACM, New York (2013). https://doi.org/10.1145/2486001.2486012
14. Jain, R., Chiu, D.M., Hawe, W.R.: A quantitative measure of fairness and discrimination for resource allocation in shared computer system, vol. 38. Eastern Research Laboratory, Digital Equipment Corporation Hudson, MA (1984)
15. Jain, S., et al.: B4: experience with a globally-deployed software defined WAN. In: Proceedings of the ACM SIGCOMM 2013 Conference on SIGCOMM, SIGCOMM 2013, pp. 3–14. ACM, New York (2013). https://doi.org/10.1145/2486001.2486019
16. Jiang, H., Dovrolis, C.: Why is the internet traffic bursty in short time scales? In: Proceedings of the 2005 ACM SIGMETRICS International Conference on Measurement and Modeling of Computer Systems, SIGMETRICS 2005, pp. 241–252. ACM, New York (2005). https://doi.org/10.1145/1064212.1064240
17. Kozu, T., Akiyama, Y., Yamaguchi, S.: Improving RTT fairness on cubic TCP. In: 2013 First International Symposium on Computing and Networking, pp. 162–167, December 2013. https://doi.org/10.1109/CANDAR.2013.30
18. Liu, H.H., et al.: Efficiently delivering online services over integrated infrastructure. In: 13th USENIX Symposium on Networked Systems Design and Implementation (NSDI 2016), pp. 77–90. USENIX Association, Santa Clara (2016). https://www.usenix.org/conference/nsdi16/technical-sessions/presentation/liu
19. Ma, S., Jiang, J., Wang, W., Li, B.: Towards RTT fairness of congestion-based congestion control. CoRR abs/1706.09115 (2017). http://arxiv.org/abs/1706.09115
20. Meng, T., Schiff, N.R., Godfrey, P.B., Schapira, M.: PCC proteus: scavenger transport and beyond. In: Proceedings of the Annual Conference of the ACM Special Interest Group on Data Communication on the Applications, Technologies, Architectures, and Protocols for Computer Communication, SIGCOMM 2020, pp. 615–631. Association for Computing Machinery, New York (2020). https://doi.org/10.1145/3387514.3405891

21. Mirza, M., Sommers, J., Barford, P., Zhu, X.: A machine learning approach to TCP throughput prediction. IEEE/ACM Trans. Netw. **18**(4), 1026–1039 (2010). https://doi.org/10.1109/TNET.2009.2037812

22. Pedregosa, F., et al.: Scikit-learn: machine learning in Python. J. Mach. Learn. Res. **12**, 2825–2830 (2011)

23. Vishwanath, K.V., Vahdat, A.: Evaluating distributed systems: does background traffic matter? In: USENIX 2008 Annual Technical Conference, ATC 2008, pp. 227–240. USENIX Association, Berkeley (2008). http://dl.acm.org/citation.cfm?id=1404014.1404031

24. Winstein, K., Balakrishnan, H.: TCP ex Machina: computer-generated congestion control. In: Proceedings of the ACM SIGCOMM 2013 Conference on SIGCOMM, SIGCOMM 2013, pp. 123–134. ACM, New York (2013). https://doi.org/10.1145/2486001.2486020

25. Yin, Q., Kaur, J.: Can machine learning benefit bandwidth estimation at ultra-high speeds? In: Karagiannis, T., Dimitropoulos, X. (eds.) PAM 2016. LNCS, vol. 9631, pp. 397–411. Springer, Cham (2016). https://doi.org/10.1007/978-3-319-30505-9_30

A Dynamic Time Warping and Deep Neural Network Ensemble for Online Signature Verification

Mandlenkosi Victor Gwetu[✉]

University of KwaZulu-Natal, Private Bag X54001, Durban 4000, South Africa
gwetum@ukzn.ac.za

Abstract. Dynamic Time Warping (DTW) is a tried and tested online signature verification technique that still finds relevance in modern studies. However, DTW operates in a writer-dependent manner and its algorithm outputs unbounded distance values. The introduction of bounded outputs offers the prospect of cross pollination with other regression models which provide normalized outputs. Writer-dependent methods are heavily influenced by the richness of the available reference signature sets. Although writer-independent methods also use reference signatures, they have the ability to learn general characteristics of genuine and forged signatures. This ability particularly gives them an edge at detecting skilled forgeries. Noting that DTW, on the other hand, has a strength at random signature verification, this study proposes a model which combines DTW and Deep Neural Networks (DNNs). When trained on a class balanced training set from the BiosecurID dataset, using a best vs 1 reference signature selection scheme, the proposed hybrid model outperforms previous methods, achieving Equal Error Rates of 5.17 and 2.64 for skilled and random signature cases, respectively.

Keywords: Normalized Dynamic Time Warping · Online signature verification · Deep Neural Network · Ensemble · Writer-independent

1 Introduction

Despite the emergence of several modern digital biometric solutions such as fingerprint and face recognition systems, signature verification is still being used as a means of authentication in banks and retail shops. While this usage can be combined with other authentication measures such as inspecting a valid identity document, in some retail shops, possession of a store card and a valid signature is enough to transact.

The benefits of signature verification include non-invasiveness and the flexibility of re-enrollment. If a user suspects that their signature may have been forged, they have the opportunity to adopt a new one and request for a signature update, on the grounds of suspected forgery. If an individual's fingerprint is forged through cast techniques [1], for example, it is not possible to alter the authentic fingerprint since it is a permanent biological trait. However, since

ⓒ Springer Nature Switzerland AG 2021
E. Renault et al. (Eds.): MLN 2020, LNCS 12629, pp. 141–153, 2021.
https://doi.org/10.1007/978-3-030-70866-5_9

online signatures can be rehearsed, verification systems need to be cognisant of hidden tendencies that can be misleading in such cases of forgery. Since humans can be subject to bias and non-scalable performance in signature verification, automated approaches are often the preferred solution. In this study we highlight a potential drawback of automated approaches - overemphasis on complex models which may be able to screen salient skilled forgeries, but be proned to breach by random forgeries.

Online signature verification compares time series pairs acquired from the same digital writing device, but at different times. It is generally considered to be more robust than offline signature verification, which merely compares global properties from static images of complete signatures [2,9]. Since it is difficult to reproduce an online signature verbatim, it is no surprise that large intra-personal variability is often observed when analyzing enrolled signatures. The availability of digital pen tablets enables easier practice of online signature imitation, giving rise to more skilled forgeries, which have close resemblance to genuine signatures.

Online signature verification is a challenging problem given that any two authentic signatures from the same user are highly unlikely to be identical and may not even be of the same length. The predominant approaches to solving this problem in previous work, have focused mainly on Dynamic Time Warping (DTW) [2,5], Hidden Markov Models (HMMs) [6,14] and Recurrent Neural Networks (RNNs) [7,12,13]. DTW is a legacy method for measuring the similarity between two signals of different lengths. Because DTW is ultimately just an instance based distance metric, it is unable to represent abstract models that support intra-user variability. HMMs support user inconsistencies through a series of chronological state transitions and observation emissions drawn from statistical distributions. Their drawbacks include the need for user-specific models and the Markovian assumption which states that the current state depends only on its previous state. RNNs are deep learning models which can capture long term ordered state dependencies but are reliant on significant amounts of training data.

Tolosana et al. [12] pioneered the use of a Siamese RNN architecture for online signature verification. Their experiments considered two scenarios of reference and query signature comparisons: 1vs1 and 4vs1. In the former, query signatures are separately matched against individual reference signatures. In the latter, a query signature's average verification score is computed using all four reference signatures. Experiments are conducted on the BioSecurID dataset, with the first 300 users forming the training set while evaluation is based on the last 100 users. Equal Error Rates (EERs) of 6.44% and 5.58% are reported for the 1vs1 and 4vs1 cases, respectively.

A subsequent extended study explored the robustness of various RNN designs such as Long Short-Term Memory (LSTM), Gated Recurrent Units (GRUs) and bidirectional schemes, in the context of skilled and random forgeries [13]. Respective EERs of 5.5% and 3% are reported when skilled and random forgeries are tested separately on the bidirectional LSTM. It is however surprising to note that the proposed models show weak performance when trained on skilled forgeries but tested on random forgeries. The best results in this scenario are EERs of 19.14% and 19.69%, achieved by the bidirectional GRU for the 1vs1 and 4vs1 cases, respectively. This reveals a major weakness in the approach, in that while it learns how to effectively detect skilled forgeries, it becomes susceptible to random/zero-effort forgeries. In comparison, the legacy DTW approach, based on 9 features chosen using the Sequential Forward Feature Selection (SFFS) algorithm; achieved random forgery EERs of 0.94% and 0.5% for the 1vs1 and 4vs1 cases, respectively [3]. It is therefore, worthwhile exploring the use of a hybrid approach that combines deep learning and DTW, to get the best of both methods in online signature verification.

This study proposes the use of DTW as an input to a fully connected Deep Neural Network (DNN) that predicts online signature similarity. Calculation of DTW distances prior to DNN training, essentially serves as a means of converting variable length online signatures to informative fixed length feature vectors. A DTW distance normalization technique that is anchored on statistics drawn from a user's reference signature set, is proposed for ensuring the that DTW output is comparable to the DNN sigmoid output. Lastly, an ensemble of DNN output and normalized DTW distance is generated, in anticipation of superior effectiveness.

The remainder of this paper is structured as follows. The next Section describes the proposed methods for achieving a DNN and DTW hybrid approach. Section 3 outlines the experimental protocol and dataset used for comparatively evaluating online signature verification performance. Results are presented and discussed in Sect. 4, before final conclusions are drawn and possible future work is proposed.

2   Proposed Methods

This study seeks to improve online signature verification through the use of a normalized DTW distance metric combined with a user-independent DNN model.

2.1   Normalized Dynamic Time Warping

Although DTW is an established algorithm that was initially applied to speech and natural language recognition [11], it has stood the test of time and still finds relevance in modern studies focusing on online signature verification [10, 13]. DTW offers the benefit of variable length signal comparison, albeit with a notable computational overhead. As an additional drawback, it returns an unbounded real number that is highly coupled to the given inputs and context, making relative comparison of outputs less objective. For example, in the case of reference and query online signature comparison, an unskilled forgery of a simple enrolled signature may result in a smaller DTW distance than an authentic reenactment of a complex enrolled signature. Likewise, the difference between any two DTW distance values does not allow for general inference. Nonetheless when DTW is applied to user specific signature verification, informative EERs can be derived.

An online reference signature R_N^V, its associated query signature Q_M^V and the warping path W_K between them, can be represented as follows.

$$R_N^V = \{r_i \in \mathbb{R}^V, 1 \leq i \leq N\}. \tag{1}$$

$$Q_M^V = \{q_i \in \mathbb{R}^V, 1 \leq j \leq M\}. \tag{2}$$

$$W_K = \{w_k = (i,j)_k, 1 \leq k \leq K\}. \tag{3}$$

Since the sequences R_N^V and Q_M^V both have V features recorded at each time stamp, DTW can compute an optimal alignment between them, even if $N \neq M$. The elements $(i,j)_k$ are index pairs such that i and j are as defined in Eqs. 1 and 2. The warping path is of length K, such that $\max(N,M) \leq K \leq N+M-1$ [10].

The DTW algorithm requires the computation of a metric such as the Euclidean distance:

$$d_{ij} = ||r_i - q_j||^2, \tag{4}$$

to compare points at positions i and j in the respective sequences. A matrix D is used to recursively compute a cumulative distance from the start of the warping path such that:

$$D_{[i,j]} = d_{ij} + \max(D_{[i-1,j]}, D_{[i-1,j-1]}, D_{[i,j-1]}), \tag{5}$$

where $D_{[1,1]} = d_{11}$ forms the base case. We introduce two extended notations:

1. $D_{[i,j]}^{(s_1, s_2)}$ for use when it beneficial to explicitly show the signatures whose similarity is being measured and
2. $D_{[i,j]}^S$ which represents the DTW between two unspecified members from the same set of sequences, S.

When the algorithm terminates, the cost associated with the optimal warping path can be represented by the value $D_{[N,M]} \in \mathbb{R}$. We seek to normalize this value such that it is scaled to the semi-open range $[0, 1)$, where 0 represents a perfect match and the concept of a total mismatch is represented by 1 but is never actually reached.

Online signature verification systems generally require users to initially enroll a set of reference signatures for future comparison against query signatures [9]. Such a set can be represented by $R = \{R_N^V\}$ and the corresponding set of all possible paired combinations by R^2. If there are $E = |R|$ enrolled reference signatures then the total number of possible reference signature pairs, $|R^2|$ is represented by the following combination:

$$_EC_2 = \frac{E!}{(E-2)! * 2}. \tag{6}$$

From the set R^2, we compute the observed range of reference signature variation for each user as:

$$range(D_{[N,M]}^R) = \max(D_{[N,M]}^R) - \min(D_{[N,M]}^R), \qquad (7)$$

where the DTW distance for any reference signatures pair in R^2 is represented by $D_{[N,M]}^R$. Likewise, we compute the mean reference signature variation for each user as:

$$mean(D_{[N,M]}^R) = \frac{\sum D_{[N,M]}^R}{{}_E C_2}. \qquad (8)$$

A normal distribution is then modelled using the parameters $\mu = mean(D_{[N,M]}^R)$ and $\sigma = \frac{range(D_{[N,M]}^R)}{2}$. Such a model is expected to yield values close to 1 for all DTW distances between reference signature pairs, since $D_{[N,M]}^R$ is within one standard deviation from the mean of the resulting normal distribution. The normalized DTW distance between a reference signature R_N^V and a query signature Q_M^V can then be derived as follows:

$$Z(R_N^V, Q_M^V) = 1 - e^{-\frac{(D_{[N,M]}^{(R_N^V, Q_M^V)} - \mu)^2}{2\sigma^2}}. \qquad (9)$$

Hence, $Z(R_N^V, R_N^V) = 0$, for a perfect match, while $0 < Z(R_N^V, Q_M^V) \leq 1$ for an imperfect match.

2.2   DTW Deep Neural Network Ensemble

The online signature verification model proposed in this study is novel in four respects:

1. When calculating DTW distances, the reference signature that has the best alignment to a given query signature is used as a representative of all the signatures in R, facilitating what we refer to as the best vs 1 scheme.
2. The DNN component learns how to discriminate forgeries using typical DTW distances from the training set, as input.
3. Instance redundancy is explored as a means of addressing class imbalance in the training set.
4. The ensemble component aggregates the DNN output with a normalized DTW measure, in an attempt to improve verification effectiveness without compromising the identification of random forgeries.

Figure 1 shows the proposed regression model for predicting query signature authenticity in a user agnostic manner. While both the DNN and normalized DTW components perform reference and query signature comparisons using a best vs 1 scheme, the latter also uses statistics drawn from all reference signature pairs of a given user. The DNN accepts several DTW outputs of the form $D_{[N,M]}^{(R_N^1, Q_M^1)}$, where DTW is performed separately for each online signature feature. The DNN is constructed and trained in line with the parameters shown in Table 1. Although several other parameter combinations were tried without any significant change in training accuracy, this study does not assume that the adopted parameters are optimal. Normalized DTW is performed by calculating $range(D^R_{[N,M]})$ and $mean(D^R_{[N,M]})$ then using them as the standard deviation σ and mean μ of a normal distribution, respectively. DTW outputs of the form $D^{(R^V_N, Q^V_M)}_{[N,M]}$ are subsequently mapped to this distribution to determine their probability of occurrence based on R^2. In this case, the Euclidean distance metric of the DTW algorithm, takes all V online signature features into account, resulting in only one output. Finally, the average of the DNN and normalized DTW outputs is used as an ensemble of the two verification models.

Table 1. Properties of the proposed fully connected DNN models.

Property	Value
Hidden Layer Activation Function	Leaky RELU
Output Layer Activation Function	Sigmoid
Layer Node Distribution [Input, ... Output]	[9, 100, 50, 9, 1]
Optimizer	Adam
Cost Function	Binary Crossentropy

3   Experimental Protocol

This study investigates the viability of a DTW and DNN hybrid model for improved online signature verification. The following research questions are formulated to guide the study towards this objective.

1. How effective is the proposed DNN model which is based on DTW outputs, in comparison to existing deep learning models that use lower level online signature features?
2. How does the proposed normalized DTW method compare against the legacy DTW algorithm, as a similarity metric?
3. How does the proposed DTW and DNN hybrid model compare against its constituent methods when they are used independently, in terms of verification effectiveness?
4. Does instance redundancy improve model verification effectiveness?

The remainder of this section describes the dataset and experiments employed to answers these questions.

3.1   Dataset

This study utilizes the BiosecurID dataset [4,8] as a benchmark for objective comparison with relevant previous studies [12,13]. The dataset captures skilled

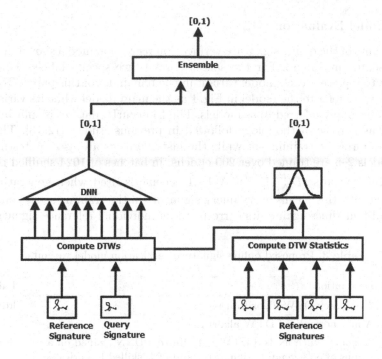

Fig. 1. Proposed online signature verification main model.

forgery and genuine attempts at impersonating online signatures of 400 users, under a controlled and supervised environment. A total of 16 genuine and 12 fake attempts are evenly captured over four separate acquisition sessions. For each user, the first four genuine signatures from the first session are used as reference signatures. For experimental purposes, the remaining 12 genuine and 12 fake attempts form the query signatures of each user. An initial set of 23 time series functions are derived from each acquired signature and recorded in a text file [12]. The dynamic information represented by these functions can be categorized as described in Table 2. The SFFS algorithm is then applied to these initial functions, resulting in the published dataset with 9 optimal features which are normalized to 0-mean and unit variance.

Table 2. Initial pool of features from the BiosecurID dataset.

Category	Description	Function #
Raw	Features extracted directly from the pen-tablet	1–3
Absolute	Non-negative functions based on raw features	5–7
Orientation	Features associated with an angle	4, 18, 20, 21
Derivatives	First and second order differentials	8–16, 19
Windowed	Features based on a sliding interval range	17, 22, 23

3.2   Model Evaluation

Ten variants of the online signature verification regression model shown in Fig. 1, are presented in Table 3. For the sake of brevity we specify labels, which are adopted to represent each model variant in the remainder of this paper. To avoid ambiguity, we refer to the model in Fig. 1 as the main model while its variants in Table 1 are simply referred to as models. The BiosecurID dataset is split into two subsets, as per the methodology followed in previous studies [12,13]. The first 300 users form the training set while the last 100 users are used as testing set. The models 2–5 are trained over 200 epochs, in batches of 1024 shuffled records of 9 DTW outputs ($D_{[N,M]}^{(R_N^1, Q_M^1)}$). A 1vs1 scheme is used when generating the DTW outputs, this means every query signature in the training set is compared a total of four times against its corresponding individual reference signatures.

Table 3. Proposed online signature verification model variants.

Model	Description	Label
0	The legacy DTW algorithm	ldtw
1	A new normalized DTW algorithm	ndtw
2	A model which feeds 9 DTW outputs to a DNN trained on a corpus of 50% genuine signatures and 50% skilled forgeries	ddnn-s
3	A model which feeds 9 DTW outputs to a DNN trained on a corpus of 50% genuine signatures and 50% random forgeries	ddnn-r
4	A model which feeds 9 DTW outputs to a DNN trained on a corpus of 33.3% genuine signatures, 33.3% skilled forgeries and 33.3% random forgeries	ddnn-a
5	A model which feeds 9 DTW outputs to a DNN trained on a corpus of 50% genuine signatures, 25% skilled forgeries and 25% random forgeries	ddnn-b
6	A hybrid model of ddnn-s and ndtw	hmdn-s
7	A hybrid model of ddnn-r and ndtw	hmdn-r
8	A hybrid model of ddnn-a and ndtw	hmdn-a
9	A hybrid model of ddnn-b and ndtw	hmdn-b

The convention in previous literature [12,13] is to use both 1vs1 and 4vs1 schemes for evaluating models. We argue that the former is not of any practical benefit since only one outcome is generally expected for each online query signature in biometric systems. Instead of a 4vs1 scheme we propose a new scheme which we refer to as best vs 1 (bvs1), for use in both training and evaluation of models 2–5. In this case, only the reference signature with the lowest DTW distance ($D_{[N,M]}^{(R_N^V, Q_M^V)}$) is used for verification. The intuition behind this scheme is to evaluate each query signature under optimal conditions, thus reducing the impact of reference signature inconsistency, on verification performance. Such a scheme is expected to give credible rejections and be robust to intra-user variability. Additionally, the computational overhead of DNN training and evaluation is reduced by a factor of E, since only one reference signature is ultimately enforced for each user.

Signature verification datasets should comprise both authentic and forged instances for effective training and realistic testing. We consider three scenarios for achieving this: combining genuine signatures with either skilled forgeries, random forgeries or all forgeries (skilled and random). This allows us to create models based on different experimental environments, in order to assess model performance under different constraints. For example, it now becomes possible to train a model on genuine and random forgery signatures but test it on skilled forgeries. This is useful in contexts where it may be combersome to attain enough skilled forgeries for training.

After noting the class imbalance exhibited by dnn-a, an additional model (dnn-b) was proposed in an attempt to reduce bias towards forgeries. In dnn-b a simple duplication of each genuine signature is used to ensure genuine and forged signatures have the same frequency in the dataset. It was not immediately evident whether this added feature in dnn-b would be enough to differentiate its performance from dnn-a. All models are evaluated on the testing set to measure their capacity for user independence and inter-session variability. Biometric systems generally use an adjustable criteria to control sensitivity, in pursuit of a simultaneous reduction in false rejection and false acceptance errors. An EER is computed from a ROC (Receiver Operating Characteristic) curve to identify the best possible balance between these two errors. It is the standard evaluation metric in online signature verification [9]. Accordingly, our model performances are compared using EERs, with lower values indicating superiority.

The model is implemented in python using scientific modules such as numpy and tensorflow. Computations are carried out on the Google Colab[1] platform using a GPU runtime environment.

4   Results

This section reports on notable observations from model training and evaluation.

4.1   Training

Figure 2 shows the training histories of DNNs which observe the parameters in Table 1. Figures 2a–d represent histories of the models: ddnn-s, ddnn-r, ddnn-a and ddnn-b, respectively. Although all 4 models demonstrate clear convergence during training, the problem space of ddnn-r seems the least complex as its initial and final training accuracies of 0.76 and 0.97, respectively. It also has the least training time as shown in Table 4. The increased total CPU time for ddnn-a and ddnn-b can be attributed to the larger training sets required by these models.

---

[1] https://colab.research.google.com/.

Evaluating all 4 models took a total of less than 2 s. It is important to note that despite the relatively fast training and evaluation times of these models, they all depend on prior DTW distance calculations. A significant computational overhead was observed in these calculations, which took an average of 1.75 seconds per signature pair.

Table 4. Model training times and final accuracies.

Model	Total CPU time (s)	Final accuracy
ddnn-s	11.3	0.88
ddnn-r	10.1	0.97
ddnn-a	13.3	0.94
ddnn-b	17.7	0.93

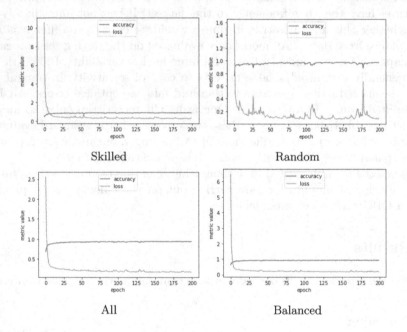

Fig. 2. Training costs and accuracies.

4.2   Evaluation

Table 5 compares the performance of models utilized in this study against results published in previous work [13], based on a similar experimental protocol. Performances are calculated on a test set of 100 users, with 4 reference signatures, 12 genuine query signatures and 12 of either skilled or random forgery signatures, provided per user. This means that each EER is informed by a ROC curve

Table 5. Model evaluations through EER. 4vs1, LSTM, GRU, BLSTM and BGRU EERs are taken from a previous study [13].

Category	Model	Skilled	Random	Average
DTW-based	4vs1	7.75	0.50	**4.13**
	ldtw	25.67	24.67	25.17
	ndtw	5.50	8.72	7.11
Train: skilled	LSTM	5.58	24.03	14.81
	GRU	6.25	28.69	17.47
	BLSTM	4.75	24.03	14.39
	BGRU	4.92	19.69	12.31
	ddnn-s	14.58	7.25	10.92
	hmdn-s	7.50	3.67	**5.59**
Train: random	LSTM	15.17	4.08	9.63
	GRU	13.92	4.25	18.17
	BLSTM	15.58	3.89	9.735
	BGRU	12.33	3.25	7.79
	ddnn-r	26.08	3.18	14.63
	hmdn-r	6.33	2.83	**4.58**
Train: all	LSTM	6.17	3.67	4.92
	GRU	5.58	3.63	4.61
	BLSTM	5.50	3.00	**4.25**
	BGRU	5.92	2.92	4.42
	ddnn-a	13.74	4.42	9.08
	hmdn-a	6.50	3.22	4.86
Train: balanced	ddnn-b	13.67	4.58	9.13
	hmdn-b	5.17	2.64	**3.91**

derived from 2400 signature verifications. This study also reports the average skilled and random forgery EER as a means of evaluating the overall robustness of a model against forgeries.

The adopted bvs1 reference signature selection method appears to favour ndtw over ldtw, with the former achieving an EER that is over 3 times smaller than the latter. Although ndtw achieved a better skilled ERR than 4vs1, the latter remains superior in terms of random EERs - recording the lowest EER of 0.5 out of all the models considered. This suggests that the 4vs1 approach may have superior generalization capability than the bvs1 greedy approach. The lowest average EERs were recorded by the hmdn models when training using equal numbers of genuine signatures and either skilled (5.59) or random (4.58) forgeries. Our proposed models demonstrate superiority against more advanced deep learning approaches from previous work, in all categories except the "Train: all", which has the challenge of class imbalance. This problem seems to be adequately addressed by hmdn-b which records the lowest average EER of 3.91, from the models considered. Despite the observed superiority of 4vs1 over bvs1, the ensemble-based hmdn-b model is seemingly able to compensate for this weakness, and ultimately achieve state of the art performance.

5   Conclusion

This study proposed a hybrid online signature verification approach, which utilizes normalized DTW and DTW-based DNNs. This section presents a summary of the observed findings and outlines possible future work.

The proposed DNN models (ddnn-s, ddnn-r, ddnn-a and ddnn-b) which use a DTW input layer was shown to provide performance which is largely inferior but comparable to existing deep learning models that use lower level online signature features. The performance of the former exceeded that of the latter only in the category of skilled forgeries, in terms of average EER. The proposed normalized DTW method outperformed the legacy DTW in the context of bvs1, for both skilled and random forgeries. This normalization is a breakthrough as it paves the way for the use of DTW in conjunction with other regression models that yield bounded values. When using a class-balanced training set, the proposed DTW and DNN hybrid model (hmdn-b) not only surpassed its constituent methods, but it also outperformed all the other models considered in this study, in terms of verification effectiveness. It yielded consistent performance for both skilled and random forgeries. The inclusion of instance redundancy seems to have had a positive effect on verification effectiveness.

Although the hmdn-b model surpassed other models considered in this study, there were cases in which the 4vs1 and legacy deep learning models excelled. Future work will extend this hybrid model through the use of the 4vs1 instead of bvs1 reference signature selection and legacy deep learning models instead of a simple DNN.

References

1. Champod, C., Espinoza, M.: Forgeries of fingerprints in forensic science. In: Marcel, S., Nixon, M.S., Li, S.Z. (eds.) Handbook of Biometric Anti-Spoofing. ACVPR, pp. 13–34. Springer, London (2014). https://doi.org/10.1007/978-1-4471-6524-8_2
2. Fahmy, M.M.: Online handwritten signature verification system based on DWT features extraction and neural network classification. Ain Shams Eng. J. 1(1), 59–70 (2010)
3. Gomez-Barrero, M., Galbally, J., Fierrez, J., Ortega-Garcia, J., Plamondon, R.: Enhanced on-line signature verification based on skilled forgery detection using sigma-lognormal features. In: 2015 International Conference on Biometrics (ICB), pp. 501–506. IEEE (2015)
4. Gomez-Barrero, M., Galbally, J., Morales, A., Fierrez, J.: Privacy-preserving comparison of variable-length data with application to biometric template protection. IEEE Access 5, 8606–8619 (2017)

5. Jaini, A.A., Sulong, G., Rehman, A.: Improved dynamic time warping (DTW) approach for online signature verification. arXiv preprint arXiv:1904.00786 (2019)
6. Kashi, R., Hu, J., Nelson, W., Turin, W.: A hidden Markov model approach to online handwritten signature verification. Int. J. Doc. Anal. Recogn. **1**(2), 102–109 (1998)
7. Lai, S., Jin, L., Yang, W.: Online signature verification using recurrent neural network and length-normalized path signature descriptor. In: 2017 14th IAPR International Conference on Document Analysis and Recognition (ICDAR), vol. 1, pp. 400–405. IEEE (2017)
8. Martinez-Diaz, M., Fierrez, J., Krish, R.P., Galbally, J.: Mobile signature verification: Feature robustness and performance comparison. IET Biometrics **3**(4), 267–277 (2014)
9. Mlaba, A., Gwetu, M., Viriri, S.: A distance-based approach to modelling reference signature for verification. In: 2017 Conference on Information Communication Technology and Society (ICTAS), pp. 1–6. IEEE (2017)
10. Rashidi, S., Fallah, A., Towhidkhah, F.: Similarity evaluation of online signatures based on modified dynamic time warping. Appl. Artif. Intell. **27**(7), 599–617 (2013)
11. Sakoe, H., Chiba, S.: Dynamic programming algorithm optimization for spoken word recognition. IEEE Trans. Acoust. Speech Signal Process. **26**(1), 43–49 (1978)
12. Tolosana, R., Vera-Rodriguez, R., Fierrez, J., Ortega-Garcia, J.: Biometric signature verification using recurrent neural networks. In: 2017 14th IAPR International Conference on Document Analysis and Recognition (ICDAR), vol. 1, pp. 652–657. IEEE (2017)
13. Tolosana, R., Vera-Rodriguez, R., Fierrez, J., Ortega-Garcia, J.: Exploring recurrent neural networks for on-line handwritten signature biometrics. IEEE Access **6**, 5128–5138 (2018)
14. Zou, J., Wang, Z.: Application of HMM to online signature verification based on segment differences. In: Sun, Z., Shan, S., Yang, G., Zhou, J., Wang, Y., Yin, Y.L. (eds.) CCBR 2013. LNCS, vol. 8232, pp. 425–432. Springer, Cham (2013). https://doi.org/10.1007/978-3-319-02961-0_53

Performance Evaluation of Some Machine Learning Algorithms for Security Intrusion Detection

Ouafae Elaeraj[1]($\boxtimes$), Cherkaoui Leghris[1], and Éric Renault[2]

[1] L@M, RTM Team, Faculty of Sciences and Techniques Mohammedia,
Hassan II University of Casablanca, Casablanca, Morocco
`ouafaeelaeraj@gmail.com`, `cherkaoui.leghris@fstm.ac.ma`
[2] LIGM, University Gustave Eiffel, CNRS, ESIEE Paris, 93162 Noisy-le-Grand, France
`eric.renault@esiee.fr`

Abstract. The growth of the Internet and the opening of systems have led to an increasing number of attacks on computer networks. Security vulnerabilities are increasing, in the design of communication protocols as well as in their implementation. On another side, the knowledge, tools and scripts, to launch attacks, become readily available and more usable. Hence, the need for an intrusion detection system (IDS) is also more apparent. This technology consists in searching for a series of words or parameters characterizing an attack in a packet flow. Intrusion Detection Systems has become an essential and critical component in an IT security architecture. An IDS should be designed as part of a global security policy. The objective of an IDS is to detect any violation of the rules according to the local security policy, it thus makes it possible to report attacks. This last multi-faceted, difficult to pin down when not handled, but most of the work done in this area remains difficult to compare, that's why the aim of our article is to analyze and compare intrusion detection techniques with several machine learning algorithms. Our research indicates which algorithm offers better overall performance than the others with the IDS field.

Keywords: IDS · Machine learning · KNN · SVM and Decision tree

1 Introduction

Information systems and therefore Internet contain very critical information for the conduct of organizational activities. It is essential to protect them against intruders and unauthorized access. In this perspective, providing a computer security system has become an essential component of the infrastructure companies. With the versatile technologies, Intrusion Detection System (IDS) can detect attacks that traditional firewalls cannot do. It analyzes the data packages to the highest layer of the OSI model and oversees why the individually run applications.

Systems with anomaly detection can also detect new patterns of flexible attacks thanks to their increasing process security of a network. However, do not believe that

© Springer Nature Switzerland AG 2021
É. Renault et al. (Eds.): MLN 2020, LNCS 12629, pp. 154–166, 2021.
https://doi.org/10.1007/978-3-030-70866-5_10

IDS software can replace the firewall, only a combination of the two safety components provides optimum protection.

As intrusion detection systems are active components of a network, they can also be a potential target for attacks, especially if the intruder knows their existence. Because of their vulnerability to DOS attack, IDS software can very soon become extinct. In addition, the hacker can also enjoy the automatic notification function of intrusion detection systems to launch DOS attacks from IDS. In particular, the detection of anomalies are a major weak point in this case if the configuration is inadequate. Indeed, if the settings are too sensitive, the amount of alert is then relatively high, even in the absence of unauthorized access. In this intent, machine learning (including supervised learning, unsupervised learning and deep learning) can be used to significantly improve the reliability of detecting the face of threats. The goal of an implementation with machine learning is to detect attacks or polymorphic and unknown threats with different algorithms in this area. Intrusion detection systems are tools that aim to detect malicious activity on the target that they monitor. An alert will be triggered when a malicious behavior is detected. Intrusion detection systems are used, in addition to traditional solutions such as Firewalls, to detect different types of malicious use of their target that cannot be detected by them.

"In 2018, the number of vulnerabilities exceeded that of the previous year, with an increase of 12% of the total number of vulnerabilities reported in 2017", says Skybox Security, the world leader in operations, analysis and reporting in the field of cybersecurity [13].

As the graph below, 16,412 new CVE (Common Vulnerabilities and Exposures) were identified by the NVD (National Vulnerability Database) in 2018 compared 14,595 in 2017, which was already a record (Fig. 1).

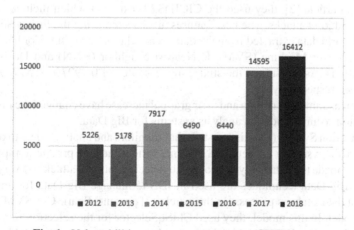

Fig. 1. Vulnerabilities and current exposures (CVE) by year.

Many researchers interested in Machine Learning to detect and predict attacks similar to known attacks, so, it becomes possible to develop better protection tools, capable of detecting novel attacks. Among these attacks, we could note:

- DoS (Denial of Service): Attacker aims to make a computer application unable to meet the requests of its users;
- U2R (User to Root): The attacker has local access to the victim machines and tries to gain super user privileges;
- R2L (Remote to Local): The attacker has no account on the victim machine, so trying to access;

This article offers a comparison of performance in intrusion detection with different machine learning algorithms. The rest of the article is organized in the following sections. In Sect. 2, we briefly describe the various existing researches used for the IDS. In Sect. 3, we study the different algorithms to compare. The comparison of these was described in Sect. 4. In Sect. 5, we summarize the article.

2 Related Works

The entire KDD CUP 99 dataset was used to analyze the effectiveness of intrusion detection with different machine learning algorithms such as Bayes Naive Bayes, J48, and Random Forest J48Graft. The experience results show that the J48, J48Graft and random forest work better than other machine learning algorithms [1]. On the whole, it is reasonable to say that the machine learning algorithms, used as classifiers for data KDD Cup 1999, does not offer much promise for the attacks and detection U2R/R2L in the context of detection abuse.

New technologies not only make life easier, but also reveal many security concerns; changing types of attacks affects many people, organizations, businesses, and more. Therefore, intrusion detection systems have been developed to avoid financial and emotional loss. In article [2], they used the CICIDS2017 dataset which includes mild and most advanced joint attacks. The best features are selected using the Fisher Score algorithm. Real world data extracted from the dataset are classified as DDoS or benign using the Support Vector Machine (SVM), K Nearest Neighbor (KNN) and Decision Tree (DT) algorithms. As a result of the study, success rates of 0.9997%, 0.5776%, 0.99% were achieved respectively.

The immense amounts of data and their gradual increase have changed the importance of information security and data analysis systems for Big Data.

The Detection System (IDS) is a system that monitors and analyzes data to detect any intrusion into the system or the network. Volume, variety and high production speed in the network have made the data analysis process to detect traditional attacks by very difficult techniques. Big Data techniques are used in IDS to manage Big Data for an accurate and efficient data analysis process. In the article [3] presents Spark-Chi-SVM intrusion detection model. In this model, they used ChiSqSelector for the selection functionality, and built an intrusion detection model using a support vector machine (SVM) classifier on the Apache Spark Big Data platform. They used KDD99 to train us and test the model. In the experiment, they introduced a comparison between the Chi-SVM classifier and the chi-logistic regression classifier. The results of the experiment showed that Spark-The Chi-SVM model has high performance, reduces training time and is effective for Big Data.

To detect and mitigate network attacks, university researchers and practitioners have developed intrusion detection systems (IDS) with automatic response systems. The response system is considered an important component of IDS, because without a timely response, IDS may not function properly to counter various attacks, especially in real time. To respond appropriately, IDSs must select the optimal response option based on the type of network attack. The article [4] provides a comprehensive study of IDS and intrusion response systems (IRS) based on our in-depth understanding of the response option for different types of network attacks. Knowing the path from the IDS to the IRS can help network administrators and network staff to understand how to combat different attacks with advanced technologies.

In the work [5], the authors proposed a method for an IDS (I-ELM) combined incremental extreme learning machine with an adaptive main component (A-PCA). In this method, the relevant characteristics of network traffic are selected adaptively, where the best detection accuracy can then be obtained by I-ELM. They used the standard dataset NSL-KDD and the standard dataset UNSW-NB15 to assess the performance of our proposed method. Thanks to the analysis of the experimental results, we can see that their proposed method has better computational capacity, greater generalization capacity and greater precision.

Various machine learning techniques have been applied to improve the performance of intrusion detection systems, among which there has been growing interest and was considered an effective method. In addition, the quality of training data is also a determinant key that can significantly improve detection capabilities. The article [6] provides an effective intrusion detection framework based on an SVM package with increased functionality. Specifically, the transformation of marginal density ratios is implemented on the original features in order to obtain better quality transformed training data; the SVM package was then used to build the intrusion detection model.

The results of the experiment show that the method proposed, in this research, can achieve good and robust performance, which has huge competitive advantages over other existing methods in terms of accuracy, detection rate, false alarm rate and training speed.

The paper [7] introduces a new clustering algorithm called Peak Density Clustering (PDC), that does not require many parameters and its iterative process is based on density. Because of its simple steps and parameters, it can have many fields of application. It's used to find a more accurate and efficient classifier, which proposes a hybrid learning model based on K-Nearest Neighbors (kNN) in order to detect attacks more efficiently and to introduce the density in KNN. In Nearest Neighbor Density Peak (NNDP), KDD-CUP 99, which is the standard data set in intrusion detection, is used for the experiment. The authors then used the data set to form and calculate some parameters that are used in this algorithm.

Finally, the DPNN classifier is used to classify the attacks. The experiment results suggest that the DPNN works better than the Support Vector Machine (SVM), K-Nearest Neighbors (kNN) and many other machine learning methods, and that it can effectively detect intrusion attacks with good performance in terms of accuracy.

Today, most intrusion detection approaches have focused on problems of feature selection or feature reduction, as some features are irrelevant and redundant, resulting in a long detection process and degrading the performance of an Intrusion Detection

System. The purpose of the study [8] is to identify important reduced input features in the construction of IDS that are efficient and computationally effective. To do this, they studied the performance of three standard feature selection methods using feature selection based on correlation, information gain and gain ratio. This paper proposes the Feature Vitality Based Reduction Method (FVBRM) to identify important reduced input features by applying one of the naive bays of the efficient classifier on reduced datasets for intrusion detection. Empirical results show that the selected reduced attributes provide better performance for designing effective and efficient IDS for network intrusion detection.

3 Learning Machine Under Surveillance Techniques

There is a Good Number of Research Papers Done on Intrusion Detection with Different Best-Known Algorithms like Support Vector Machine (SVM), Decision Arbe and K-Nearest Neighbor (KNN) in This Regard the Goal of Our Article is to Compare Between These to Detect the Most Efficient Algorithm in the Field of Intrusion Detection. The Comparison Will Be Made Based on Accuracy and Test Time of Each Algorithm.

3.1 Support Vector Machine (SVM)

Support Vector Machine (SVM) is an automatic supervised learning algorithm that can be used for classification or regression challenges. However, it is mainly used in classification problems. This technique is a method of classification in two classes that attempts to separate the positive examples of negative ones. The method then seeks the hyperplane that separates the positive examples of negative ones, ensuring that the margin between the closest of either maximum positive and negative. This ensures a generalization of the principle because new examples cannot be too similar to those used to find the hyperplane but be located on one side or the other of the border. The advantage of this method is the selection of support vectors that represent the discriminant vectors by which is determined the hyperplane.

Jamal Hussain [9] proposes a model based on SVM linear kernel and nonlinear operating RSO NSL-KDD data as input to classification methods followed by evaluation of the performance. He used 10 times the cross-validation technique to validate the data set for different SVM classifiers such as linear SVM, SVM Quadratic, SVM Cubic, SVM Gaussian Fine, medium and coarse Gaussian SVM where used for classification with NSL-KDD datasets.

To measure the experimental results, the different performance metrics are used, such as: (i) Overall Accuracy, (ii) specificity, (iii) sensitivity, (iv) G-average and (v) the ROC curve.

$$Overall\ Accuracy = (TP + TN) / (TP + TN + FP + FN);$$

$$Sensitivity = TP / TP + FN :$$

$$Specificity = TN / TN + FP :$$

$$G - mean = \sqrt{sensitivity * specificity};$$

Where: TP = True Positive; FP = False Positive; TN = True Negative; FN = False Negative.

The information visualizes different classifiers such as prediction speed (for binary and multiclass) memory usage, intelligibility and modern flexibility for all classification methods after performing the experiments. It also expresses the parameters and the formation time of the SVMs of the linear and nonlinear kernel. Finally, it is deduced that coarse Gaussian SVM takes the optimal time of 00:00:31 to form the data set (Table 1).

Quadratic and cubic SVMs display the best results with 100% accuracy. The SVM Fine Gaussian- SVM and Medium Gaussian- SVM come in second degree showed the best result among all the classifiers with 99.40% of the precision, the Linear- SVM gave the result with 97.90% of the precision, then the Coarse Gaussian SVM comes last with a result of 97.50%.

The ROC (Receiver Operating Characteristic) curve is generally used to measure the performance of the method. The latter is a plot of the intrusion detection precision compared to the probability of the false positive. It can be obtained by varying the detection threshold.

The operating characteristic curve receiver (ROC) and its surface under the curve (AUC) are used for the analysis of the classifier performance as appears in Fig. 2. The ROC graph is the compromise between benefits and costs, the fine Gaussian SVM gave the maximum result, ie, 99.98% (ROC curve).

Table 1. Accuracy and processing time of SVM algorithm with different methods.

Methods	Linear-SVM	Quadratic-SVM	Cubic-SVM	Fine Gaussian-SVM	Medium Gaussian-SVM	Coarse Gaussian SVM
Accuracy (%)	97.90	100	100	99.40	99.40	97.50
Training Time	02 :57	01 :39	13 :33	01 :09	00 :39	00 :31
ROC curve (AUC) (%)	99,43	99,93	99,90	99,98	99,95	99,49

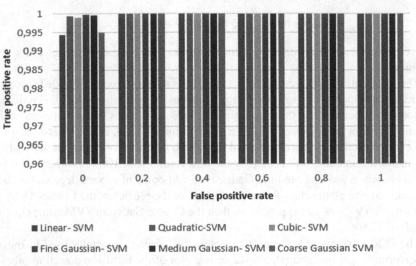

Fig. 2. Curve ROC pour SVM algorithm with different methods.

3.2 Decision Tree

Decision trees are used to visually represent decisions and show or inform decision making. When working with machine learning and data mining, decision trees are used as a predictive model. These models map data observations and draw conclusions about the target value of the data.

The goal of the decision tree learning is to create a model that will predict the value of a target based on inputs variables. In the predictive model, the attributes of the data that are determined by the observation are represented by the branches, while conclusions about the target value of the data are represented in the leaves. When learning a tree, the source data is divided into subsets based on an attribute value test, which is repeated recursively on each of the derived subsets. Once the subset of a node reaches the value equivalent to its target value, the recursion process is completed.

Kajal Rai [10] proposed a decision tree based on intrusion detection. Following the C4.5 decision tree approach, a decision tree algorithm is developed in the proposed system. In the latter, the algorithm is designed to solve two problems such as feature selection and shared value.

Using the information gain, the most relevant features are selected so that the classifier is not biased towards the most frequent values, the split value is selected. The DTS (Decision Tree Split) algorithm can be used for signature-based intrusion detection. It is compared with the Classification, Regression and AD Tree. The experiment is performed on the NSL-KDD (Knowledge Network Security Laboratory Discovery and Data Mining) feature dataset. The time taken by the classifier to build the model and the accuracy obtained is analyzed.

ROC curves of the AD Tree, C4.5, CART and DTS algorithm without selection of features on the NSL-KDD test data are plotted as shown in Fig. 3, as observed from Table 2 and Fig. 3, that the true positive rate of DTS is better than C4.5 technique;

however CART (Classification And Regression Trees) show the best performance in terms of positive real rates. But, if we compare the results in terms of delay to build the model, we can see that CART is very high time compared to other techniques. We can see from the results that the technical proposed, with all the features, get good accuracy with fewer features selected using the information gain. However, CART shows the best performance in terms of positive real rates.

Based on the experiment results (Table 2), it is concluded that the proposed algorithm for signature-based intrusion detection is more effective in detecting attacks in the network with a lower number of features and it takes less time to build the model. It is also concluded that the efficiency depends on the size of the data set and the number of functionalities used to build the decision tree.

Table 2. Accuracy and processing time of decision tree algorithm with different methods.

Methods	AD Tree	CART	C 4.5	DTS
Accuracy (%)	78.45	82, 5	75.6	79.52
Training Time	00:02:10	00:05:09	00:00:07	00:01:24

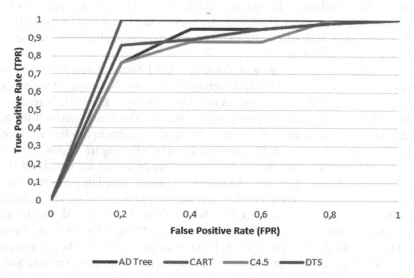

Fig. 3. Comparison of ROC Curve of proposed algorithm with various other classifiers.

3.3 K-Nearest Neighbor (KNN)

The KNN algorithm is a simple supervised machine learning algorithm that can be used to solve classification and regression problems. The KNN algorithm is one of the most basic algorithms in Machine Learning to such an extent that it is considered by some to be a statistical algorithm. It will allow highlighting the membership of a point in an observation group. It is also a data classification algorithm that tries to determine in which group a data point is located by examining the data points surrounding it.

The article [11] proposes a new intrusion detection model for an ad hoc mobile network using the CP-KNN (Conformal Prediction K-Nearest Neighbor) algorithm, which is used to calculate the resemblance between new individuals and other samples of the class using the K-nearest neighbor method, this work aims to classify the audit data for the detection of anomalies and also make it possible to efficiently detect various anomalies with a high rate of true positives, low false positive rates and a high confidence rate. The proposed method is robust, effective and also retains its good detection performance to avoid abnormal activity. In Fig. 4. The graph is based on the classifier which consumes time for its classification at several levels. In this experiment, the accuracy of the nearest K-neighbor is 90% and the accuracy of the nearest K-conformal prediction is 94% as seen in Fig. 4.

In another article [12], which discusses a new technique for learning the behaviour of the program in intrusion detection. Their approach uses the nearest k-neighbour (k NN) classifier to classify each new program behaviour in normal or intrusive class. The frequencies of system calls used by a program, instead of their local order, are used to characterize the behaviour of the program. Each system call is treated as a "word" and each process, that is to say program execution, as a "document". Then the k The NN algorithm, which has been successful in text categorization applications, can be easily adapted to intrusion detection.

Their preliminary experiences with DARPA BSM 1998 audit data have shown that this approach is capable of effectively detecting the intrusive behaviour of the program.

They formed a test data set to assess the performance of kNN. Their results also show that a low rate of false positives can be achieved. The performance of the KNN also depends on the value of k, the number of closest neighbours to the test process. Usually, the optimal value of k is determined empirically. Figure 6 shows the ROC curves for 3 different k values from 5 to 25 when the processes are transformed with the TF-IDF (term frequency-inverse document frequency) weighting method. For this particular data set, $k = 10$ is a better choice than the other values since the attack detection rate reaches 100% faster. For $k = 10$, the kNN can detect 10 of the 35 attacks with a zero false positive rate. And the detection rate reaches 100% quickly when the threshold is raised to 0.72 and the false positive rate remains as low as 0.44%. They also used the frequency weighting method to transform the processes of the same training and test data sets.

Thus, for the frequency weighting method, $k = 15$ provides the lowest false positive rate of 0.87% (46 false alarms out of 5285 normal processes) when the attack the detection rate reaches 100% with the threshold value of 0.99. The reason for the high threshold value is that some attack instances are very similar to normal processes with frequency weight. Therefore, a lower threshold value is necessary, and a better false positive rate can be obtained with the TF-IDF weighting method.

Although this result cannot stand in the way of a more sophisticated data set, text categorization techniques seem well applicable to the field of intrusion detection.

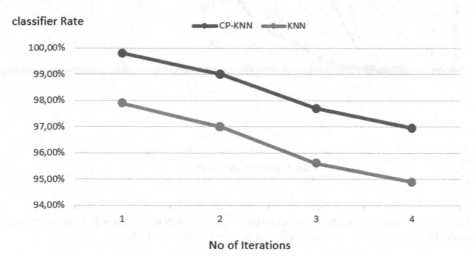

Fig. 4. Classifier Rate for CP-KNN vs KNN.

In Fig. 5 shows a graph based on the classifier who consumes time for its classification at several levels. The proposed CP-KNN reduces the classification period for iteration to several levels, as noted above.

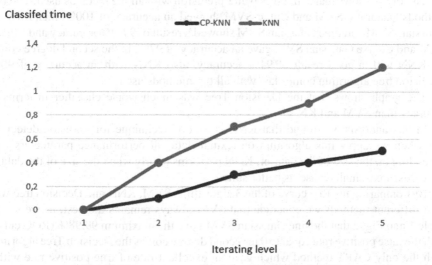

Fig. 5. Time Accuracy Graph for CP-KNN and KNN.

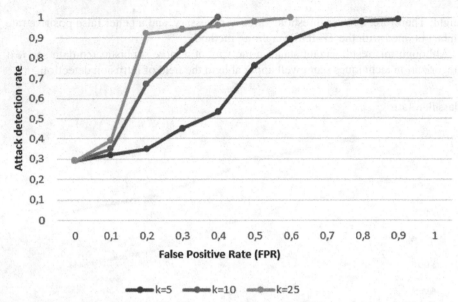

Fig. 6. Performance of the kNN classifier method expressed in ROC curves for the tf•idf weighting method.False positive rate vs attack detection rate for k = 5, 10 and 25.

4   Technical Comparison in IDS Area

To evaluate the performance of the intrusion detection of the different algorithms (SVM, KNN and Decision Tree), we compared the latter in terms of accuracy and test time and we summarized the results obtained in Figs. 4 and 5.

The Fig. 7 shows that svm gives better precision with all the methods used. The two methods Quadratic-SVM and Cubic-SVM showed an accuracy of 100%, then Medium Gaussian-SVM and Fine Gaussian SVM showed a result of 99, 4% accuracy and Linear-SVM and Coarse Gaussian SVM gave an accuracy of 97%. In the second degree comes CP-KNN with a good result of 94% accuracy, also KNN with an accuracy of 90%. 'Decision tree algorithm comes last with all the methods used.

The graph shows that the Decision Tree was much worse classifier in terms of accuracy than SVM and KNN.

In this analysis, we found that KNN is a good technique for intrusion detection. But when we apply this algorithm on textual data, all performance parameters vary depending on the size of the dataset; KNN performs poorly when the size of the dataset increases, so, a small dataset is better suited.

By comparing the Roc curve of the 3 algorithms SVM, KNN and Decision tree with their different methods we conclude that SVM always remains the best, as we see on Table 1 and Fig. 2 that the fine Gaussian SVM gave the maximum 99.98% (ROC curve) with the false positive rate equal to 0 in second degree comes the Decision Tree algorithm with the only CART method which gave an excellent rate of true positive rate with a minimum of false positive rate which equals 0.2 and lastly comes the KNN algorithm with k = 10, the kNN can detect 10 of the 35 attacks with a false positive rate 0.4.

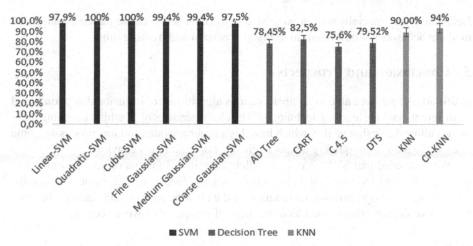

Fig. 7. Precision comparison of the three algorithms with different methods.

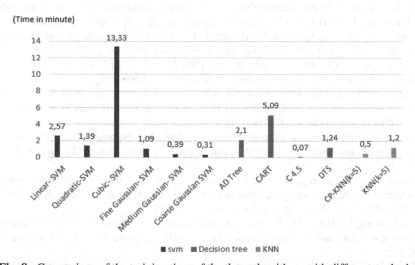

Fig. 8. Comparison of the training time of the three algorithms with different methods.

SVM is a complex classifier; we found that its accuracy does not depend too much on the size of the dataset, but on all the factors that depend on the number of training cycles. It is the best classifier that shows excellent accuracy and its training time is slightly better than KNN and Decision Tree.

From the results we can conclude that Support Vector Machine performs better than KNN and Decision Tree, we can say that KNN works well with small training data. The results also show that SVM test time is better than KNN and Decision Tree especially with a large size in the form of results we can conclude that Support Vector Machine performs better than KNN and Decision Tree, we can say that KNN works well with small training data. The results also show that SVM training time is better than KNN and

Decision Tree, especially with a large size in the data set, because the Gaussian SVM medium achieves the best result in terms of precision and training time.

5 Conclusion and Prospects

In this article, we were able to compare various algorithms for intrusion detection based on different machine learning techniques. The characteristics of Machine Learning techniques allow the design of IDS which have high accuracy rates and low processing time rates while the system quickly adapts to changes in malicious behavior.

We concluded that SVM was an excellent algorithm in terms of accuracy and processing time and ROC curves than KNN and Decision Tree. Our future work will consist of applying SVM optimization techniques for the classification of IDS data sets in order to improve the detection of attacks in the field of computer network security.

References

1. Modi, U., Jain, A.: An improved method to detect intrusion using machine learning algorithms. Inf. Eng. Int. J. (IEIJ) **4**(2), 17–29 (2016). https://doi.org/10.5121/ieij.2016.4203
2. Aksu, D., Üstebay, S., Aydin, M.A., Atmaca, T.: Intrusion detection with comparative analysis of supervised learning techniques and fisher score feature selection algorithm. In: Czachórski, T., Gelenbe, E., Grochla, K., Lent, R. (eds.) ISCIS 2018. CCIS, vol. 935, pp. 141–149. Springer, Cham (2018). https://doi.org/10.1007/978-3-030-00840-6_16
3. Othman, S.M., Ba-Alwi, F.M., Alsohybe, N.T., Al-Hashida, A.Y.: Intrusion detection model using machine learning algorithm on Big Data environment. J. Big Data **5**(1), 1–12 (2018). https://doi.org/10.1186/s40537-018-0145-4
4. Anwar, S., et al.: From intrusion detection to an intrusion response system: fundamentals, requirements, and future directions. Algorithms **10**(2), 39 (2017). https://doi.org/10.3390/a10 020039
5. Gao, J., Chai, S., Zhang, B., Xia, Y.: Research on network intrusion detection based on incremental extreme learning machine and adaptive principal component analysis. Energies **12**(7), 1223 (2019). https://doi.org/10.3390/en12071223
6. Gu, J., Wang, L., Wang, H., Wang, S.: A novel approach to intrusion detection using SVM together with increased feature. Comput. Secur. (2019). https://doi.org/10.1016/j.cose.2019. 05.022
7. Zhang, H., Peng, H., Yang, Y.: Nearest neighbors based approach to density peaks intrusion detection. Chaos, Solitons Fractals **110**, 33–40 (2018). https://doi.org/10.1016/j.chaos.2018. 03.010
8. Mukherjee, S., Sharma, N.: Intrusion detection using naive bayes classify with feature reduction. Procedia Technol. **4**, 119–128 (2012). https://doi.org/10.1016/j.protcy.2012.05.017
9. Jamal, H., Mishra, A.: An actual intrusion detection based on support vector framework machine using NSL - KDD dataset. Indian J. Comput. Sci. Eng. (IJCSE) **8**(6), 703–713, December 2017–January 2018. e-ISSN: 0976–5166
10. Rai, K., Devi, M.S., Guleria, A.: Decision tree-based algorithm for intrusion detection. Int. J. Adv. Netw. Appl. **7**, 2828–2834 (2016)
11. Mani, L., Vidya, P.: A novel intrusion detection model for mobile ad -hoc networks using CP – KNN. Int. J. Comput. Netw. Commun. **6** (2014). https://doi.org/10.5121/ijcnc.2014.651
12. Liao, Y., Rao, V.: Use of K-nearest neighbor classifier for intrusion detection. Comput. Secur. **21**, 439–448 (2002). https://doi.org/10.1016/S0167-4048(02)00514-X
13. Charles, B.: Skybox security. In: 2019 Vulnerability and Threat Trends, 29 March 2019

Three Quantum Machine Learning Approaches for Mobile User Indoor-Outdoor Detection

Frank Phillipson[(✉)], Robert S. Wezeman, and Irina Chiscop

TNO, PO Box 96800, 2509 JE The Hague, The Netherlands
{frank.phillipson,robert.wezeman,irina.chiscop}@tno.nl

Abstract. There is a growing trend in using machine learning techniques for detecting environmental context in communication networks. Machine learning is one of the promising candidate areas where quantum computing can show a quantum advantage over their classical algorithmic counterpart on near term Noisy Intermediate-Scale Quantum (NISQ) devices. The goal of this paper is to give a practical overview of (supervised) quantum machine learning techniques to be used for indoor-outdoor detection. Due to the small number of qubits in current quantum hardware, real application is not yet feasible. Our work is intended to be a starting point for further explorations of quantum machine learning techniques for indoor-outdoor detection.

Keywords: Quantum machine learning · Mobile devices · Indoor-outdoor detection · Hybrid quantum-classical · Variational quantum classifier · Quantum classification · Quantum SVM

1 Introduction

The environmental context of a mobile device is important for operators of mobile communication networks. They can use this information to determine how the device is used, to optimise the device and the network for greater efficiency and usability. A lot of research has been published recently on improving functionality for location based services and service delivery by optimising network dimensioning and provisioning using machine learning techniques. These machine learning methods can learn to classify the environment the mobile device is in, for example indoor-outdoor detection or more detailed environments detection, like home, office, transportation, etcetera. There is a wide variety of solutions, regarding the machine learning techniques and used data. Examples of these machine learning approaches are: Random forest and AdaBoost classifiers are used on mobile device sensor data to classify the environment, Esmaeili Kelishomi et al. [9]; Support Vector Machine (SVM) and Deep Learning (DL) techniques are used in combination with a hybrid semi-supervised learning system to identify the indoor-outdoor environment using large and real collected

© Springer Nature Switzerland AG 2021
É. Renault et al. (Eds.): MLN 2020, LNCS 12629, pp. 167–183, 2021.
https://doi.org/10.1007/978-3-030-70866-5_11

3GPP (3rd Generation Partnership Project) signals measurements, Saffar et al. [28]. The same authors extended their work using deep learning, based on radio signals, time related features and mobility indicators for a more complex environment classification, with multiple environments in [27]. Other examples of machine learning for indoor-outdoor detection use Ensemble Learning Schemes, Zhang et al. [38], a semi-supervised learning algorithm, Bejarano-Luque et al. [3], and an ensemble model based on stacking and filtering the detection results with a hidden Markov model, Zhu et al. [39]. In an overview paper, Wang et al. [34] investigated a wide range of machine learning algorithms for indoor-outdoor classification including Decision Tree, Random Forest, Support Vector Machine, K-Nearest Neighbour, Logistic Regression, Naive Bayesian, and Neural Network (NN).

In general, machine learning approaches are improving and getting better results at the cost of using more and more computational power. Where the classical computer reaches the end of Moore's law [16], a new computing paradigm appears at the horizon: quantum computing. Also for real-time machine learning application quantum computing can be helpful. Quantum computers make use of quantum-mechanical phenomena, such as super-position (the system can be described by a linear combination of distinct quantum states) and entanglement (measuring the state of a qubit is always correlated to the state of the qubit it is entangled with), to perform operations on data. Where classical computers require the data to be encoded into binary digits (bits), each of which is always in one of two definite states (0 or 1), quantum computation uses quantum bits, which can be in a superposition of multiple definite states. These computers would, at least theoretically, be able to solve certain problems much quicker than any classical computer that use even the best currently known algorithms, thus giving a quantum advantage to quantum computers [26]. One of the promising candidate areas to show a useful quantum advantage on near-term devices, the so called noisy intermediate-scale quantum (NISQ) devices, is believed to be machine learning. Introduction in quantum machine learning can be found in [7,22,23,32]. A recent overview on quantum machine learning can be found in Abohashima et al. [2]. An indication for the speed up attained for various quantum machine learning techniques/algorithms is given by Biamonte et al. [4]. Specifically, for Bayesian Inference [18], Online Perceptron [15], and Reinforcement Learning [7,21] quadratic speedup is expected. Whereas an exponential speed up is expected for Principal Component Analysis (PCA) [17], SVM [24] and Least Square Fitting [36] compared to their classical counterpart. These algorithms with an exponential speed up all use the famous HHL [13] algorithm as subroutine for solving linear systems of equations. There are however caveats to the applicability of the HHL method [1], which may limit the applicability of these quantum machine learning methods. Here the main problems are that the data has to be encoded in a quantum state and the output is a quantum state that has to be translated into classical output, which both are non-trivial to solve without losing the expected exponential gain.

Goal of this paper is to give a practicable overview of (supervised) quantum machine learning techniques to be used for indoor-outdoor detection. It is meant as a starting point for further exploration as current applications are still limited by the number of qubits in current quantum hardware. It is expected that the number of qubits will grow rapidly in the next 5–10 years, making a real application of quantum computing for machine learning available within that time span. Therefore, we do not use real data, nor show the precise performance of the machine learning approaches in terms of calculation times, accuracy, F1 score, loss, etcetera.

In Sect. 2, we give a short introduction to quantum computing. The data we use in the examples is generated from a simple theoretical propagation model, that is explained in Sect. 3. In Sect. 4, we use the generated data to elaborate three quantum machine learning approaches: a hybrid quantum variational classifier, a quantum distance-based classifier and quantum annealing-based support vector machine. In Sect. 5, we end with some conclusions, recommendations and ideas for further research.

2   Quantum Computing

Quantum computing is the technique of using quantum mechanical phenomena such as superposition, entanglement and interference for doing computational operations. The type of devices which are capable of doing such quantum operations are still being actively developed and named quantum computers. We distinguish between two paradigms of quantum computing devices: gate-based and quantum annealers. A practically usable quantum computer is expected to be developed in the next few years. In less than ten years quantum computers will begin to outperform everyday computers, leading to breakthroughs in artificial intelligence, the discovery of new pharmaceuticals and beyond. Currently, various parties, such as Google, IBM, Intel, Rigetti, QuTech, D-Wave and IonQ, are developing quantum chips, which are the basis of the quantum computer [25]. The size of these computers is limited, with the state-of-the-art being around 70 qubits for gate-based quantum computers and 5000 qubits for quantum annealers. In the meantime, progress is being made on algorithms that can be executed on those quantum computers and on the software (stack) to enable the execution of quantum algorithms on quantum hardware.

2.1   Gate-Based Quantum Computing

Quantum bits, unlike classical bits, can be in a superposition between the states which correspond to what we classically think of as a 0 and 1. The state of a qubit, denoted by $|\psi\rangle$, can be described by two complex numbers α and β

$$|\psi\rangle = \alpha |0\rangle + \beta |1\rangle. \tag{1}$$

When a measurement is performed on a qubit in this state, either the state $|0\rangle$ or the state $|1\rangle$ will be observed with probability $|\alpha|^2$ and $|\beta|^2$, respectively. Hence, we have the constraint $|\alpha|^2 + |\beta|^2 = 1$. We can parameterise any state by the three real valued parameters $\theta \in [0, \pi]$ and $\phi, \gamma \in [0, 2\pi]$

$$|\psi\rangle = e^{i\gamma} \left(\cos\left(\frac{\theta}{2}\right) |0\rangle + e^{i\phi} \sin\left(\frac{\theta}{2}\right) |1\rangle \right). \tag{2}$$

The global phase factor $e^{i\gamma}$ has no physical consequences on measurement probabilities, and hence is often chosen to be one. All possible states which result in a different physical state can be visualised by the surface of a 3-dimensional sphere, better known as the Bloch sphere shown in Fig. 1.

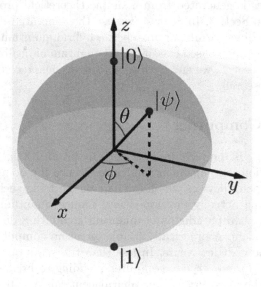

Fig. 1. Bloch sphere

Once a qubit is prepared in a specific state, we can apply quantum gates on it, which correspond to a rotation on the Bloch sphere. For example, the Pauli-X gate, which is often thought of as the quantum equivalent of the classical NOT gate, corresponds to a 180° rotation around the x-axis. This quantum gate indeed rotates a qubit in the state $|0\rangle$ into the state $|1\rangle$ and vice versa.

Once we start combining multiple qubits, the Bloch sphere becomes less visual intuitive. A multi-qubit gate acts on multiple qubits simultaneously. The CNOT (controlled NOT) gate for example acts on two qubits. This gate applies a Pauli-X gate to the target qubit only if the control qubit is in the $|1\rangle$ state.

It is the challenging task to design a quantum circuit, a combination of multiple quantum gates acting on, possibly multiple, qubits, such that after the circuit is applied the resulting quantum state is 'useful'. In particular, one looks to create quantum algorithms such that, due to their quantum properties, the resulting state can be used to solve a problem that outperforms the classical algorithmic counterpart in terms of complexity or efficiency. A famous example of such an algorithm is Grover's search algorithm [11], which finds with high probability the unique input to a black box function with domain of size N. The quantum algorithm only needs $\mathcal{O}(\sqrt{N})$ function calls, whereas the classical algorithm requires $\mathcal{O}(N)$ function calls.

For quantum machine learning algorithms, the data encoding phase is usually the bottleneck of the algorithm which could even nullify the gained speedup. In Fig. 2, a small example is shown of a quantum circuit which can be used to encode two data points in two qubits.

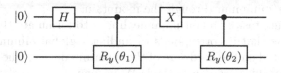

Fig. 2. Amplitude encoding: two-qubit circuit for data encoding of two normalised two-dimensional data points.

The figure should be read from left to right, where each line represents a qubit. In this case we look at a two-qubit system. We first apply a Hadamard gate on the first qubit which brings it in an equal superposition between the states $|0\rangle$ and $|1\rangle$. The two data points $x_i = (x_{i0}, x_{i1})$ for $i = 0, 1$ can be encoded in the amplitudes of the second qubit by using controlled rotations, where the angles of the rotations θ_i are chosen such that $\alpha_i = x_{i0}$ and $\beta_i = x_{i1}$. The circuit rotates the initial state $|0\rangle |0\rangle$ to the desired state given by

$$|0\rangle |0\rangle \rightarrow \frac{1}{\sqrt{2}} \big(|0\rangle |x_1\rangle + |1\rangle |x_2\rangle \big), \tag{3}$$

here the first qubit acts as a counter, while the two features of the data point are encoded in the second qubit. Alternatively, using just two qubits, it is also possible to encode a single data point with four features.

Such circuit diagrams have to be translated to a specific physical lay-out of quantum hardware, when executed. In this way, a gated quantum computer can be programmed. To create these circuits, quantum programming languages can be used, like QASM, PyQuil, QCL and Q#. These quantum programming languages only focus on a part of the quantum software stack [6], tools for other layers are also in development.

Making gate-based quantum computers is hard for multiple reasons. One of the reasons is that qubits are very sensitive and can quickly become useless due to small interactions with the environment. State-of-the-art devices therefore typically have in the order of 70 qubits: IBM presented the Hummingbird chip with 65 qubits in 2020, Google has its Bristlecone chip with 72 qubits.

2.2   Annealing-Based Quantum Computing

The quantum computing techniques, as described until now, fall in the gate-based quantum computing category. There is, however, also a different form of quantum computing which is known as quantum annealing, based on the work of Kadowaki and Nishimori [14]. The idea of quantum annealing is to create an equal superposition over all possible states. Then, by slowly turning on a problem-specific magnetic field, the qubits interact with each other and move towards the state with the lowest energy. This can be thought of as a ball rolling down a mountain and finally ending up at the valley. Classically, this would often result in the ball ending up in a local minimum rather than the global minimum. Quantum effects make it possible that sometimes the ball rolls out of a local minimum and hence still ends up in the global minimum or use a tunnel to go to an other minimum through the mountain.

The challenging task of quantum annealing is to formulate the desired problem in such terms that it corresponds to finding a global minimum, which can also be implemented on the hardware of the quantum device. The most advanced implementation of this paradigm is the D-Wave quantum annealer. This machine accepts a problem formulated as an Ising Hamiltonian, or rewritten as its binary equivalent, in QUBO formulation. The QUBO, Quadratic Unconstrained Binary Optimisation problem [10], is expressed by the optimisation problem:

$$\text{QUBO:} \quad \min_{x \in \{0,1\}^n} y = \pm x^t Q x, \tag{4}$$

where $x \in \{0,1\}^n$ are the decision variables and Q is a $n \times n$ coefficient matrix. QUBO problems belong to the class of NP-hard problems. For a large number of combinatorial optimisation problems the QUBO representation is known [10,19]. Many constrained integer programming problems can easily be transformed to a QUBO representation.

Next, this formulation needs to be embedded on the hardware. In the most advanced D-Wave 2000Q version of the system, the 2048 qubits are placed in a Chimera architecture [20]: a 16×16 matrix of unit cells, each consisting of 8 qubits. This allows every qubit to be connected to at most 5 or 6 other qubits. With this limited hardware structure and connectivity, fully embedding a problem on the QPU can sometimes be difficult or simply not possible. In such cases, the D-Wave system employs built-in routines to decompose the problem into smaller sub-problems that are sent to the QPU, and in the end reconstructs the complete solution vector from all sub-sample solutions. The first decomposition algorithm introduced by D-Wave was *qbsolv* [5], which gave a first possibility to solve larger scale problems on the QPU. Although *qbsolv* was the main decomposition approach on the D-Wave system, it did not enable customisations, and therefore is not particularly suited for all kinds of problems. The new decomposition approaches D-Wave offers are D-Wave Hybrid and the Hybrid Solver Service, offering more customability.

3   Generating Data

We propose to look at a UMTS (Universal Mobile Telecommunications System, also called third generation, 3G, of wireless mobile telecommunications technology) network and have two features for the machine learning step: the received power level of the active connection and the received pilot signal to interference ratio. In order to describe the statistics of the received power we can use the classical propagation model that includes the characteristics of the radio channel. We use the simplified model from [8] and add extra loss to emulate the indoor environment. We can describe the propagation model in the classical form as follows:

$$P_R = \frac{P_T}{L(d)\beta\Gamma},\tag{5}$$

or in logaritmic (dB) terms:

$$P_R^{dB} = P_T^{dB} - L(d)^{dB} - \beta^{dB} - \Gamma^{dB},\tag{6}$$

where

- P_R: Received power level,
- P_T: Transmitted power level,
- $L(d)$: Path loss/Attenuation,
- β: Slow fading/Shadowing,
- Γ: Fast fading/Multipath effects.

The level of the slow-faded power (β) fluctuates with a log-normal distribution around the received mean power. If the linear valued β is log-normally distributed, then the logarithmic valued $\beta^{dB} = 10^{0.1\beta}$ is normally distributed and expressed in dB, with parameters (μ, σ), where $\mu = 0$. Now, for a user at a certain location x the receiving signal from cell c_i, neglecting the fast fading effect, we obtain (again in dB):

$$P_{R,i}^{dB}(x) = P_{T,i}^{dB} - L_i^{dB}(x) - \beta_i^{dB},\tag{7}$$

which is the first feature. Now we can define the second feature, the signal-to-interference-ratio SIR [12,33] of cell c_i as

$$SIR_i(x) = \frac{P_{T,i}L_i(x)\beta_i}{\sum_{j=1,j\neq i}^{M} P_{T,j}L_j(x)\beta_j},\tag{8}$$

$$SIR_i^{dB}(x) = 10\log_{10} SIR_i(x).\tag{9}$$

Using this model, we generated 300 data points, using $P_{T,i}^{dB} - L_i^{dB}(x) = 90$ dB, $\sigma = 8$ dB and an additional path loss for indoor settings of 15 dB. As data preparation, we scaled the data to the $[0, 1]$ interval and inversed the SIR axis. The original data is shown in Fig. 3, and the prepared data in Fig. 4.

4   Quantum Machine Learning Approaches

In this section we describe three different quantum machine learning algorithms. The first two algorithms use gate-based quantum computers while the last one uses a quantum annealing approach. To execute the algorithms, we rely mainly on simulators, classical machines which are capable to simulate quantum machines up to a small number of error free qubits. The last algorithm is also executed on the D-Wave 2000Q machine.

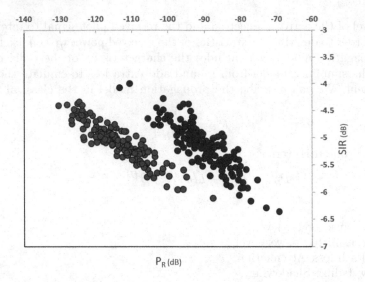

Fig. 3. Original generated data points. Red dots are the outdoor measurements, blue dots are the indoor measurements. (Color figure online)

4.1 Quantum Variational Classifier

The first approach we present is a variational classifier, where we used the implementation by pennylane.ai[1], inspired on [31]. Generally, in quantum variational approaches, a circuit is built using quantum gates performing rotations, and the corresponding parameters, the angles of the rotations, are learned classically, such that the output of the circuit fits best the expected output. This approach is very similar to conventional neural networks, where weights are learned with the same purpose.

The quantum circuit acts on a quantum state that represents the input via amplitude encoding. This requires a state preparation step. After applying the state preparation as well as the model circuit, the prediction is retrieved from the measurement of a single qubit. This means that there are three main steps: 1. state preparation, 2. model circuit, 3. measurement. The circuit model can be understood as a black box routine that executes the inference step of the machine learning algorithm on a small-scale quantum computer. Each layer consists of a generic single qubit gate rotating on all angles for each qubit and a controlled gate. Training of the angle parameters per qubit and layer is done by a classical gradient method. The data was already rescaled and is now also padded to have four features and then normalised, see Fig. 5. Working on the positive subspace makes encoding of the data into the qubits much easier, as proposed in [30]. The used cost function is the standard square loss that measures the distance between the target labels and model predictions. An example of a 2-layer model circuit, including the data encoding is:

---

[1] https://pennylane.ai/qml/demos/tutorial_variational_classifier.html.

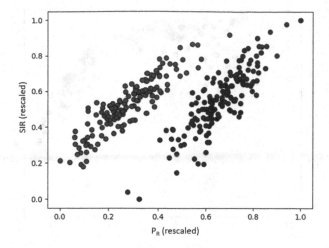

Fig. 4. Prepared artificial data points. Red dots are the outdoor measurements, blue dots are the indoor measurements. (Color figure online)

Where $ROT(\phi, \omega, \psi) = R_z(\psi)R_y(\omega)R_z(\phi)$. Note that only one single data point with four features is encoded. This is different from the example in Sect. 2.1 where we gave an example of encoding two data points with two features. The required number of qubits in this model only depend on the number of features and not the size of the data set.

For our data we obtained the best results when using a 3-layer model on a quantum computer simulator. We use 75% of the data points as training data and the remaining 25% as validation set. The classical gradient used for the optimisation of the angles used 200 optimisation steps. The result is shown in Fig. 6.

The two main benefits of this approach, following [31], are: First, the hybrid classical-quantum approach makes it possible to store the parameters for later use, outside the coherence times; Second, this classifier will work on a small quantum computer even for a large number of features and in the absence of strong error correction. An example in [31] shows that an eight qubit model uses 100 parameters to classify inputs of $2^8 = 256$ dimensions, which is much more compact than a conventional feed-forward neural network.

4.2 Quantum Distance-Based Classifier

The second approach we present is a classification algorithm, which classifies a data point based on its distance to the training data. The algorithm takes as input M training data points $\boldsymbol{x}_i \in \mathbb{R}^N$ with binary labels $y_i \in \{-1, 1\}$. The goal of the algorithm is to assign a label $\tilde{y}$ to a new data point $\tilde{\boldsymbol{x}}$. Classification is based on the classification function

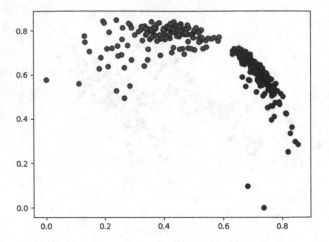

Fig. 5. Padded and normalised data.

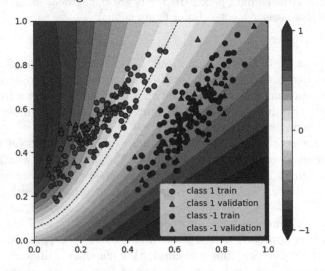

Fig. 6. Resulting assignments of the variational classifier.

$$\tilde{y} = \text{sgn}\left(\sum_{m=0}^{M-1} y_m \left[1 - \frac{1}{4M}|\tilde{\boldsymbol{x}} - \boldsymbol{x}_m|^2\right]\right).\tag{10}$$

The nature of qubits restrict the input data to lie on a unit sphere and hence an additional data preprocessing step is required, where the data points are first projected on the unit sphere. For the specific details how to implement this classifier on a quantum machine we refer to [35]. The quantum algorithm can be implemented using $2 \cdot (log_2(F) + log_2(N) + 1)$ qubits, where N is the number of training data points and F is the number of features of the data. In Fig. 7 the resulting assignments are shown for $N = 256$ and $F = 2$, which can be implemented on a quantum device using 20 qubits. In the figure, the projection on the unit sphere has been reverted before being plotted. Due to this projection on the unit sphere, it can be seen that classification of a data point depends only on the angle between the different features.

This approach is especially powerful if the initial quantum state is given, for instance as a result from a quantum process or using a quantum RAM as was also proposed by [29]. In that case, the complexity is constant $\mathcal{O}(1)$, independent on the number of features and the number of data points. If the quantum state has to be generated explicitly, the complexity is $\mathcal{O}(NF)$. Due to the probabilistic nature of quantum computing, the obtained result is probabilistic and multiple evaluations of the quantum state are required.

4.3   Quantum Annealing-Based SVM

Kernel-based support vector machines (SVMs) are supervised machine learning algorithms for classification and regression problems. We will use the approach as proposed by Willsch et al. [37], who created and ran an SVM on the D-Wave quantum annealer. For this, a QUBO formulation is required that can be embedded on the quantum annealer. For a labelled dataset containing N data points (x_n, y_n) with x_n a feature vector and $y_n \in \{-1, 1\}$ the corresponding label. The SVM can be formulated as a quadratic programming problem:

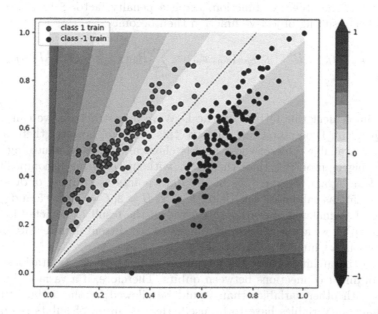

Fig. 7. Resulting assignments of the distance-based classifier with 256 training data points.

$$\text{minimise} \quad \left(\frac{1}{2} \sum_{n=1}^{N} \sum_{m=1}^{N} \alpha_n \alpha_m y_n y_m k(x_n, x_m) - \sum_n \alpha_n \right) \tag{11}$$

$$\text{subject to } 0 \leq \alpha_n \leq C \quad \forall n = 1, ..., N, \tag{12}$$

$$\sum_{n=1}^{N} \alpha_n y_n = 0 \tag{13}$$

where α_n is a coefficient, C a regularisation parameter and $k(\cdot, \cdot)$ the kernel function. A commonly used kernel is the Gaussian kernel $k(x_n, x_m) = e^{-\gamma ||x_n - x_m||^2}$, using a hyperparameter $\gamma > 0$. The coefficients define a decision boundary that separates the vector space in two regions, corresponding to the predicted class labels. To translate this quadratic programming formulation to a QUBO formulation there are two main steps. First, real number input has to be translated to binary input, using the encoding:

$$\alpha_n = \sum_{k=0}^{K-1} B^k a_{K_{n+k}} \tag{14}$$

with $a_{K_{n+k}} \in \{0,1\}$ binary variables, K the number of binary variables to encode α_n and B the base used for the encoding, usually $B = 2$ or $B = 10$. More details about the choice for K can be found in [37]. The second step is to translate the constraints to the objective function, using a penalty factor ξ for a quadratic penalty. The resulting objective function then becomes:

$$\frac{1}{2} \sum_{n,m,k,j} a_{K_{n+k}} a_{K_{m+j}} B^{k+j} y_n y_m k(x_n, x_m) - \sum_{n,k} B_k a_{K_{n+k}} + \xi \Big(\sum_{n,k} B^k a_{K_{n+k}} y_n \Big)^2 \tag{15}$$

If we implement this QUBO on the D-Wave infrastructure, we can use the QPU (Quantum Processor Unit) to solve this problem, however there are also other tools available. One of those tool is classical simulated annealing, which can solve bigger problems than the QPU, without requiring decomposition of the problem. Again, we use 75% of the data points as training data and the remaining 25% as validation set. Using $K = 2$, $B = 2$, $C = 3$, $\xi = 5$ and $\gamma = 16$, and using the simulated annealing solver we get the result as depicted in Fig. 8. Using the QPU, we are restricted by the size of the current architecture, namely 2000 qubits in a Chimera architecture.

Due to the architecture of the quantum chip, there are limitations to the number of direct connections between qubits. Therefore, if a variable has more relations with other variables than would be allowed by the architecture, so-called chains of variables have to be used. Here, a group of qubits is linked to represent a single variable, thereby allowing for implementing more constraints with more different variables. The number of available qubits and the limitations in connectivity, therefore quickly leads to embedding problems. Then problem decomposition techniques have to be used to cut the problem in pieces, solve each

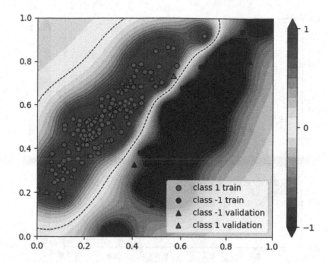

Fig. 8. Resulting assignments of the simulated annealing-based SVM.

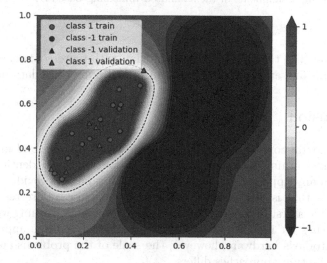

Fig. 9. Resulting assignments of the quantum annealing-based SVM, trained on 25 data points.

piece separately and than combine the pieces back together. However, this is at the expense of the quality of the solution. We will not use any decomposition techniques here and show the solution for the largest problem that could be embedded, using 25 data points for learning and 8 for validation. The QPU uses less than 10 ms annealing time for solving this problem. The result for this problem can be found in Fig. 9. For reference, also the result from the simulated annealing approach for the same data and the same parameters is presented in

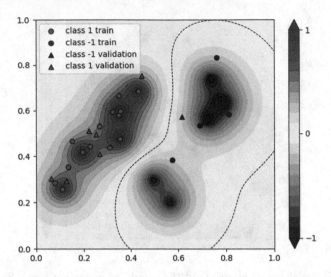

Fig. 10. Resulting assignments of the simulated annealing-based SVM, trained on 25 data points.

Fig. 10. It looks like the QPU creates a solution with more confidence than the simulated annealing approach with the same small number of data points.

5   Conclusion

The environmental context of a mobile device is getting more and more important. Machine learning techniques used for indoor-outdoor identification are proposed in many approaches. Quantum computing might provide the computational power that is needed when conventional computers are reaching their borders. To give a first idea of the use of quantum machine learning we proposed three approaches for indoor-outdoor classification. All of these approaches are usable using today's hardware, however, the scale of the problems and the benefits of the quantum approaches differs.

The Quantum Variational Classifier only uses a small number of qubits and is able to manage a huge number of datapoints. However, the gain here is not directly the calculation time, but the compactness of the used feed-forward neural network. This will lead to more efficient learning of a model with multiple features in the quantum approach than in the conventional classical approach. The Quantum Distance-based Classifier brings the complexity back to $\mathcal{O}(1)$, independent on the number of features and the number of data points. However, the data management is a big problem at this moment. The number of qubits needed to store big data input and the time it takes to encode the data are still under (scientific) investigation. The quantum annealing-based SVM is a straightforward implementation of the conventional SVM on a quantum computer. It is much easier to understand than the gate-based approaches. The scale of the current hardware is currently the limiting factor, leading to less optimal values for bigger problems. Here, the problem has to be decomposed, which has its cost. The hope is that when larger hardware is available, the computation time will grow less quickly than in the conventional classical approach. It is remarkable that the quantum annealing results show a solution with more confidence than the simulated annealing solution. Further research is needed on this point.

Overall, we believe that the quantum era has begun. It is important to monitor the developments on quantum hardware and applications on machine learning closely. Applications in real time learning for telecommunication will need the power this will produce.

References

1. Aaronson, S.: Read the fine print. Nat. Phys. **11**(4), 291–293 (2015)
2. Abohashima, Z., Elhosen, M., Houssein, E.H., Mohamed, W.M.: Classification with quantum machine learning: a survey. arXiv preprint arXiv:2006.12270 (2020)
3. Bejarano-Luque, J.L., Toril, M., Fernandez-Navarro, M., Acedo-Hernández, R., Luna-Ramírez, S.: A data-driven algorithm for indoor/outdoor detection based on connection traces in a LTE network. IEEE Access **7**, 65877–65888 (2019)
4. Biamonte, J., Wittek, P., Pancotti, N., Rebentrost, P., Wiebe, N., Lloyd, S.: Quantum machine learning. Nature **549**(7671), 195–202 (2017)
5. Booth, M., Reinhardt, S.P., Roy, A.: Partitioning optimization problems for hybrid classical/quantum execution. Technical report, D-Wave Systems, September 2017
6. van den Brink, R.F., Phillipson, F., Neumann, N.M.P.: Vision on next level quantum software tooling. In: Computation Tools (2019)
7. Dunjko, V., Taylor, J.M., Briegel, H.J.: Quantum-enhanced machine learning. Phys. Rev. Lett. **117**(13), 130501 (2016)
8. Erdbrink, R.: Analysis of UMTS cell assignment probabilities. Master's thesis, VU University Amsterdam, The Netherlands (2005)
9. Esmaeili Kelishomi, A., Garmabaki, A., Bahaghighat, M., Dong, J.: Mobile user indoor-outdoor detection through physical daily activities. Sensors **19**(3), 511 (2019)
10. Glover, F., Kochenberger, G., Du, Y.: A tutorial on formulating and using QUBO models. arXiv preprint arXiv:1811.11538 (2018)
11. Grover, L.K.: A fast quantum mechanical algorithm for database search. In: Proceedings of the Twenty-Eighth Annual ACM Symposium on Theory of Computing, STOC 1996, pp. 212–219. Association for Computing Machinery, New York (1996). https://doi.org/10.1145/237814.237866
12. Haenggi, M., Andrews, J.G., Baccelli, F., Dousse, O., Franceschetti, M.: Stochastic geometry and random graphs for the analysis and design of wireless networks. IEEE J. Sel. Areas Commun. **27**(7), 1029–1046 (2009)
13. Harrow, A.W., Hassidim, A., Lloyd, S.: Quantum algorithm for linear systems of equations. Phys. Rev. Lett. **103**(15), 150502 (2009)
14. Kadowaki, T., Nishimori, H.: Quantum annealing in the transverse Ising model. Phys. Rev. E **58**(5), 5355 (1998)
15. Kapoor, A., Wiebe, N., Svore, K.: Quantum perceptron models. In: Advances in Neural Information Processing Systems, pp. 3999–4007 (2016)

16. Leiserson, C.E., et al.: There's plenty of room at the top: what will drive computer performance after Moore's law? Science **368**(6495) (2020). https://doi.org/10.1126/SCIENCE.AAM974. Review summary in print version on page 1079: Computer Science
17. Lloyd, S., Mohseni, M., Rebentrost, P.: Quantum principal component analysis. Nat. Phys. **10**(9), 631–633 (2014)
18. Low, G.H., Yoder, T.J., Chuang, I.L.: Quantum inference on Bayesian networks. Phys. Rev. A **89**(6), 062315 (2014)
19. Lucas, A.: Ising formulations of many NP problems. Front. Phys. **2**, 5 (2014)
20. McGeoch, C.C.: Adiabatic quantum computation and quantum annealing: theory and practice. Synth. Lect. Quant. Comput. **5**(2), 1–93 (2014)
21. Neumann, N.M.P., de Heer, P.B.U.L., Chiscop, I., Phillipson, F.: Multi-agent reinforcement learning using simulated quantum annealing. In: Krzhizhanovskaya, V.V., et al. (eds.) ICCS 2020. LNCS, vol. 12142, pp. 562–575. Springer, Cham (2020). https://doi.org/10.1007/978-3-030-50433-5_43
22. Neumann, N.M.P., Phillipson, F., Versluis, R.: Machine learning in the quantum era. Digitale Welt **3**(2), 24–29 (2019)
23. Phillipson, F.: Quantum machine learning: benefits and practical examples. In: QANSWER, pp. 51–56 (2020)
24. Rebentrost, P., Mohseni, M., Lloyd, S.: Quantum support vector machine for big data classification. Phys. Rev. Lett. **113**(13), 130503 (2014)
25. Resch, S., Karpuzcu, U.R.: Quantum computing: an overview across the system stack. arXiv preprint arXiv:1905.07240 (2019)
26. Rønnow, T.F., et al.: Defining and detecting quantum speedup. Science **345**(6195), 420–424 (2014)
27. Saffar, I., Morel, M.L.A., Amara, M., Singh, K.D., Viho, C.: Mobile user environment detection using deep learning based multi-output classification. In: 2019 12th IFIP Wireless and Mobile Networking Conference (WMNC), pp. 16–23. IEEE (2019)
28. Saffar, I., Morel, M.L.A., Singh, K.D., Viho, C.: Machine learning with partially labeled data for indoor outdoor detection. In: 2019 16th IEEE Annual Consumer Communications & Networking Conference (CCNC), pp. 1–8. IEEE (2019)
29. Schuld, M., Fingerhuth, M., Petruccione, F.: Implementing a distance-based classifier with a quantum interference circuit. EPL (Europhys. Lett.) **119**(6), 60002 (2017)
30. Schuld, M., Petruccione, F.: Supervised Learning with Quantum Computers. QST. Springer, Cham (2018). https://doi.org/10.1007/978-3-319-96424-9
31. Schuld, M., Bocharov, A., Svore, K.M., Wiebe, N.: Circuit-centric quantum classifiers. Phys. Rev. A **101**(3), 032308 (2020)
32. Schuld, M., Sinayskiy, I., Petruccione, F.: An introduction to quantum machine learning. Contemp. Phys. **56**(2), 172–185 (2015)
33. Tsalaile, T., Sameni, R., Sanei, S., Jutten, C., Chambers, J., et al.: Sequential blind source extraction for quasi-periodic signals with time-varying period. IEEE Trans. Biomed. Eng. **56**(3), 646–655 (2008)
34. Wang, W., Chang, Q., Li, Q., Shi, Z., Chen, W.: Indoor-outdoor detection using a smart phone sensor. Sensors **16**(10), 1563 (2016)
35. Wezeman, R., Neumann, N., Phillipson, F.: Distance-based classifier on the quantum inspire. Digitale Welt **4**, 85–91 (2020). https://doi.org/10.1007/s42354-019-0240-5
36. Wiebe, N., Braun, D., Lloyd, S.: Quantum algorithm for data fitting. Phys. Rev. Lett. **109**(5), 050505 (2012)

37. Willsch, D., Willsch, M., De Raedt, H., Michielsen, K.: Support vector machines on the D-wave quantum annealer. Comput. Phys. Commun. **248**, 107006 (2020)
38. Zhang, L., Ni, Q., Zhai, M., Moreno, J., Briso, C.: An ensemble learning scheme for indoor-outdoor classification based on KPIs of LTE network. IEEE Access **7**, 63057–63065 (2019)
39. Zhu, Y., et al.: A fast indoor/outdoor transition detection algorithm based on machine learning. Sensors **19**(4), 786 (2019)

Learning Resource Allocation Algorithms for Cellular Networks

Thi Thuy Nga Nguyen[1,2]($\boxtimes$), Olivier Brun[2], and Balakrishna J. Prabhu[2]

[1] Continental Digital Service in France, Toulouse, France
Thi.Thuy.Nga.Nguyen@continental-corporation.com
[2] LAAS-CNRS, Université de Toulouse, CNRS, Toulouse, France
{brun,Balakrishna.Prabhu}@laas.fr

Abstract. Resource allocation algorithms in wireless networks can require solving complex optimization problems at every decision epoch. For large scale networks, when decisions need to be taken on time scales of milliseconds, using standard convex optimization solvers for computing the optimum can be a time-consuming affair that may impair real-time decision making. In this paper, we propose to use Deep Feedforward Neural Networks (DFNN) for learning the relation between inputs and the outputs of two such resource allocation algorithms that were proposed in [18,19]. On numerical examples with realistic mobility patterns, we show that the learning algorithm yields an approximate yet satisfactory solution with much less computation time.

Keywords: Scheduling · Deep Feedforward Neural Networks · Supervise learning

1 Introduction

In cellular wireless networks, a central scheduling problem is to choose one among several concurrent users (for example, mobile phones) to which the scheduler (henceforth also referred to as the base station) must send data to. This scheduling decision is taken every time-slot which is of the order of 2 ms [16] and is based on what are called as channel conditions of the users. Roughly, the channel condition of a user determines the data rate at which the base station can communicate with this user. In wireless networks, these conditions can vary randomly on short as well as on long time scales. Also called fading and shadowing, these random variations are a consequence of the interference patterns induced by the different obstacles (building, trees, etc.) in the path of the radio waves used for wireless communications [23]. However, the decision is not as easy as scheduling the user with the best channel condition as such a policy could lead to unfairness between users. Imagine a user with a direct line of sight path with the base station and another who is inside a building or an underground metro station. Quite possibly, the former user will always have a better channel which would starve the latter user of any communication.

© Springer Nature Switzerland AG 2021
E. Renault et al. (Eds.): MLN 2020, LNCS 12629, pp. 184–203, 2021.
https://doi.org/10.1007/978-3-030-70866-5_12

To avoid these unfair allocations, a typical solution is to define a utility which is a usually a concave function of the throughput[1] of the users and then compute the scheduling decision as the one that maximizes the sum of the utilities of the users. For example, the utility function could be the logarithm of the average data rates of the users. These solutions fall under the umbrella of the network utility maximization problem [27].

In an ideal scenario, the base station will solve this utility maximization problem every 2 ms. However, there are two practical issues that make this infeasible. First, the throughput of a user depends upon future channel conditions which are unknown to the scheduler. Second, the integer scheduling problem is known to be NP-complete [16]. Solving such an optimization problem over a time horizon of seconds (thousands of time-slots) and with hundreds of users can be unrealistic in time-slots of 2 ms (which are actually becoming shorter as the technology progresses). To overcome these practical issues, several heuristics have been proposed that are based on an estimation of the future data rates [16,18,19]. The heuristics (called STO1 and STO2) in [18,19] use the estimated future data rates as an input to a relaxed version of the original problem restricted to a shorter time horizon thereby reducing the dimensionality of the problem. These heuristics are thus much faster to solve than the original problem but they still require solving rather frequently a large-scale concave optimization problem which can be time consuming. It was shown numerically in [18,19] that STO1 and STO2 performed better than the one in [16] as well as the popular PF algorithm [13] that does not use future estimated rates. Therefore, in this paper, we shall address the problem of improving the speed of these heuristics without compromising on their superior total utility.

1.1 Contributions

We propose a machine learning based solution to speed up the operations of STO1[2]. The key idea is to use a Deep Feedforward Neural Network (DFNN) to approximate the output of the relaxed optimization problem in the heuristics. It will be shown on numerical examples that once the DFNN is trained, it takes much less time to generate a reasonable accurate solution compared to using the specialized Python package CVXPY [4] that uses the solver MOSEK [1] for solving of a concave optimization problem. We compare different DFNN architectures and different loss functions to find the most appropriate ones for our problem. We then compare the behavior of the learning algorithm with STO1 and with other existing algorithms as well. The comparison shall be done on scenarios created by SUMO [15] which can generate realistic mobility patterns on

---

[1] We use data rate and the throughput to denote two different but related quantities. The throughput only takes into account the data rate of the time-slots in which a user is served. It is thus no more than the total data rate of this user.

[2] STO2 is similar to STO1 but solves the optimization problem less frequently. In this paper we shall focus on STO1 but the ideas developed here can be applied to STO2 as well.

road networks with vehicular mobility. Based on numerical results, the learning algorithm is shown to perform close to STO1 with much less computation time.

1.2   Related Works

The theory of approximation for DFNN has been studied in many papers. Motivated by Komogorov's superposition theorem [12] in 1957, many approximation results have proven the approximation capabilities of feed-forward neural networks for the class of continuous functions such as [3,8,17]. In his theorem, Komogorov proved that any continuous function can be represented as a superposition of continuous functions of one variable. In [3] (1989), Cybenko proved that any multivariate continuous function with support in a hypercube can be uniformly approximated by a linear finite combinations of compositions of a sigmoidal functions and a set of affine functions. This representation is in fact a feed-forward neural networks with sigmoidal activation functions. Independently with the work of Cybenko, Hornik [8] (1989) also proved a similar result. Two years later, Hornik [7] showed that multi-layer feed-forward neural networks with arbitrary bounded and non-constant activation function can approximate arbitrary well real-valued continuous functions on compact subsets of $\mathbf{R}^n$ as long as sufficiently many hidden layers are available. The word "deep" in "deep learning" thus simply means many layers.

Learning an algorithm to produce an approximate algorithm in order to reduce computation time has been proposed in several recent research papers [6,21]. In [6], the authors consider a Sparse Coding problem which is used for extracting features from raw data. The problem is that Sparse Coding is often too slow for real-time processing in several applications such as pattern recognition. The authors propose a method using a non linear, feed-forward function to learn Sparse Coding to produce an approximate algorithm with 10 times less computation.

Learning an algorithm for wireless resource management has been proposed in [21]. In that work, the authors used DFNN to learn an algorithm for the interference channel power control problem. They obtain an almost real time algorithm, since passing the input through a DFNN to get the output only requires a small number of simple operations as compared to an iterative optimization algorithm. They show that, by choosing an appropriate initialization, the initial power control algorithm performs a continuous mapping which can be efficiently learnt.

In this paper, we use DFNNs for learning a channel allocation algorithm maximizing the proportional fairness between vehicular users. The proposed method is however potentially applicable to other convex optimization problems.

1.3   Organization

In Sect. 2, we recall the resource allocation problem in the case of a single Base Station (BS). We also remind the reader of the STO1 algorithm and state the learning problem we address in this paper. In Sect. 3, we formally define the input-output relationship for the DFNN model. Numerical results are presented in Sect. 4. We compare the computing times of the DFNN-based prediction algorithm against those of the original algorithm in Sect. 5 to evaluate the reduction in computing times. Finally, in Sect. 6 we discuss several research directions that can be followed in future work.

2 Problem Formulation

Consider the following downlink discrete-time channel allocation problem for a single Base Station (BS) (see [13] and references therein):

$$
\begin{cases}
\text{maximize} \quad O(\alpha) = \sum_{i=1}^{K} \log \left(\sum_{j=1}^{T} \alpha_{ij} r_{ij} \right) \\
\text{subject to} \\
\qquad \sum_{i=1}^{K} \alpha_{ij} \leq 1, \quad j = 1, \ldots, T; \\
\qquad \alpha_{ij} \in \{0,1\}, \quad j = 1, \ldots, T, \ i = 1, \ldots, K.
\end{cases}
\tag{I}
$$

Here $r_{ij} \geq 0$ is the data rate of user i in time-slot j, and α_{ij} is the corresponding allocation in this slot. The constraints impose that the BS can choose at most one user in each time-slot. The objective of the BS is to maximize the sum of the individual user utilities which are defined as the logarithm of the total throughput of the user over a time horizon of T. As mentioned in Sect. 1, solving this problem is not practical because the data rates $r_{i,j}$ become known to the scheduler only in slot j. Further, users arrive and leave and it is not possible to know in advance which users will be present in the network in the future. Hence, the algorithms proposed use either no information on the future rate (see [13] and references therein) or an estimation of the future rates [16, 18].

A more fine-grained scheduling problem on the downlink involves joint power control and channel allocation which allows the BS to vary the transmit power in addition to choosing how much of the channel (or bandwidth) it can allocate to the users [9], and is defined as:

$$
\begin{cases}
\text{maximize} \quad \sum_{i=1}^{K} \log \left(\sum_{j=1}^{T} x_{ij} \log \left(1 + \frac{p_{ij}\gamma_{ij}}{x_{ij}} \right) \right) \\
\text{subject to} \quad \sum_{i=1}^{K} x_{ij} \leq 1, \ \forall j; \quad x_{ij} \geq 0, \forall i, j \\
\qquad \frac{1}{T} \sum_{j} \sum_{i} p_{ij} \leq \bar{P}; \quad \sum_{i} p_{ij} \leq P_{max}, \ \forall j.
\end{cases}
\tag{P}
$$

Here p_{ij} (resp. x_{ij}) is the power (resp. fraction of bandwidth) allocated by the BS to user i in slot j. The parameter γ_{ij} represents the channel condition of the user i in slot j and $x_{ij} \log \left(1 + \frac{p_{ij}\gamma_{ij}}{x_{ij}} \right)$ is the Shannon rate obtained by this user when it is allocated power p_{ij} and fraction of bandwidth x_{ij}. The utility of the user is again the logarithm of its total throughput and the objective of the BS is to maximize the sum utility. Note that there are constraints on the power allocation which involve both the maximum power that can be expended in a time slot as well as the average power spend over the whole horizon. For this joint power and channel allocation problem, an adapted version of STO1 was proposed in [19].

Remark 1 (Joint power control and channel allocation). For brevity, we shall explain the STO1 algorithm and its associated machine learning solution only for the channel allocation problem (I). A remark shall be made wherever the treatment of this problem differs from that of the joint power control and channel allocation problem (P).

2.1   The STO1 Algorithm

STO1 is a sequential algorithm which computes the allocation on time-slot j based on the current and the past data rates, and the estimated future data rates. It operates on two time-scales (shown in Fig. 1) in order to reduce its complexity. Define big-slots as a certain number (order of hundreds) of time-slots (or small-slots) that reflect the duration over which the distance of each user to the BS does not change much, from that the data rate of each user in each time-slot can change a certain amount but the sum of its data rate over big-slot does not change much by law of large number. For example, while the scheduling time-slots are 2 ms in length, one big slot can be equal to 100 time-slots, and one user can move at most by a few meters in a big slot. Big slots is defined to reduce dimension of the original optimization problem as explained below. It shall be assumed that estimations are available for users' future positions at the granularity of big-slots, and that the mean of future rates based on the future positions are estimated.

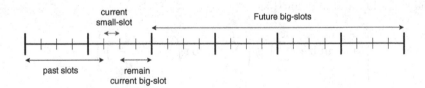

Fig. 1. The different types of time slots in the STO1 algorithm.

Before giving a formal definition of the STO1 algorithm, we give an intuitive explanation. In each small-slot j, STO1 does two steps. In the first step, it solves the problem (I) but with several restrictions: *(i)* the horizon is shortened to J big-slots; *(ii)* the future allocations are computed only on the aggregated level of big-slots; and *(iii)* the integer constraints on α_{ij} are relaxed. In the second step, the fractional allocation for the current small-slot is projected onto the set of the feasible integral allocations. The first step reduces the number of variables and hence the dimensionality of the problem as the future allocations are computed only for big-slots. Note that the second step is optional and is relevant only to problems with integral constraints.

A more formal definition is as follows. Let δ be the size of the small-slot, and let Δ be the size of the big-slot in absolute time units. Let $m = \Delta/\delta$ be the number of small slots in a big-slot (see Fig. 1). Denote by $\bar{r}_{ij}$ the mean rate in slot j for user i. At each small-slot t, with a slight abuse of notation, we shall denote by $\bar{\rho}_{i,0} = \sum_{j=t+1}^{(m-(t \bmod m))+t} \bar{r}_{ij}$ the total rate for user i in the remaining channel allocation slots of the current big-slot $\tau = 0$, where $t \bmod m$ denotes the remainder when dividing t by m. We also define $\bar{\alpha}_{i,0}$ as the corresponding allocation for the current big-slot $\tau = 0$.

Denote by $\bar{\rho}_{i\tau} = \sum_{j=(\tau-1)m+A+1}^{\tau m + A} \bar{r}_{ij}$ where $A = \left(\lfloor \frac{t}{m} \rfloor + 1 \right) m$, is the total average data rate that user i will get in the future big-slot τ ($\tau = 1, 2, ..., J-1$), where big slot τ starts after the current big-slot, and J is the short time horizon in term of big slots over which we can estimate the mean future rate. We also define $\bar{\alpha}_{i\tau}$ as the corresponding allocation for user i in future big slot τ. These allocations $\bar{\alpha}_{i\tau}$ can be interpreted as the fraction of small slots that user i will be allocated in the big-slot τ.

Note that this definition is slightly different from the definition in [18]. The differences are as follows: in [18], there is no current big slot and the future big-slot starts just after the current small slots, but it does not change much. The above definition of two time slot types corresponds in fact to the ones introduced in [19].

Denote by $a_i(t) = \sum_{j=1}^{t} \alpha_{ij} r_{ij}$ the total throughput allocated to user i up to time slot t, and let $K(t)$ be the number of users inside the coverage range of the BS at time t.

The algorithm STO1 contains two steps which are as follows:

- **Step 1** – solve the following optimization problem over a short-term horizon of J big-slots:

$$
\begin{cases}
\text{maximize} \sum_{i=1}^{K(t)} \log \left(a_i(t-1) + \alpha_{it} r_{it} + \sum_{\tau=1}^{J} \bar{\alpha}_{i\tau} \bar{\rho}_{i\tau} \right) \\
\text{subject to} \\
\qquad \sum_{i=1}^{K(t)} \alpha_{it} = 1, \\
\qquad \sum_{i=1}^{K(t)} \bar{\alpha}_{i\tau} = 1, \quad \tau = 0, ..., J-1, \\
\qquad \alpha_{it}, \bar{\alpha}_{i\tau} \in [0,1], , \quad \tau = 0, ..., J-1, \; i = 1, ..., K(t).
\end{cases}
$$
$$\text{(STO1-Opt)}$$

The decision variables in Problem (STO1-Opt) are the channel allocations in the current small slot, α_{it}, and the channel allocations in the current and future big-slots, $\bar{\alpha}_{i\tau}$. Since the future allocations are only computed on the time-scale of big-slots, there is a reduction by factor m in the number of variables in (STO1-Opt).

- **Step 2** – obtain a feasible integral allocation from the fractional one. For example, set $\alpha_{i^*j} = 1$ with $i^* = \text{argmax}_i\alpha_{ij}$ and set it to 0 for the other users.

Remark 2. (STO1-Opt) is solved thanks to the python package CVXPY [4] and the solver MOSEK [1]. In [18], this optimization problem was solved using a projected gradient algorithm, since it allows to iteratively solve a convex optimization problem when the feasible set is a simplex or a Cartesian product of simplices, no matter how complex the objective function is as long as it is smooth and convex. The advantage of CVXPY [4] is that it can be used to generalize this idea to other convex optimization problems, with more complex constraints such as power control constraints or others QoS constraints (e.g. delay constraints).

2.2   Learning STO1 with DFNN

As presented in [18] and [19], STO1 and STO2 are better than the other existing algorithms. However they have to solve an optimization problem with a large number of variables and constraints frequently (every small slot for STO1 or every big slot for STO2), so even if their performance is good, their computations are heavy. When the system is large and requires many more QoS contraints, they may not be able to run in real time. In addition, even when they are able to run in real time, it is good to reduce the computation time without reducing too much the quality of the allocation.

Therefore in this paper, our objective is to learn the STO1 algorithm using Deep Feedforward Neural Networks to obtain a new algorithm that behaves like STO1 but with a significantly reduced computation time. In other words, we want to learn the input-output relationship of STO1, by approximating the input-output mapping of STO1 with a DFNN (see [5] for background material on supervised learning with DFNN). After getting the approximation function (that is, the DFNN), the output can be computed by feeding the DFNN with the input value, instead of solving an optimization problem. This simpler method is expected to work faster than the original algorithm.

The main idea is to train a DFNN to predict an approximate solution to (STO1-Opt) instead of using a specialized convex optimization package. Obviously, the same idea could be used for other problems in order to obtain an approximate method which performs almost as well as the original algorithm but requires much less computing time. In short, the approach is as follows:

1. Design a well-performing algorithm based on the best available information;
2. Learn it by an approximate algorithm which behaves closely to the original one and has less computation.

The basic idea can be summarized as follows. Supervised learning is a learning task that amounts to learning an input-output relationship from a bunch of examples of input-output pairs which is called training data. The relationship can be derived from those examples by analyzing the training data and an inferred function will then be produced. A learning algorithm is considered as good if it can generalize the input-output relationship, i.e, it is able to determine outputs of unseen inputs with small error. So the objective here is to find the function representing as well as possible the input-output relationship.

Mathematically, let F be the true function that represents the relation between input and output. However, if the true function is unknown, our task is to find an approximate function $\hat{F}$ that is inferred from examples taken from training data $(x_n, y_n)_{n=1}^{N}$. The STO1 algorithm can be seen as a function F that maps an input $x_n \in \mathcal{X}$ (a problem instance) to an output $y_n \in \mathcal{Y}$ (a channel allocation), where $\mathcal{X}$ and $\mathcal{Y}$ denote the input and output spaces of STO1, respectively. Unfortunately, STO1 is too complicated to get the exact formula of F. Therefore here we want to approximate it by another function $\hat{F} : \mathcal{X} \to \mathcal{Y}$ which is in the form of a DFNN (an example of DFNN is illustrated in Fig. 2). Our objective is thus to find $\hat{F}$ such that it minimizes the empirical risk

$$R_{\mathrm{erm}}(F) = \frac{1}{N} \sum_{n=1}^{N} l(y_n, F(x_n)),$$

where $l(\cdot)$ is a loss function which measures how far y_n is from $F(x_n)$. Note that $l(\cdot)$ is not necessarily a mathematical distance but it is similar to a distance in the sense that it is small if the difference between y_n and $f(x_n)$ is small.

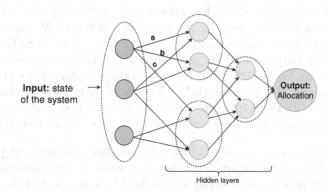

Fig. 2. An example of Deep Feed-forward Neural Network.

A DFNN is composed of many linear functions (sum of matrix multiplications and bias vectors) and non-linear functions (relu, sigmoid, softmax, etc.), and inside linear functions there are many parameters. So finding a good DFNN function means finding a good architecture, that is: the way the linear and non-linear functions are combined, the linear and non-linear functions in each layer, the size (number of units in each hidden layer) and then their parameters. Finding a good architecture is in general not an easy task [22]. In this paper, we shall empirically compare some architectures through experiments presented in Sect. 4.2. After fixing the architecture, we have to find appropriate parameters by minimizing the empirical risk defined above.

3   System Setup for Learning

Recall that, in STO1, we have two types of time scales: one is the big slot Δ and the other one is the small slot δ. In STO1, we solve problem (STO1-Opt) with variables of size $(1 + J) * K$ every small slot, where J is the time horizon in terms of big slots and K is the number of users in the system. The size of the allocation vector for each user is equal to $1 + J$ since it contains the allocation for the current small slot, α_{it}, and the average allocation $\bar{\alpha}_{i\tau}$ for the subsequent J big slots (including the current big slot). The input of STO1 is a data rate vector of size $(1 + J) * K$ and the total allocated throughput for the K users. The output of STO1 is the current allocation vector $(\alpha_{it})_{i=1,...,K}$ which is of size K, since we shall only use current allocation for making decision.

As it is defined at the moment, STO1 is not well suited to be modelled as a learning problem for the two reasons stated below.

Firstly, since K can vary over time, the dimension of the input vector will also vary. To circumvent this problem and to properly define STO1 as a function, we have to fix the size of the state. To do that, we extend the real state of the system by adding some pseudo users. Let us assume that there are at most K_M users inside the system. We will then add $K_M - K$ pseudo users, where K is the number of real users in the system at time t. We will actually learn an extended version of STO1 which is STO1 when we restrict it to K users. There are many ways to extend STO1, but here we try to define an extended version that preserves as much as possible the continuity of STO1. When we mention "learning STO1", it means "learning the extended function" of STO1.

Secondly, the output of STO1 as defined above is the solution of an optimization problem. So in fact STO1 is a set-valued mapping since the solution need not be unique. But by using the CVXPY package to solve the convex optimization problem (STO1-Opt), we agree with the way it determines one of the solutions. This makes STO1 becomes a function (instead of set-valued mapping).

Remark 3 (Joint power control and channel allocation). For the joint power control and channel allocation problem, the state needs to be augmented by the remaining total power. The output of the DFNN will now give the transmit power to each user as well as the fraction of the channel it gets allocated.

3.1   State

We define a state as a matrix of size $(2 + J) \times K_M$, where K_M is the maximum number of users in the system. There are thus $2 + J$ rows, and each row has K_M elements. The interpretation is as follows:

- The first row gives the current rates of the K users. We fill in the K positions on the left hand side with these current rates, and the remaining $K_M - K$ positions are filled with -1.
- The next J rows (from $2, .., J + 1$) give the average rates of the users in the next J big slots. For pseudo users (the $K_M - K$ columns on the right hand

side), we use the value $(-1) \cdot (\Delta/\delta - (t \mod \Delta/\delta))$ for the current big slot and the value $(-1) \cdot \Delta/\delta$ for the other big slots.

- The last row gives the total allocated throughput of the K users. For pseudo users, we use a large enough value which is significantly greater than the total allocated throughput of real users.

By observing how STO1 works, we remark, as expected, that STO1 gives priority to users with a low allocated throughput and a high current rate. Therefore, the way we define the state (that is, by using negative values for the current and future rates of pseudo users, and extremely large values for their allocated throughput) is intended to help the model ignore quickly the pseudo users.

Remark 4. Remark that there are K real users in the system at present time. Therefore, the K places of the real users in the first row which give the current rates of those users have to be strictly positive. The future rates (from the second row to the $(J + 1)$-th row) can be zero.

3.2 Target

We remind the reader that we want to learn only the current allocation, not the future allocation. Therefore, the target will be a vector of size K_M, where the first K positions represent the fractional allocation α_{it} of the K users as computed by STO1, and the last positions are filled with zero. Since in the optimization problem, the sum of allocation should be equal to 1, when there is no user in the system (all positions correspond to pseudo users), the allocation vector will be set to $(1/K_M, ..., 1/K_M)$ by convention.

Figure 3 illustrates the input and output of the DFNN model as described above.

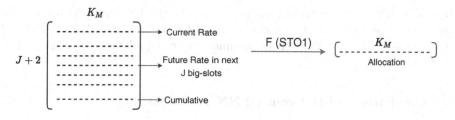

Fig. 3. Input and output of the DFNN model.

3.3 Loss, DFNN Architecture, Initial Parameters and Optimizer

We will try several different loss functions and architectures and compare them in the numerical section. The initial parameters (weights) of the DFNN will be chosen as proposed in [14], which allows the initial parameters to be not too big and not too small. The optimizer is Adam, which was first introduced in [11] and is a stochastic first-order gradient-based algorithm. The convergence of Adam is proven in [20].

4   Numerical Comparisons

In this section, we do simulations to evaluate the influence of many factors on the behavior of the DFNN model (loss functions, architecture of the DFNN). We use the keras library [2] to implement our code.

There are actually a lot of factors that can have an impact on the behavior of the learning procedure such as the initial learning rate, the learning rate decay, the optimizer, the initial weight, the number of parameters, the activation functions in layers... Here we are not able to justify all our choices, but we focus on the factors which have the most significant impact on the learning algorithm in our opinion. The initial learning rate is chosen equal to 0.0015 and after each epoch, this learning rate decays by a factor 0.998.

4.1   An Unified Data Generator for Comparison

To support the comparisons in this section, the data (both for training and validation) is generated as follows. The number of users is generated randomly from 0 to $K_M = 10$. The sojourn time of each user is generated in $(0, 400)$ seconds. This value could of course be increased, but here in order to reduce the learning time and be able to make many comparisons, we consider only small scenarios. At each learning epoch, the model will go through 1600 samples (that is, input-output pairs). The transmission rate in each small slot is generated randomly between 0 and $5 * \delta/\Delta$. The rate we use for evaluating in SUMO scenarios are given by

$$r(x) = \eta \left(1 + \kappa\, e^{-d(x, BS)/\sigma} \right), \tag{1}$$

where $d(x, BS)$ is the distance from position x to the BS, and η represents the noise level. For the SUMO scenarios in Sect. 4.4, we use $\kappa = 3$, $\sigma = 100$ and $\eta \sim$ Uniform$(0.7, 1.3)$. The others parameters are equal to $J = 10, \Delta = 1$ s, $\delta = 2$ ms.

4.2   Comparison of Different DFNN Architectures

In this part, we will consider 4 different architectures of the DFNN model and compare their performances. For the 4 models, the activation function used in hidden layers is the relu function, whereas the output layer uses the softmax function since we want the sum of the allocations to be equal to 1. In this comparison, we use the same loss function for all models, the huber loss [26].

Model 1. The first model used in this section contains 2 layers which are 1 hidden layer and 1 output layer. The hidden layer contains 500 units, and in total the model has $67,510$ parameters. The architecture of this model is illustrated in Fig. 4.

```
Layer (type)                  Output Shape            Param #
=================================================================
input_1 (InputLayer)          (None, 120)                0

dense_1 (Dense)               (None, 500)              60500

activation_1 (Activation)     (None, 500)                0

batch_normalization_1 (Batch  (None, 500)              2000

dense_2 (Dense)               (None, 10)               5010

activation_2 (Activation)     (None, 10)                 0
=================================================================
Total params: 67,510
Trainable params: 66,510
Non-trainable params: 1,000
```

Fig. 4. Model 1 architecture.

Model 2. As the first model, the second model contains 2 layers: 1 hidden layer and 1 output layer. However, the hidden layer contains 1000 units, and in total the model has 135, 010 parameters. We take the same number of layers as in model 1 (but with more units in hidden layers) in order to compare whether it is better to have more parameters.

Model 3. As the two previous models, the third model contains 2 layers (1 hidden layer and 1 output layer). The hidden layer contains 100 units, and we have 13, 510 parameters in total. We take the same number of layers as in model 1 (but fewer units in the hidden layer) to compare whether it is better to have fewer parameters.

Model 4. The last model contains 10 layers which are 9 hidden layers and 1 output layer. Each hidden layer contains 82 units, and in total the model contains 67, 496 parameters. We take a model that has almost the same number of parameters as Model 1, to compare whether it is better to have more layers or fewer layers.

Remark 5 (Joint power control and channel allocation). For the joint power control and channel allocation problem, we still compare the four above models except that the output layer of each model will be modified since it includes not only the channel allocation but also the power.

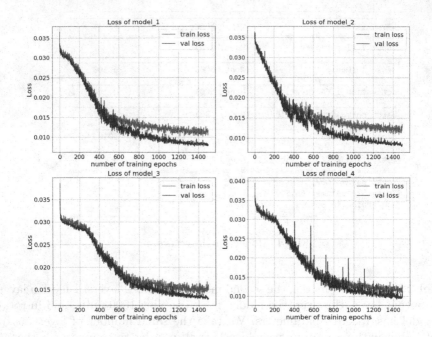

(a) Loss on training set and validation set of each model

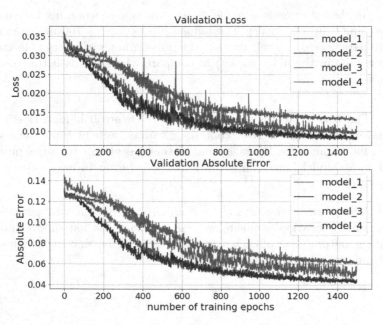

(b) Plot on same axis for loss and absolute error on validation set of all the four models

Fig. 5. Comparison of the 4 DFNN models.

Figure 5a illustrates the loss of the 4 models on training and validation data. Figure 5b plots loss and absolute error of the 4 models on the same axis on validation set. The same quantities but for the problem of joint power control and channel allocation are shown in Fig. 6.

From these figures, we observe that for the model without power control:

- Having almost the same number of parameters, Model 1 with fewer layers is better than Model 4.
- Having the same layers, Model 1 and Model 2 with more parameters are better than Model 3.
- Model 1 and Model 2 behave similarly and have the same number of layers. However Model 1 has less parameters than Model 2 so it is less costly from a computational point of view. Therefore from now on we shall use Model 1 for other comparisons in the sequel.

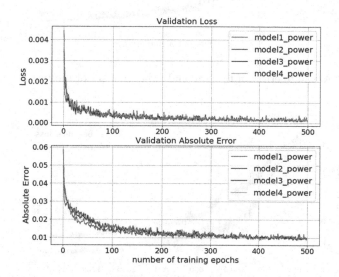

Fig. 6. Comparison of the 4 DFNN models for the joint power control and channel allocation problem.

For the model with power control:

- Having almost the same number of parameters, Model 1 with few layers is slightly better than Model 4, but the difference is quite small in this case.
- Having the same layers, models 1, 2 and 3 are almost the same but Model 3 has fewer parameters.

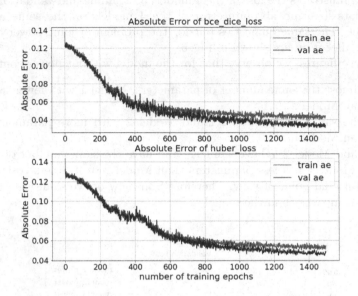

(a) Absolute error on training and validation set of the two losses

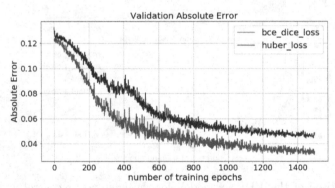

(b) Plot on same axis of the absolute error on validation of the two losses

Fig. 7. Comparison of loss functions.

4.3 Comparisons of Different Loss Functions

To compare the quality of the learning model obtained using different loss functions, we use the same model, that is Model 1. Figure 7 presents the results obtained with the huber loss [26], with the sum of binary cross-entropy [24] and dice loss (which equals to 1− dice coefficient [25]). The second loss function is denoted by bce_dice loss. From the figures, the bce_dice loss function is better in this case. So for the next comparisons, we shall use Model 1 and bce_dice loss.

4.4   Performance Evaluation on SUMO Scenarios

In this section, we shall use a mobility simulation software for comparisons of the algorithms, that is Simulation of Urban Mobility application (SUMO) [15]. SUMO is an open source software designed for simulating mobility of moving users (vehicles, bus, truck, bicycle, pedestrian, ...) in large road traffic networks. It allows to import maps of different cities and simulate realistic mobility traces. This application is used to simulate the complex moving dynamic systems in several specific regions of Toulouse city to compare our heuristics against existing algorithms in realistic scenarios. The performance evaluation of the heuristic is done in two steps: firstly, SUMO is used for generating the mobility traces of vehicles; finally, these traces are then fed to a Python script which implements the algorithms and computes the value of the objective function of those algorithms.

We shall compare the learning-based allocation scheme with STO1 and other existing algorithms on two different scenarios created with SUMO. The first scenario contains 244 users and lasts 61.7 min. The map of this scenario is shown in Fig. 8. The results obtained with the different learning schemes on this scenario are shown in Fig. 10. The second scenario contains 214 users and lasts 62.4 min. The map of this scenario is shown in Fig. 9. The data for BS location can be found on the website[3] of the French Frequency Agency (ANFR), which manages all radio frequencies in France. The results obtained with the different learning schemes on this scenario are shown in Fig. 11. We also simulate two existing algorithms, $(PS)^2S$ [16] and PF [10] (which are also used in [18] for comparisons) in order to show that the approximation algorithm performs better than the existing algorithms. As mentioned above, for the learning algorithm, we use Model 1 and bce_dice loss.

When the number of learning epochs is large enough, the learning-based scheme performs well compared to STO1 and other algorithms.

5   Computing Times

The computing time of STO1 depends on the convex optimization solver used, whereas the learning algorithm has only to feed the DFNN model with the input matrix. We consider the same setting as in Sect. 4.2, that is $K = 10$ (there are 10 users in the system) and the short term horizon is $J = 10$ s. For these values, the average computing time of Mosek is around 43.7 ms, whereas the prediction with the DFNN model (Model 1, which contains 67,510 parameters) takes only 0.65 ms on average. When adding power control, Mosek solves the optimization problem in around 113.4 ms, while the prediction of the DFNN model (almost

---

[3] https://data.anfr.fr/anfr/portail.

Fig. 8. The Carmes borough in Toulouse, with one BS (Free Mobile type LTE1800). The actual size is 200 m × 400 m.

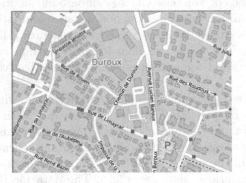

Fig. 9. Duroux, one BS type LTE1800, operator SFR. The actual size is around 350 × 500 m.

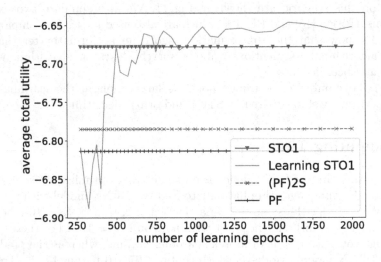

Fig. 10. Comparisons of evaluated on Carmes scenario created by SUMO.

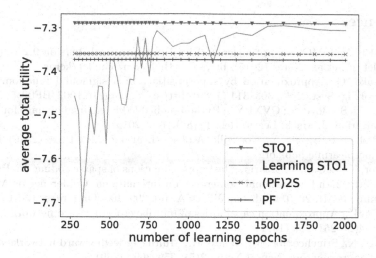

Fig. 11. Comparisons of evaluated on Duroux scenario created by SUMO.

the same with Model 1 but the output layer contains power vector in addition, which contains 73,020 parameters) takes 0.68 ms on average. These computing times are averaged over 10000 samples, all are measured on a machine using GPU (graphics processing unit) which allows computing many calculations in parallel.

From the above measurements, we can conclude that the computing time with a solver can vary widely with the number of QoS constraints for the same network (number of users, time horizon). In contrast, the prediction time of the learning-based algorithm with DFNN is almost insensitive to such changes.

6   Summary and Discussion

We have proposed to use DFNN for learning the channel allocation obtained with one of the heuristics (STO1) introduced in [18] and [19]. Numerical results on SUMO scenarios show that the learning-based method yields approximate yet satisfactory channel allocations with much less computation time as long as there are enough learning epochs. The state of the DFNN is defined in such a way that the model is not restricted to a particular scenario, that is, it can learn the channel allocation for a general network.

There are several directions of research that can be investigated to improve the learning algorithm, such as a better generator of data, a better loss function, a better architecture of the DFNN model, and other things such as the optimizer, the learning rate, etc.

Acknowledgements. This work was partially funded by a contract with Continental Digital Services France.

References

1. MOSEK ApS: MOSEK optimizer API for Python manual. Version 9.2.21 (2019)
2. Chollet, F., et al.: Keras (2015). https://github.com/fchollet/keras
3. Cybenko, G.: Approximation by superpositions of a sigmoidal function. Math. Control Sig. Syst. **2**(4), 303–314 (1989). https://doi.org/10.1007/BF02551274
4. Diamond, S., Boyd, S.: CVXPY: a Python-embedded modeling language for convex optimization. J. Mach. Learn. Res. **17**(83), 1–5 (2016)
5. Goodfellow, I., Bengio, Y., Courville, A.: Deep Learning. MIT Press (2016). http://www.deeplearningbook.org
6. Gregor, K., LeCun, Y.: Learning fast approximations of sparse coding. In: Proceedings of the 27th International Conference on International Conference on Machine Learning, ICML 2010, Madison, WI, USA, pp. 399–406. Omnipress (2010)
7. Hornik, K.: Approximation capabilities of multilayer feedforward networks. Neural Netw. **4**(2), 251–257 (1991)
8. Hornik, K., Stinchcombe, M., White, H.: Multilayer feedforward networks are universal approximators. Neural Netw. **2**(5), 359–366 (1989)
9. Huang, J., Subramanian, V., Agrawal, R., Berry, R.: Downlink scheduling and resource allocation for OFDM systems. IEEE Trans. Wirel. Commun. **8**, 288–296 (2009)
10. Kelly, F.: Charging and rate control for elastic traffic. Eur. Trans. Telecommun. **8**(1), 33–37 (1997)
11. Kingma, D., Ba, J.: Adam: a method for stochastic optimization. In: International Conference on Learning Representations, December 2014
12. Kolmogorov, A.: On the representation of continuous functions of many variables by superposition of continuous functions of one variable and addition. Dokl. Akad. Nauk SSSR **114**(5), 953–956 (1957)
13. Kushner, H.J., Whiting, P.A.: Convergence of proportional-fair sharing algorithms under general conditions. IEEE Trans. Wirel. Commun. **3**(4), 1250–1259 (2004)
14. Lecun, Y., Bottou, L., Orr, G., Müller, K.-R.: Efficient backprop, August 2000
15. Lopez, P.A., et al.: Microscopic traffic simulation using sumo. In: The 21st IEEE International Conference on Intelligent Transportation Systems. IEEE (2018)
16. Margolies, R., et al.: Exploiting mobility in proportional fair cellular scheduling: measurements and algorithms. IEEE/ACM Trans. Netw. **24**(1), 355–367 (2016)
17. Montanelli, H., Yang, H.: Error bounds for deep RELU networks using the Kolmogorov-Arnold superposition theorem (2019)
18. Nguyen, T.T.N., Brun, O., Prabhu, B.J.: An algorithm for improved proportional-fair utility for vehicular users. In: Gribaudo, M., Sopin, E., Kochetkova, I. (eds.) ASMTA 2019. LNCS, vol. 12023, pp. 115–130. Springer, Cham (2020). https://doi.org/10.1007/978-3-030-62885-7_9
19. Nguyen, N., Brun, O., Prabhu, B.: Joint downlink power control and channel allocation based on a partial view of future channel conditions. In: The 15th Workshop on Resource Allocation, Cooperation and Competition in Wireless Networks, June 2020
20. Reddi, S.J., Kale, S., Kumar, S.: On the convergence of Adam and beyond. CoRR, abs/1904.09237 (2019)
21. Sun, H., Chen, X., Shi, Q., Hong, M., Fu, X., Sidiropoulos, N.D.: Learning to optimize: training deep neural networks for wireless resource management. CoRR, abs/1705.09412 (2017)

22. Sun, S., Chen, W., Wang, L., Liu, X., Liu, T.-Y.: On the depth of deep neural networks: a theoretical view. In: Proceedings of the Thirtieth AAAI Conference on Artificial Intelligence, AAAI 2016, pp. 2066–2072. AAAI Press (2016)
23. Tse, D., Viswanath, P.: Fundamentals of Wireless Communication. Cambridge University Press, Cambridge (2005)
24. Wikipedia: Cross entropy, 20 June 2020. https://en.wikipedia.org/wiki/Cross_entropy
25. Wikipedia: Sorensen dice coefficient, 28 July 2020. https://en.wikipedia.org/wiki/S%C3%B8rensen%E2%80%93Dice_coefficient
26. Wikipedia: Huber loss, 29 May 2020. https://en.wikipedia.org/wiki/Huber_loss
27. Yi, Y., Chiang, M.: Stochastic network utility maximisation-a tribute to Kelly's paper published in this journal a decade ago. Eur. Trans. Telecommun. **19**(4), 421–442 (2008)

Enhanced Pub/Sub Communications for Massive IoT Traffic with SARSA Reinforcement Learning

Carlos E. Arruda[2], Pedro F. Moraes[2], Nazim Agoulmine[1] (iD),

and Joberto S. B. Martins[2(✉)] (iD)

[1] University of Paris-Saclay/IBISC Laboratory, Paris, France
nazim.agoulmine@univ-evry.fr
[2] Salvador University - UNIFACS, Salvador, Brazil
arruda.ceas@gmail.com, pmoraes_@hotmail.com, joberto.martins@gmail.com

Abstract. Sensors are being extensively deployed and are expected to expand at significant rates in the coming years. They typically generate a large volume of data on the internet of things (IoT) application areas like smart cities, intelligent traffic systems, smart grid, and e-health. Cloud, edge and fog computing are potential and competitive strategies for collecting, processing, and distributing IoT data. However, cloud, edge, and fog-based solutions need to tackle the distribution of a high volume of IoT data efficiently through constrained and limited resource network infrastructures. This paper addresses the issue of conveying a massive volume of IoT data through a network with limited communications resources (bandwidth) using a cognitive communications resource allocation based on Reinforcement Learning (RL) with SARSA algorithm. The proposed network infrastructure (PSIoTRL) uses a Publish/Subscribe architecture to access massive and highly distributed IoT data. It is demonstrated that the PSIoTRL bandwidth allocation for buffer flushing based on SARSA enhances the IoT aggregator buffer occupation and network link utilization. The PSIoTRL dynamically adapts the IoT aggregator traffic flushing according to the Pub/Sub topic's priority and network constraint requirements.

Keywords: Publish/Subscribe · Reinforcement learning · SARSA · IoT · Massive IoT traffic · Resource allocation · Network communications

1 Introduction

Sensors are being extensively deployed and are expected to expand at significant rates in the coming years in the internet of things (IoT) application areas like smart city, intelligent traffic systems (ITS), connected and autonomous vehicles

Work supported by CAPES and Salvador University (UNIFACS).

(CAV), video analytics in public safety (VAPS) and mobile e-health [31,37], to mention some.

Cloud, edge and fog computing are currently the most prevailing and competing used strategies to convey IoT data between producers and consumers. They use several different alternatives to decide where to process, how to distribute, and where to store IoT data [23]. Nevertheless, cloud, edge, and fog-based solutions need to tackle the distribution of a high volume of IoT data efficiently.

IoT data distribution, whatever data handling, processing, or storing strategy used, does require efficient network communications infrastructures. In fact, there is a research challenge on enhancing network communications and providing efficient resource allocation based on the fact that, in many applications like IoT, networks have limited and constrained resources [5].

Machine learning and reinforcement learning with distinct algorithms like Q-Learning [33], deep reinforcement learning [11], and SARSA (State-Action-Reward-State-Action) [1] are being applied to support an efficient allocation of resources in networks for application areas like 5G, cognitive radio, and mobile edge computing, to mention some [7,28].

The Publish/Subscribe (Pub/Sub) paradigm is extensively used as an enabler for data distribution in various scenarios like information-centric networking, data-centric systems, smart grid, and other general group-based communications [2,10,25]. The Pub/Sub paradigm is an alternative for IoT data distribution between producers and consumers in general [10,22].

This work addresses the issue of enhancing the utilization of a communication channel (MPLS label switched path - LSP, physical link, fiber optics slot, others) deployed in net- work infrastructures with limited resources (bandwidth) using the SARSA reinforcement learning algorithm. The target application area is the massive exchange of IoT data using the Pub/Sub paradigm. The framework integrating the SARSA module with the Pub/Sub message exchange deployment (PSIoT described in [22]) is the PSIoTRL.

Differently from other proposals for network infrastructure resource allocation and optimization that consider, for instance, the quality of service for the entire set of network nodes and Pub/Sub aggregators, the SARSA-based PSIoTRL framework provides a simple data ingress based solution. In fact, it controls the quality of Pub/Sub topics data distribution in an aggregator by keeping Pub/Sub buffer topics occupation below a defined threshold. This indirectly means that a certain amount of bandwidth is available for the Pub/Sub topic and, consequently, a level of quality is allocated for the communication channel. Since the Pub/Sub data transfers are dynamically requested, SARSA deploys a dynamic control of bandwidth allocation for buffer data flushing.

The contribution of this work is multi-fold. Firstly, we modeled the SARSA agent communications with a generic buffer occupation metric. The adopted metric results in a limited amount of SARSA states to allow the allocation of bandwidth preserving Pub/Sub topic priorities without using extensive computational resources. Moreover, we demonstrate that Pub/Sub communications with SARSA can be enhanced by dynamically adjusting IoT aggregators queues occupation to Pub/Sub topics priority and network resources availability.

The remainder of the paper is organized as follows. Section 2 presents the related work on IoT data processing deployments with cognitive communications approaches. Section 3 is a background section about reinforcement learning with SARSA. Sections 4 and 5 describe the basic PSIoT Pub/Sub framework, discuss the RL applicability for the constrained communication problem and present the PSIoTRL framework components including the intelligent orchestrator for IoT traffic management. In Sect. 6, we present a proof of concept for the PSIoTRL and evaluate how it enhances the IoT network resource management. Finally, Sect. 7 concludes with an overview of the main highlights, contributions and future work.

2   Related Work

Architectural and system-wide studies about IoT massive data processing and data flow in smart cities are presented in Al-Fuqaha [20], Rathore [27] and Martins [17]. Al-Fuqaha [20] introduces the concept of cognitive smart city where IoT and artificial intelligence are merged in a three-level model with different requirements. In relation to the communication level, the basic approach assumes a fog cloud computing (fog-CC) communication without considering any specific IoT massive traffic requirement. In fact, the proposal assumes that the aggregation with edge processing reduces the required bandwidth for fog-CC communication to a minimum. Rathore [27] explores the issue of big data analytics for smart cities. The paper proposes an edge-based aggregation and processing strategy for raw data. Processed data is forwarded through gateways to smart city applications using Internet and assuming bandwidth reduction at edge-level. Martins [17] discusses the potential benefits and impacts of how some technological enablers, like software-defined networking [12] and machine learning [36], are integrated and aim at cognitive management [28]. IoT data edge-processed aggregation, network communication and service deployments towards an efficient overall smart city solution are discussed and the relevance of intelligent communication resource provisioning and deployment are highlighted.

Resource provisioning from the perspective of IoT services deployment for smart city is considered by Santos [31]. The paper proposes a container-based micro-services approach, using Kubernetes, for service deployment that aims to off-load IoT processing with fog computing. In relation to this paper discussion, the proposal endorses fog processing offloading using a edge-computed approach and does not consider the network communication resource provisioning necessary to distribute the outcomes of the edge processing.

Edge intelligence for service deployment is discussed in Zhang [37]. The proposed solutions defines a framework capable to support the deployment of AI algorithms like RL on common edge aggregators like Raspberry. The approach assumed is to enable the execution of AI algorithms in the edge for those applications that require near real time edge processing like voice recognition and on-board autonomous vehicles processing.

Machine learning in communication networks is broadly addressed by Cote [7] and Boutaba et al. in [5].

Intelligent network communication resources are considered by Zhao [38]. The work presents a smart routing solution for crowdsourcing data with mobile edge computing (MEC) in smart cities using reinforcement learning (RL). The solution defines routes, differently from ours that optimizes the bandwidth allocated for IoT flow flushing.

In summary, this work advances on existing studies that propose service provisioning by adding an intelligent component for the allocation of communication resources between IoT data aggregators and IoT data consumers. From the architectural point of view, this work adopts an edge-based processing approach coupled with an efficient communication resource allocation for massive IoT data transfers.

3   Reinforcement Learning and SARSA Algorithm

Reinforcement learning (RL) is a largely used machine learning (ML) technique in which a trial-and-error learning process is executed by an agent that acts on a system aiming to maximize its rewards. The RL algorithm is expected to learn how to reach or approach a certain objective by interacting with a system through a feedback loop [13, 19, 33].

In RL, a reward value r is received by the agent for the transitions from one state to another. The overall objective of the agent is to find a policy π which maximizes the expect future sum of rewards received, each of them, subjected to a discount policy γ.

The *value function* in RL is a prediction of the return available from each state, as indicated in Eq. 1.

$$V(x_t) \leftarrow E\{\sum_{k=o}^{\infty} \gamma^k r_{t+k}\} \tag{1}$$

Where r_t is the reward received for the transition from state x_t to x_{t+1} and γ is the discount factor $(0 \leq \gamma \leq 1)$. The value function $V(x_t)$ represents the discounted sum of rewards received from step t onward. Therefore, it will depend on the sequence of actions taken and on the policy adopted to take these actions.

Two well-known and somehow similar reinforcement learning algorithms are Q-learning [8, 33] and SARSA [1, 14].

Q-learning is an off-policy RL algorithm in which the agent finds an optimal policy that maximizes the total discount expected reward for executing a particular action at a particular state. Fundamentally, Q-learning finds the optimal policy in a step-by-step manner. Q-learning is off-policy because the next state and action are uncertain when the algorithm updates the value function.

In Q-learning, the value function termed Q-function is learnt. It is a prediction of the return associated with each action $a \in A$ (set of actions). This prediction can be updated with respect to the predicted return of the next state visited (Eq. 4).

$$Q(x_t, a_t) \leftarrow r_t + \gamma V(x_{t+1}) \tag{2}$$

Since the overall objective is to maximize the reward received, the current estimate of $V(x_t)$ becomes:

$$Q(x_t, a_t) \leftarrow r_t + \gamma \max_{a \in A} Q(x_{t+1}, a) \tag{3}$$

The Q-function is shown to converge for markovian decision processes (MDP) [26]. In Q-learning, the agent maintains a lookup table of $Q(X, A)$ and $Q(x, a)$ represents the current estimate of the optimal action value function. Once the Q-function has converged, the optimal policy π is to take the action in each state with the highest predicted return (greedy policy) [29].

3.1 SARSA Algorithm

SARSA (State-Action-Reward-State-Action) algorithm is the reinforcement learning approach used by PSIoTRL framework [1,14].

SARSA is a temporal-difference (TD) on-policy algorithm that learns the Q-values based on the action performed by the current policy. SARSA algorithm differs from Q-learning by the way it sets up the future reward. In SARSA the agent uses the action and the state at time $t + 1$ to update the Q-value. The SARSA tuple of main elements involved in the interaction process are:

$$< x_t, a_t, r_{t+1}, x_{t+1}, a_{t+1} > \tag{4}$$

Where:

- x_t, a_t are the current state and action;
- r_{t+1} is the reward; and
- x_{t+1}, a_{t+1} are the next state and action reached using the policy $(\epsilon - Greedy)$.

SARSA Q-values are therefore updated based on the Eqs. 5 or 6.

$$Q(x_t, a_t) \leftarrow Q(x_t, a_t) + \alpha[r_{t+1} + \gamma Q(x_{t+1}, a_{t+1}) - Q(x_t, a_t)] \tag{5}$$

$$\Delta Q(x_{t-1}, a_{t-1}) = \alpha(t)[r_{t-1} + \gamma Q(x_t, a_t) - Q(x_{t-1}, a_{t-1})] \tag{6}$$

Where:

- α is the learning rate $(0 \leq \alpha \leq 1)$; and
- γ is the discount factor $(0 \leq \gamma \leq 1)$.

3.2   Network Communications Resource Allocation with SARSA

Reinforcement learning has achieved superhuman performance in games like Go [32] and also obtained a critical result bridging the divide between high-dimensional sensory inputs and actions with ATARI [18].

Video games and communication networks have imperfect but interesting operation similarities. As discussed in Cote [7], video games and networks are closed systems with a finite number of states. In games, actions include pressing buttons and moving the joystick, the image pixels define the state, and the cost function is the game score. In networks, actions correspond mainly to network configuration parameters (bandwidth allocation, fiber allocation, others) or network routing configuration. The network state (snapshot) defines the state of the system at each iteration, and the cost function corresponds to the performance of the network such like utilization rate, throughput, or number of dropped packets.

Hence, it is reasonable to consider that SARSA may have a parallel success for allocating network communications resources (bandwidth) in the PSIoTRL.

Reinforcement learning with SARSA has proved efficient and obtained substantial success in resource allocation [1,30], cloud computing [4] and computational offloading [3,24].

An evaluation of SARSA and various other model-free algorithms is presented in Dafazio [9]. The evaluation considered diverse and difficult problems within a consistent environment (Arcade [15]) involving configuration aspects like Epsilon-greedy policy, exploration versus exploitation and state space with SARSA outperforming algorithms like Q-learning, ETTR (Expected Time-to-Rendezvous) [34], R-learning [16], GQ algorithm [35] and Actor-Critic algorithm [6].

SARSA algorithm features and potential advantages within the PSIoTRL environment include:

- Being an on-policy algorithm, SARSA attempts to evaluate or improve the policy that is used to make decision;
- SARSA avoids the state explosion issue common with the Q-learning algorithm;
- PSIoTRL has a reduced state space and, consequently, SARSA behaves effectively by exploring all states at least once; and
- Key issues that RL algorithms experience like bad initial performance and large training time are minimized in the PSIoTRL environment since SARSA addresses the allocation of network bandwidth locally with a reduced state-space.

4   The Publish/Subscribe Framework for IoT (PSIoT)

The objective of this work is to enhance the Publish/Subscribe (Pub/Sub) communications used in the context of massive IoT data transfers (Fig. 1).

The Pub/Sub communications is supported by the PSIoT framework described in [22]. The PSIoT framework aims to efficiently handle network resources and IoT QoS requirements over the network between the IoT devices and consumer IoT applications [22]. Consumers are applications executed in servers located beyond the backbone network, cloud computing infrastructure accessed by the network or any other scheme that makes use of the managed network infrastructure for communications (Fig. 1).

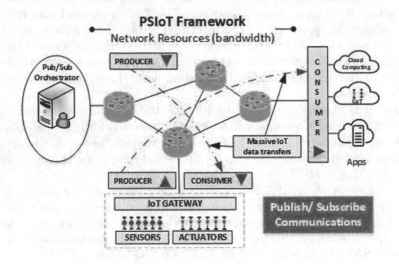

Fig. 1. PSIoT framework functional view [22].

The PSIoT framework was developed to manage IoT traffic in a network and to provide QoS based on IoT data characteristics and network-wide specifications, e.g. total network use, realtime network traffic, routing and bandwidth constraints.

IoT data generated from sensors and devices can be aggregated and processed in the cloud or at the network edge, where each of these points are considered as *Aggregators* or *Producers* in the PSIoT framework. Whereas each client that receives IoT data, both end-users applications and even other aggregation nodes, are denoted as *Consumers*.

The PSIoT framework was designed to orchestrate massive IoT traffic and allocate network resources between producers and consumers. Besides, the PSIoT framework can schedule the data flow, based on Quality of Service (QoS) requirements, and allocates bandwidth on a backbone with limited resources. For this, the PSIoT was modeled to operate with four main components: producers, consumers, a backbone with limited resources; and the orchestrator (Fig. 1) [22].

Aggregators are elements that gather the data obtained by connected sensors to send them (in an opportune moment) to their consumers (clients) using a

Pub/Sub-style communication channel. In the opposite direction, aggregators deliver data received from producers to send them to actuators.

This exchange of information is performed through the Pub/Sub architecture, whose characteristic of asynchronous, use of topics and the use of a broker, makes it attractive for IoT applications [10].

PSIoT implements QoS when forwarding the data gathered by the aggregator to output buffers, according to the following characteristics [22]:

- b0 (priority) - high transmission rates and low delay application requirements for health care and data from critical industrial sensors, as examples;
- b1 (sensitive) - commercial data and security sensors; and
- b2 (insensitive) - best effort.

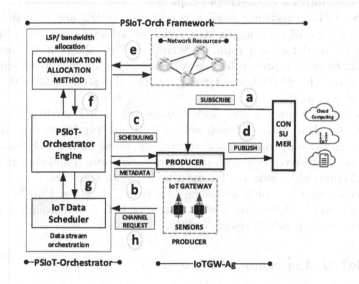

Fig. 2. PSIoT Pub/Sub message model [22].

The Pub/Sub message model adopted by the PSIoT is illustrated in Fig. 2. In summary, the message flow is as follows [22]:

1. A consumer subscribes to a particular topic and specify the requested QoS level with a particular aggregator (*a*);
2. The aggregator sends to the orchestrator relevant metadata such as number of subscribers and their associated QoS levels and buffer allocation (*b*);
3. The orchestrator notifies the aggregator with an amount of bandwidth that can be consumed by each level of QoS (*c*); and
4. The aggregator publishes the data to the consumer according to bandwidth and data availability in the buffer (*d*).

The PSIoT uses fixed rule scheduling for IoT data consumption.

5   PSIoTRL Framework - Architectural Components, Communication Model and SARSA Algorithm

The PSIoTRL architectural components are illustrated in Fig. 3. Its main proposed modules are:

- The SARSA agent;
- The PSIoTRL orchestrator module;
- Aggregators (at least one for each cluster where IoT data will be consumed);
- The network infrastructure (backbone); and
- Producers and consumers.

The SARSA agent is integrated in the PSIoT framework [22] and, by demand, allocates bandwidth for the aggregator queues.

The PSIoTRL orchestrator module is described in [22] and [21]. It basically has the knowledge of each aggregator Pub/Sub subscriptions, as well as the QoS levels required for each topic subscription. This allows it to control the transmission emptying rates of IoT data from each bu er within the aggregator in the network. The SARSA agent computes the allocated bandwidth and the orchestrator deploys it.

The aggregators are Fog-like nodes connected to IoT devices that act as aggregators for their data and also act as Pub/Sub producers regarding IoT applications subscribing to topics and consuming corresponding IoT data [10,21].

The network interconnects data producers (aggregators) and consumers and, has limited bandwidth resources for massive IoT data transfers.

Producers and consumers exchange IoT data. They use Pub/Sub [10] to communicate and exchange massive amounts of IoT data through a network with constrained bandwidth resources [21,22].

5.1   PSIoTRL Aggregator Configuration

The PSIoTRL framework uses the Pub/Sub paradigm to produce and consume data. Consumers request IoT data on the aggregator's queues, and the aggregator empties its buffers according to consumer requests. The aggregator queues are deployed with the following configuration:

- Three IoT data queues (buffers) are configured per aggregator (one Pub/Sub topic per queue): $B1$, $B2$ and $B3$;
- Initial buffer transmission rates[1] are: $T1$, $T2$ and $T3$; and
- Buffer priorities are: $p1$, $p2$ and $p3$ with $p1 > p2 > p3$.

For this evaluation we consider one aggregator Ag_i located at network node n_z that delivers IoT data to a set of consumers C_i, with $i = 1, 2, ..., n$ located at network node n_y. Between nodes n_z and n_y there is a communication resource

---

[1] By initial buffer transmission rate, we mean the configured initial transmission rate to empty buffers without any buffer over-utilization.

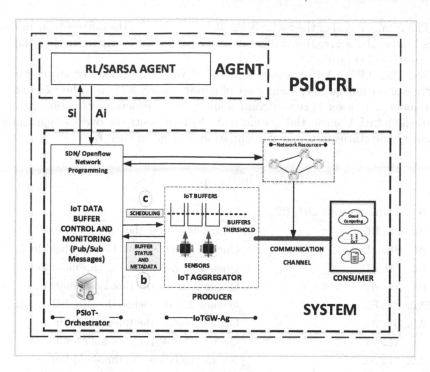

Fig. 3. PSIoTRL components and communication - orchestrator, aggregator, buffers and communication chanel.

(MPLS LSP, physical link, fiber slot, other) with a limited bandwidth BW_{nz}. The communication resource interconnecting Pub/Sub producer and consumers is then constrained as follows:

$$BW_{zy} \geq \sum_{j=1}^{n} TB_j \qquad (7)$$

Where TB_j is the currently transmission rate to empty a buffer j.

The aggregator monitors buffer occupation and occupation above a defined threshold limit is signaled to the orchestrator (Fig. 3).

5.2   The SARSA Agent Communication Model

The PSIoTRL deployment and operation require SARSA agent modeling for the buffer bandwidth allocation problem and the definition of basic SARSA algorithm configuration parameters.

The primary goal of the SARSA agent is to arbitrate the output transmission rates among the aggregator buffers in order to efficiently distribute IoT data to the consumers (Fig. 3).

The PSIoTRL SARSA agent is modeled with a finite set of states for the aggregator output queues, a finite set of configuration actions to be executed on each queue, and a set of reward values for each state/action transition pair.

In the SARSA agent, the system state S_i represents the overall aggregator's output queue status in terms of occupation at a given time t_i.

Table 1. Buffer (queue) states.

Buffer (queue) state (B1, B2, B3)	Description
BL, BL, BL	All queues below the threshold limit
BL, BL, AL	Queues 1 and 2 below the threshold limit
BL, AL, BL	Queues 1 an 3 below the threshold limit
BL, AL, AL	Queue 1 below the threshold limit
AL, BL, BL	Queues 2 and 3 below the threshold limit
AL, BL, AL	Queue 2 below the threshold limit
AL, AL, BL	Queue 1 below the threshold limit
AL, AL, AL	No queues below the threshold limit

† Legend: BL - Below Limit; AL - Above Limit.

Table 2. SARSA agent actions.

Buffer (queue)	Actions
B1	T+, T−, N
B2	T+, T−, N
B3	T+, T−, N

† T+: increase transmission rate; T−: decrease transmission rate; N: null action.

The SARSA agent states are indicated in Table 1. Each queue has two states either below (BL) or above (AL), a preconfigured threshold. Having two states for each queue allows, in this context, sufficient information for bandwidth allocation and contributes to avoid the common state explosion issue existing in many RL deployments.

A discrete and small number of states for the queues can be reasonably assumed because it mainly works as a threshold to indicate that queues require attention to enforce priorities or maximize throughput. The first case eventually happens when queued Pub/Sub messages require more capacity to be transferred than the actually allocated one. The second case corresponds to the situation where some queues have plenty of data to transmit, while others have unused capacity.

The utilization of the buffer threshold results in having a kind of agnostic Pub/Sub implementation. In fact, the threshold-based cognitive actions allow:

- To tune the PSIoTRL to have a faster or proactive reaction to adjust the queue's bandwidth; and
- To tune the Pub/Sub dynamic reconfiguration capability according to IoT data transfer sensitivity.

The PSIoTRL SARSA agent actions are illustrated in Table 2. Three actions are defined for each buffer: increase capacity (transmission rate), reduce capacity (transmission rate) and do nothing. Therefore, twenty seven actions are possible for each agent state. The amount of bandwidth increased or reduced per queue is a SARSA configuration parameter. A set of rewards is also defined for each executed state/action pair.

5.3   The SARSA ϵ-Greedy Policy

The SARSA agent uses an ϵ-greedy policy since it must exploit as much as possible the acquired knowledge but also explores new possibilities of enhancing the allocation of band-width for queue communications. The SARSA agent will take new actions at random with the ϵ-greedy policy defining the probability of choosing random actions. No value change or on-the-fly fine-tuning of ϵ was considered in this solution.

The ϵ-greedy policy matches adequately the inherent need of a Pub/Sub IoT message delivery framework. Random demands from consumers consume Pub/Sub data with different priorities. These demands generate a random volume of data to be transferred from aggregator queues to consumers. The per-queue bandwidth distribution must be dynamically computed among IoT data queues considering the existing constraint (bandwidth limitation) of the available communication channel.

5.4   The SARSA Algorithm for Bandwidth Allocation

The SARSA agent manages the aggregator's output queues transmission rate according to IoT data demands, IoT data priority, and communication resource constraints (bandwidth limitation). The bandwidth allocation algorithm pseudo-code is presented in Algorithm 1.

Algorithm 1. PSIoTRL-SARSA Bandwidth Allocation Algorithm Pseudo-code

1: **procedure** PSIoTRL-SARSA($Q(S_t, A_t, r, S_{t+1}, A_{t+1})$)   ▹ SARSA states and reward
2:    **for** each pair (Q_t, A_t) **do**   ▹ Initialization
3:        Initialize Q-values - $Q(S_t, A_t) = 0$ ▹ *Tabula rasa* approach
4:    **repeat**   ▹ Forever
5:        Gets current PSIoTRL buffers state S_t ▹ Buffer state is the trigger event
6:        **repeat**   ▹ Finishes upon terminal condition
7:            Choose action A_t using the ϵ-greedy policy   ▹ Exploration and exploitation
8:            Execute action A_t on the PSIoTRL system   ▹ T+, T- or do nothing on queues
9:            Get immediate reward r_t
10:           Collect new state S_{t+1}
11:           Choose new action A_{t+1} using the ϵ-greedy policy
12:           Update Q-value - $Q(S_t, A_t)$ using SARSA equation 5   ▹ SARSA algorithm
13:       **until** Terminal condition reached
14:   **until** forever

The algorithm procedure is as follows:

1. The algorithm starts initializing the Q-values in table, $Q(s, a)$;
2. The current state, s is captured;
3. An action a is chosen for the current state using the greedy policy;
4. The agent triggers the action, and observes the immediate reward, r, as well as the new reached state s';
5. The Q-value for the state s' is updated using the observed reward and the maximum reward possible for the next state; and
6. Finally for the current cycle, set the state to the new state, and repeat the process until the end condition is reached.

The end condition for the PSIoTRL bandwidth allocation process is the following:

- Bandwidth constraint limit is reached (Eq. 7); or
- Current buffers transmission rates corresponding priorities are ($T1 > T2 > T3$); or
- Maximum number of attempts is reached.

6   Proof of Concept

The purpose of this proof of concept is to validate that the agent behavior enhanced the allocation of bandwidth when one or more buffers exceeded the defined buffer occupation limit (50% - AL condition). When this event occurs, the aggregator detects the problem and sends an alarm containing metadata to the orchestrator. After that, the SARSA agent allocates a new percentage of bandwidth among the aggregator buffer.

Four initial buffer conditions were used:

- One queue exceeds the occupation threshold (AL, BL, BL);
- Two queues exceed the occupation threshold (AL, AL, BL);

- Three queues exceed the occupation threshold (AL, AL, AL); and
- One queue exceeds the link bandwidth capacity and the two other queues exceed the occupation threshold $(+AL, AL, AL)$.

The considered performance parameters are the following:

- Queue (buffer) Occupation;
- Link Occupation; and
- Packet Loss.

The allocation of bandwidth to the aggregator queues is evaluated in two scenarios. In scenario 1, a predefined simple algorithm is used to allocate the bandwidth. In scenario 2, the SARSA agent does the same task. Table 3 presents a summary of the evaluation scenarios.

Table 3. Summary of the proof of concept scenarios.

Aggregator				Orchestrator	
Queues States (B1, B2,B3)	T1i	T2i	T3i	Scenario 1	Scenario 2(SARSA)
AL, BL, BL					
AL, AL, BL	35%	25%	15%	T1 = T1i * factor;	SARSA module
AL, AL, AL				T2 = (100 − T1)/2;	
+AL, BL, BL				T3 = (100 − T1)/2;	

† +AL: 120% buffer occupation; dropping packets; factor: 1.15 or 1.25 or 1.50.

6.1   Scenario 1 - Bandwidth Allocation with Fixed Rule

In this scenario, the orchestrator has a simple fixed rule to increase or decrease the bandwidth for buffer flushing. $T1$ bandwidth is adjusted by a fixed factor (15%, 25% or 50%) and the remaining queues get half of the remaining bandwidth as follows:

- $T1 = T1_i * \text{factor}$ where $T1_i = T1$ initial state;
- $T2 = (100 - T1)/2$; and
- $T3 = (100 - T1)/2$.

Buffer transmission emptying initial rates are respectively 35%, 25% and 15% for T1, T2 and T3. Tables 4, 5 and 6 present the results obtained with the fixed rule bandwidth allocation method. In these Tables, packets loss for queue B_i is P_i.

Table 4. Bandwidth allocation with fixed rule - 15% bandwidth adjustment.

Initial QS (B1i, B2i, B3i)	Final QS (B1, B2, B3)	Final TR (T1, T2, T3)	Packet loss (P1, P2, P3)	Link occupation
AL, BL, BL	(72, 16, 0)	(40, 30, 30)	(0, 0, 0)	100
AL, AL, BL	(72, 64, 0)	(40, 30, 30)	(0, 0, 0)	100
AL, AL, AL	(72, 64, 0)	(40, 30, 30)	(0, 0, 0)	100
+AL, BL, BL	(108, 16, 0)	(40, 30, 30)	(8, 0, 0)	100

† QS: Queue State; TR: Transmission Rate; AL: 80%; BL: 20%; +AL: 120%.

Table 5. Bandwidth allocation with fixed rule - 25% bandwidth adjustment.

Initial QS (B1i, B2i, B3i)	Final QS (B1, B2, B3)	Final TR (T1, T2, T3)	Packet loss (P1, P2, P3)	Link occupation
AL, BL, BL	(56, 18, 2)	(44, 28, 28)	(0, 0, 0)	100
AL, AL, BL	(56, 72, 2)	(44, 28, 28)	(0, 0, 0)	100
AL, AL, AL	(56, 72, 8)	(44, 28, 28)	(0, 0, 0)	100
+AL, BL, BL	(84, 18, 2)	(44, 28, 28)	(0, 0, 0)	100

† QS: Queue State; TR: Transmission Rate; AL: 80%; BL: 20%; +AL: 120%.

Table 6. Bandwidth allocation with fixed rule - 50% bandwidth adjustment.

Initial QS (B1i, B2i, B3i)	Final QS (B1, B2, B3)	Final TR (T1, T2, T3)	Packet loss (P1, P2, P3)	Link occupation
AL, BL, BL	(40, 20, 10)	(53, 24, 23)	(0, 0, 0)	100
AL, AL, BL	(40, 80, 10)	(53, 24, 23)	(0, 0, 0)	100
AL, AL, AL	(40, 80, 40)	(53, 24, 23)	(0, 0, 0)	100
+AL, BL, BL	(60, 20, 10)	(53, 24, 23)	(0, 0, 0)	100

† Legend: QS: Queue State; TR: Transmission Rate; AL: 80%; BL: 20%; +AL: 120%.

6.2   Scenario 2 - Enhanced Communications with SARSA Agent

The parameters and initial conditions used in this scenario of the proof-of-concept are the following:

- SARSA Agent configuration parameters:
  - Epsilon-greedy policy ϵ of 2%;
  - Learning rate α of 20%; and
  - Discount factor γ of 80%
- Pub/Sub queue threshold limit (triggers agent action) = 50%
- Agent actions: bandwidth increased or reduced by 10%
- Maximum number of attempts = 400

The SARSA configuration parameters are fixed for all evaluations in this scenario 2. Typical values were used and the impact of the variation of the values of these parameters was not considered in this evaluation and left for future works.

In terms of the PSIoTRL operation and SARSA evaluation, the agent action in the orchestrator is triggered by the aggregator anytime one of its queues occupation goes beyond the configured threshold limit (Fig. 3).

The initial condition with SARSA are the same as those applied for the scenario 1 as indicated in Table 3. Ten runs were executed for each of the four initial states and the obtained results are summarized in Table 7 with a confidence interval of 5%.

Table 7. Queue occupation, packet loss and link occupation with SARSA.

Initial QS (B1i, B2i, B3i)	Final QS (B1, B2, B3)	Final TR (T1, T2, T3)	Packet loss (P1, P2, P3)	Link occupation
AL, BL, BL	(48, 12, 22)	(49, 35, 13)	(0, 0, 0)	97
AL, AL, BL	(47, 47, 23)	(50, 35, 12)	(0, 0, 0)	97
AL, AL, AL	(48, 48, 87)	(49, 35, 14)	(0, 0, 0)	98
+AL, BL, BL	(24, 24, 26)	(63, 20, 11)	(28, 0, 0)	94

† QS: Queue State; TR: Transmission Rate; AL: 80%; BL: 20%; +AL: 120%.

6.3 Bandwidth Allocation with SARSA for Buffer Flushing - Results

First of all, it is essential to highlight that this analysis's objective is to verify that the SARSA algorithm enhances bandwidth allocation for an IoT aggregator using specifically a Pub/Sub method to exchange data. In this context, enhanced bandwidth allocation means that the constrained bandwidth resources can be tuned and, consequently, better used for attained QoS performance parameters like delay, jitter, and packet loss.

The evaluated parameters are queue occupation, link occupation and packet loss.

The algorithm's ability to bring the queue state back to a threshold limit is not an effective performance parameter but demonstrates that the algorithm enhances the usage of the constrained bandwidth resources. For example, keeping the occupation of queues 1 and 2 below 50% would imply having the delay and jitter for the IoT data belonging to the specific Pub/Sub topics within its application's requirements.

The link occupation performance parameter evaluated in the runs indicates the efficient use or not of the bandwidth available, and the packet loss performance parameter is relevant to Pub/Sub topics sensitive to data loss.

The Figs. 4, 5, 6 and 7 present the final buffer occupation with bandwidth allocation for buffer flushing done by the orchestrator using fixed rule (case 1) and SARSA (case 2).

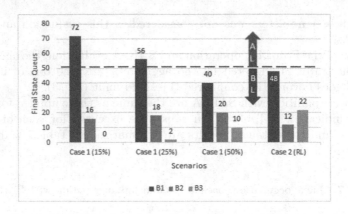

Fig. 4. Final queue occupation - initial state = *AL, BL, BL*.

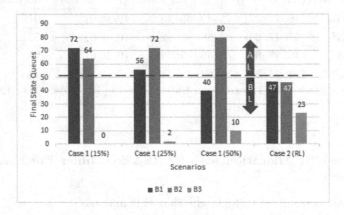

Fig. 5. Final queue occupation - initial state = *AL, AL, BL*.

For all initial states **(AL, BL, BL;   AL, AL, BL;   AL, AL, AL and   + AL, BL, BL)** bandwidth allocation with SARSA performed better that the fixed rule using increments of 15%, 25% and 50% in the buffer flushing bandwidth. SARSA algorithm brought all buffers to the final condition *BL, BL, BL*.

In Fig. 5, the case 1 (50%) was the unique fixed rule option that succeeded to bring one of the queue occupation (queue 3) below the threshold limit.

In Fig. 6, the bandwidth allocation with SARSA performed again better than the fixed rule algorithm. However, while the two priority buffers *B1* and *B2* are below the limit, buffer *B3* (least effort) was set above the limit due to its lower priority.

Figure 7 corresponds to the evaluation scenario in which we want to observe the behavior of the fixed rule bandwidth allocation and SARSA when buffer *B1* is already experiencing packet loss (above 100% capacity). Figure 7 shows that the SARSA algorithm is again the most efficient approach to bring all buffers below the threshold limit.

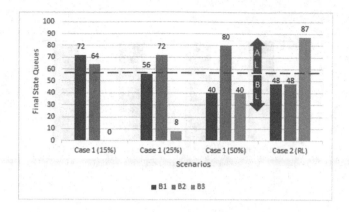

Fig. 6. Final queue occupation - initial state $= AL, AL, AL$.

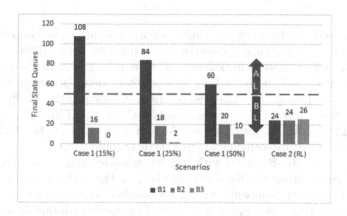

Fig. 7. Final queue occupation - initial state $= +AL, BL, BL$.

6.4 Allocated Bandwidth per Queue, Link Occupation and Packet Loss

In this experimentation, the objective is to use the minimum amount of link bandwidth per buffer to transmit data. Buffer priority should be preserved, and the aimed final condition is to bring all buffers to the BL state after bandwidth allocation.

Figure 8 illustrates two related aspects of link occupation. Firstly, it shows that the SARSA allocation of bandwidth per buffer is aligned with buffer priorities. In fact, more bandwidth is allocated to $B1$ than to $B2$, and, in turn, $B2$ gets more bandwidth than $B3$. This result is fully consistent with the defined buffer priorities. The fixed rule method allocates bandwidth to buffer $B1$ and splits the remaining bandwidth among the other buffers.

A second result illustrated in Fig. 8 is link occupation. The result indicates that the SARSA algorithm is more efficient than the fixed rule allocation method.

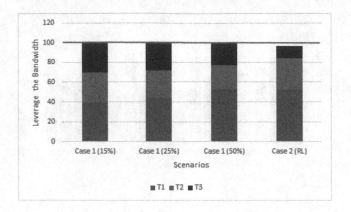

Fig. 8. Allocated bandwidth per queue and link occupation.

SARSA succeeds in bringing all buffer occupation levels bellow the defined threshold and, concomitantly, preserves link utilization below 100%.

Packet loss is indicated in Tables 4, 5, 6 and 7. The allocation of chunks of bandwidth for IoT data flushing in queues is done proactively once the 50% buffer occupancy threshold is reached. Consequently, no packet loss occurs for three out of four cases considered in the PSIoTRL evaluation *(AL, BL, BL; AL, AL, BL and AL, AL, AL)*. It happens that the buffer flushing bandwidth is adjusted before any packet loss could occur.

The case *+AL, BL, BL* starts with buffer overloaded and, as such, packet loss does occur since either the fixed rule method or the SARSA algorithm takes some time to allocate bandwidth to the overloaded buffer and fix the problem. As expected, SARSA takes more time to allocate bandwidth due to the learning process inherent to the algorithm and, consequently, presents a more significant packet loss.

7   Final Considerations

This work has presented a solution to address the problem of network resources allocation in the context of massive IoT data distribution. It has presented a PSIoTRL framework that introduces an architecture that aim to manage the allocation of limited network resources to aggregators.

As highlighted in the simulations, the SARSA agent in PSIoTRL was able to reconfigure the queues' flushing transmission rate efficiently, providing in 100% of the tested cases, that these queues reached the BL (below threshold limit) final condition with less link bandwidth usage. The PSIoTRL demonstrates that a solution based on SARSA is an efficient reinforcement learning approach for enhancing resource (bandwidth) allocation in the Pub/Sub-based massive IoT data exchange context.

Future works will address SARSA computation scalability concerning the granularity for allocating new chunks of bandwidth to flush IoT data and adjust the Pub/Sub queue occupancy. The impact of the SARSA configuration parameters (epsilon-greedy policy, learning rate, and discount factor), in the algorithm time response, with a focus on the learning rate, will also be evaluated.

Acknowledgments. Authors want to thanks CAPES (Coordination for the Improvement of Higher Education Personnel) for the master's scholarship support granted.

References

1. Alfakih, T., Hassan, M.M., Gumaei, A., Savaglio, C., Fortino, G.: Task offloading and resource allocation for mobile edge computing by deep reinforcement learning based on SARSA. IEEE Access **8**, 54074–54084 (2020)
2. An, K., Gokhale, A., Tambe, S., Kuroda, T.: Wide Area Network-Scale Discovery and Data Dissemination in Data-Centric Publish/Subscribe Systems, ACM Press, pp. 1–2 (2015)
3. Asghari, A., Sohrabi, M.K., Yaghmaee, F.: Task scheduling, resource provisioning, and load balancing on scientific workflows using parallel SARSA reinforcement learning agents and genetic algorithm. J. Supercomput. **77**, 2800–2828 (2021). https://doi.org/10.1007/s11227-020-03364-1
4. Bibal Benifa, J.V., Dejey, D.: RLPAS: reinforcement learning-based proactive autoscaler for resource provisioning in cloud environment. Mob. Netw. Appl. **24**(4), 1348–1363 (2018). https://doi.org/10.1007/s11036-018-0996-0
5. Boutaba, R., et al.: A comprehensive survey on machine learning for networking: evolution, applications and research opportunities. J. Internet Serv. Appl. **9**(1), 16 (2018)
6. Ciosek, K., Vuong, Q., Loftin, R., Hofmann, K.: Better exploration with optimistic actor critic, pp. 1787–1798 (2019)
7. Cote, D.: Using machine learning in communication networks. IEEE/OSA J. Opt. Commun. Networking **10**(10), D100–D109 (2018)
8. Dabbaghjamanesh, M., Moeini, A., Kavousi-Fard, A.: Reinforcement learning-based load forecasting of electric vehicle charging station using Q-learning technique. IEEE Trans. Indust. Inform. **V1**, 1–9 (2020)
9. Defazio, A., Graepel, T.: A comparison of learning algorithms on the arcade learning environment. arXiv:1410.8620 [cs], October 2014
10. Happ, D., Wolisz, A.: Limitations of the Pub/Sub pattern for cloud based IoT and their implications. In: Cloudification of the Internet of Things (CIoT), Paris, pp. 1–6. IEEE, November 2016
11. Koo, J., Menditta, V.B., Rahman, M.R., Walid, A.: Deep reinforcement learning for network slicing with heterogeneous resource requirements and time varying traffic dynamics. In: 2019 15th International Conference on Network and Service Management, pp. 1–5, October 2019
12. Kreutz, D., Ramos, F.M.V., Verissimo, P., Rothenberg, C.E., Azodolmolky, S., Uhlig, S.: Software-defined networking: a comprehensive survey. Proc. IEEE **103**(1), 14–76 (2014)
13. Latah, M., Toker, L.: Artificial intelligence enabled software-defined networking: a comprehensive overview. IET Netw. **8**(2), 79–99 (2019)
14. Liao, X., Wu, D., Wang, Y.: Dynamic spectrum access based on improved SARSA algorithm. IOP Conf. Ser. Mater. Sci. Eng. **768**(7), 072015 (2020)

15. Machado, M.C., Bellemare, M.G., Talvitie, E., Veness, J., Hausknecht, M., Bowling, M.: Revisiting the arcade learning environment: evaluation protocols and open problems for general agents. J. Artif. Intell. Res. **61**, 523–562 (2018)
16. Mahadevan, S.: Average reward reinforcement learning: foundations, algorithms, and empirical results. Mach. Learn. **22**(1), 159–195 (1996)
17. Martins, J.S.B.: Towards smart city innovation under the perspective of software-defined networking, artificial intelligence and big data. Revista de Tecnologia da Informação e Comunicação **8**(2), 1–7 (2018)
18. Mnih, V., et al.: Human-level control through deep reinforcement learning. Nature **518**(7540), 529–533 (2015)
19. Moerland, T.M., Broekens, J., Jonker, C.M.: A framework for reinforcement learning and planning. Ph.D. thesis, TU Delft. June 2020
20. Mohammadi, M., Al-Fuqaha, A.: Enabling cognitive smart cities using big data and machine learning: approaches and challenges. IEEE Commun. Mag. **56**(2), 94–101 (2018)
21. Moraes, P.F., Martins, J.S.B.: A Pub/Sub SDN-integrated framework for IoT traffic orchestration. In: Proceedings of the 3rd International Conference on Future Networks and Distributed Systems, ICFNDS 2019, Paris, France, pp. 1–9 (2019)
22. Moraes, P.F., Reale, R.F., Martins, J.S.B.: A publish/subscribe QoS-aware framework for massive IoT traffic orchestration. In: Proceedings of the 6th International Workshop on ADVANCEs in ICT Infrastructures and Services (ADVANCE), Santiago, pp. 1–14, January 2018
23. Mukherjee, M., Shu, L., Wang, D.: Survey of fog computing: fundamental, network applications, and research challenges. IEEE Commun. Surv. Tutorials **20**(3), 1826–1857 (2018)
24. Nassar, A., Yilmaz, Y.: Reinforcement learning for adaptive resource allocation in fog RAN for IoT with heterogeneous latency requirements. IEEE Access **7**, 128014–128025 (2019)
25. Nour, B., Sharif, K., Li, F., Yang, S., Moungla, H., Wang, Y.: ICN publisher-subscriber models: challenges and group-based communication. IEEE Netw. **33**(6), 156–163 (2019)
26. Ramani, D.: A short survey on memory based reinforcement learning. arXiv:1904.06736 [cs], April 2019
27. Rathore, M.M., Ahmad, A., Paul, A., Rho, S.: Urban planning and building smart cities based on the Internet of Things using big data analytics. Comput. Netw. **101**, 63–80 (2016)
28. Rendon, O.M.C., et al.: Machine learning for cognitive network management. IEEE Commun. Mag. 1–9 (2018)
29. Rummery, G.A., Niranjan, M.: On-line Q-learning using connectionist systems. Technical report, TR 166, Cambridge University Engineering Department, Cambridge, England (1994)
30. Sampaio, L.S.R., Faustini, P.H.A., Silva, A.S., Granville, L.Z., Schaeffer-Filho, A.: Using NFV and reinforcement learning for anomalies detection and mitigation in SDN. In: 2018 IEEE Symposium on Computers and Communications (ISCC), pp. 00432–00437, June 2018
31. Santos, J., Wauters, T., Volckaert, B., De Turck, F.: Resource provisioning in fog computing: from theory to practice. Sensors (Basel, Switzerland) **19**(10), 2238 (2019)
32. Silver, D., et al.: Mastering the game of go without human knowledge. Nature **550**(7676), 354–359 (2017)

33. Sutton, R.S., Barto, A.G.: Introduction to Reinforcement Learning, 1st edn. MIT Press, Cambridge (1998)
34. Wang, J.H., Lu, P.E., Chang, C.S., Lee, D.S.: A reinforcement learning approach for the multichannel rendezvous problem. In: 2019 IEEE Globecom Workshops (GC Wkshps), pp. 1–5, December 2019
35. Wang, Y., Zou, S.: Finite-sample analysis of Greedy-GQ with linear function approximation under Markovian noise. In: Proceedings of Machine Learning Research. Proceedings of the 36th Conference on Uncertainty in Artificial Intelligence (UAI), vol. 124, pp. 1–26 (2020)
36. Xie, J., et al.: A survey of machine learning techniques applied to software defined networking (SDN): research issues and challenges. IEEE Commun. Surv. Tutorials **21**(1), 393–430 (2019)
37. Zhang, X., Wang, Y., Lu, S., Liu, L., Xu, L., Shi, W.: OpenEI: an open framework for edge intelligence. In: 39th IEEE International Conference on Distributed Computing Systems (ICDCS), Dallas, US, pp. 1–12, July 2019
38. Zhao, L., Wang, J., Liu, J., Kato, N.: Routing for crowd management in smart cities: a deep reinforcement learning perspective. IEEE Commun. Mag. **57**(4), 88–93 (2019)

Deep Learning-Aided Spatial Multiplexing with Index Modulation

Merve Turhan[1]([⊠])(iD), Ersin Öztürk[2](iD), and Hakan Ali Çırpan[3](iD)

[1] Ericsson Research, Kista, Sweden
merve.turhan@ericsson.com
[2] Netas, Department of Research and Development, Pendik, 34912 Istanbul, Turkey
[3] Faculty of Electrical and Electronics Engineering, Istanbul Technical University,
34469 Maslak, Istanbul, Turkey

Abstract. In this paper, deep learning (DL)-aided data detection of spatial multiplexing (SMX) multiple-input multiple-output (MIMO) transmission with index modulation (IM) (Deep-SMX-IM) has been proposed. Deep-SMX-IM has been constructed by combining a zero-forcing (ZF) detector and DL technique. The proposed method uses the significant advantages of DL techniques to learn transmission characteristics of the frequency and spatial domains. Furthermore, thanks to using subblock-based detection provided by IM, Deep-SMX-IM is a straightforward method, which eventually reveals reduced complexity. It has been shown that Deep-SMX-IM has significant error performance gains compared to ZF detector without increasing computational complexity for different system configurations.

Keywords: GFDM · OFDM · Deep learning · Spatial multiplexing · Index modulation

1 Introduction

The demand for wireless communications continues to increase and expand rapidly with new applications. In order to carry out the requested demand, orthogonal frequency division multiplexing (OFDM) has been proposed by Third Generation Partnership Project (3GPP) [2,3]. In spite of its proven advantages, OFDM has some drawbacks such as high out-of band (OOB) emission and high peak-to-average power ratio (PAPR) [22]. In this sense, generalized frequency division multiplexing (GFDM) [11] came into prominence in terms of reduced latency, high spectral efficiency, and low OOB emission. Also, spatial multiplexing (SMX) is an effective method to improve spectral efficiency. On the other hand, index modulation (IM) techniques [4] provide spectral and energy efficiency with using transmissions entities to convey digital information innovatively. Taking account of GFDM, MIMO and IM efficiencies, their tight integration has been introduced in [12–17]. In the past decade, deep neural networks

---

M. Turhan—Work done while author was afliated with Istanbul Technical University.

© Springer Nature Switzerland AG 2021
É. Renault et al. (Eds.): MLN 2020, LNCS 12629, pp. 226–236, 2021.
https://doi.org/10.1007/978-3-030-70866-5_14

(DNNs) has been widely applied in miscellaneous areas, e.g. speech recognition, object detection, natural language processing [7]. Also, it has become an important area for communication systems, especially for physical layer problems [9,18]. In [5,8,19], the use of DNN for MIMO detection has been examined. In [10,20,21], deep learning (DL)-aided data detection scheme has been presented for GFDM, OFDM with IM (OFDM-IM) and GFDM with IM (GFDM-IM), respectively. In this article, a DNN-aided detector is proposed for the combined application of SMX transmission, GFDM, and IM for the purpose of improving error performance without increasing complexity. The main contribution of this article is to adapt a convolutional neural network (CNN) and a fully connected neural network (FCNN) to learn the transmission characteristics of spatial and frequency multiplexing, respectively. Note that, a CNN approach provides a flexible structure for SMX transmission thanks to supporting multi-channel operation and preserves the spatial dependence. Besides, IM scheme enables to implement subblock-based and parallel data detection, which simplifies the DL model and reduces the complexity as well as the processing delay significantly. As far as we know, the proposed method would be the first attempt to implement DL-aided SMX with IM (SMX-IM) detection. It has been shown that the proposed method has a significant bit error rate (BER) gain compared to ZF detector with the same complexity in terms of order of magnitude of complex multiplication (CM).

2 System Model

Consider a GFDM-based SMX system with T transmit and R receive antennas. Note that the considered system covers the OFDM-based SMX system. The transmitter gets PT information bits as input. A GFDM symbol, each composing of M subsymbols with K subcarriers, is split into L IM groups. Each group is consisted of $u = MK/L$ subcarrier locations, and v out of u subcarrier locations are activated and used to transmit quadrature amplitude modulation (QAM) symbols. Hence, an IM group carries a p-bit binary message $s_t^l = \left[s_t^l(1), s_t^l(2), \ldots, s_t^l(p) \right]$, for $l = 1, \ldots, L$, $t = 1, \ldots, T$ and $P = pL$. Each p-bit binary message consist of p_i and p_q bits. While $p_q = v \log_2(Q)$ are mapped by Q-array mapper, $p_i = \lfloor \log_2(C(u,v)) \rfloor$ are executed to determine sub-carrier locations with reference to lookup table [4]. Therefore, $\alpha = 2^{p_i}$ possible realizations are obtained. Here, $C(\mu, \nu)$ represents the binomial coefficient and $\lfloor \cdot \rfloor$ is the floor function. As a consequence, IM groups $d_t^l = \left[d_t^l(1), d_t^l(2), \ldots, d_t^l(u) \right]^T$, where $d_t^l(\gamma) \in \{0, S\}$, is built as using mapping operation with p input bits [12]. Here, S denotes Q-ary constellation. The resulting IM groups are then concatenated to form the GFDM-IM symbol

$$\mathbf{d_t} = [d_{t,0,0}, \ldots, d_{t,K-1,0}, d_{t,0,1}, \ldots, d_{t,K-1,1}, \ldots, d_{t,K-1,M-1}] \tag{1}$$

where $d_{t,k,m} \in \{0, S\}$, for $m = 0, \ldots, M-1$, $k = 0, \ldots, K-1$, $t = 1, \ldots, T$, is the data symbol of k-th subcarrier on m-th subsymbol of a GFDM symbol belonging to t-th transmit antenna. Then, the GFDM-IM symbol $\mathbf{d}_t$ is modulated by a GFDM modulator and the resulting GFDM transmit signal can be written in linear form as

$$\mathbf{x}_t = \mathbf{A}\mathbf{d}_t, \tag{2}$$

where $\mathbf{A}$ represents an $MK \times MK$ GFDM transmitter matrix [11]. Eventually, a cyclic prefix (CP) with length N_{CP} is attached to $\mathbf{x}_t$ and the resulting vector $\hat{\mathbf{x}}_t = \left[\mathbf{x}_t \left(MK - N_{\mathrm{CP}} + 1 : MK \right)^T, \mathbf{x}_t{}^T \right]^T$ is transmitted over a frequency-selective channel.

At the receiver side, under the assumption that CP is longer than the maximum delay spread of the channel (N_{Ch}), the whole received signal can be obtained as

$$\underbrace{\begin{bmatrix} \mathbf{y}_1 \\ \vdots \\ \mathbf{y}_R \end{bmatrix}}_{\mathbf{y}} = \underbrace{\begin{bmatrix} \mathbf{H}_{1,1}\mathbf{A} \dots \mathbf{H}_{T,1}\mathbf{A} \\ \vdots \quad \ddots \quad \vdots \\ \mathbf{H}_{R,1}\mathbf{A} \dots \mathbf{H}_{R,T}\mathbf{A} \end{bmatrix}}_{\widetilde{\mathbf{H}}} \underbrace{\begin{bmatrix} \mathbf{d}_1 \\ \vdots \\ \mathbf{d}_T \end{bmatrix}}_{\mathbf{d}} + \underbrace{\begin{bmatrix} \mathbf{n}_1 \\ \vdots \\ \mathbf{n}_R \end{bmatrix}}_{\mathbf{n}} \tag{3}$$

after the removal of CP. Here, $\mathbf{y}_r = [y_r(0), y_r(1), \dots, y_r(N-1)]^T$ is the vector of the received signals, $\mathbf{H}_{r,t}$, for $t = 1, \dots, T, r = 1, \dots, R$, indicates the $N \times N$ circular convolution matrix formed from the channel impulse response coefficients given by $\mathbf{h}_{r,t} = [h_{r,t}(1), h_{r,t}(2), \dots, h_{r,t}(N_{\mathrm{Ch}})]^T$, and $\mathbf{n}_r$ denotes an $N \times 1$ additive white Gaussian noise (AWGN) vector. The elements of $\mathbf{h}_{r,t}$ and $\mathbf{n}_r$ follow $\mathcal{CN}(0,1)$ and $\mathcal{CN}(0,\sigma_n^2)$ distributions, respectively. Here, $\mathcal{CN}(\mu, \sigma^2)$ represents the distribution of a circularly symmetric complex Gaussian random variable with mean μ and variance σ^2. Equation 3 can be rewritten in a more succinct form as

$$\mathbf{y} = \widetilde{\mathbf{H}}\mathbf{d} + \mathbf{n}. \tag{4}$$

3   Deep Detector

The block diagram of the proposed deep learning-aided data detection of spatial multiplexing scheme, termed as Deep-SMX-IM, is given in Fig 1. The channel information is assumed to be perfectly known at the receiver. The proposed detector is built as two stages, namely coarse detector and fine detector. It also has two intermediate steps for regularizing coarse detector output and fine detector outputs, which are SMX-IM Block Splitter and Combiner, respectively. As a first stage, coarse detector applies ZF detection to handle channel and GFDM modulation effects together and the coarse detector's output can be given by

$$\begin{bmatrix} \psi_1 \\ \psi_2 \\ \vdots \\ \psi_T \end{bmatrix} = \left(\widetilde{\mathbf{H}}^H \widetilde{\mathbf{H}} \right)^{-1} \widetilde{\mathbf{H}}^H \mathbf{y}, \tag{5}$$

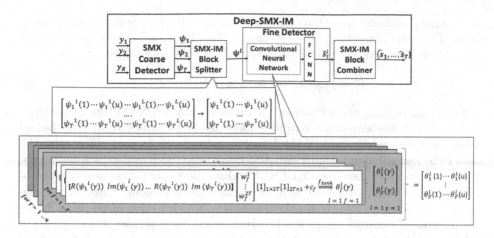

Fig. 1. Block diagram of the Deep-SMX-IM.

where $\boldsymbol{\psi}_t = [\boldsymbol{\psi}_t^{1^T}, \boldsymbol{\psi}_t^{2^T}, ..., \boldsymbol{\psi}_t^{L^T}]^T$, for $t = 1, ...T$ and for $l = 1, ...L$, $\boldsymbol{\psi}_t^l$ denotes a $u \times 1$ vector. After that, coarse detector output $\boldsymbol{\psi}_t^l$ is subdivided into IM subblocks by SMX-IM Block Splitter and the resulting matrix can be expressed as

$$\boldsymbol{\Psi}^l = \begin{bmatrix} \psi_1^l(1) \ \dots \ \psi_1^l(u) \\ \vdots \ \ddots \ \vdots \\ \psi_T^l(1) \ \dots \ \psi_T^l(u) \end{bmatrix}. \tag{6}$$

The fine detector stage of Deep-SMX-IM is built by using CNN and FCNN. The fine detector's CNN part convolves the IM subblock $\boldsymbol{\Psi}^l$ with the kernels $\mathbf{w}^f = [w_1^f, ..., w_{2T}^f]$, adds bias $\mathbf{c}_f$ for $f = 1, ...F$ and stride 1, and the modified received IM subblock is obtained as

$$\theta_f^l(\gamma) = f_{tanh}(\Re(\psi_1^l(\gamma) * w_1^f) + \Im(\psi_1^l(\gamma)) * w_2^f + \dots$$
$$+ \Re(\psi_T^l(\gamma) * w_{2T-1}^f) + \Im(\psi_T^l(\gamma)) * w_{2T}^f + c_f) \tag{7}$$

for $f = 1, ..., F$ and $\gamma = 1, ..., u$. Note that since complex numbers are not supported by any DL framework yet, real and imaginary parts of the received signals are processed separately. The CNN part repeats the convolution operation for F different kernel filters. The output from CNN can be obtained as

$$\boldsymbol{\Theta}^l = \begin{bmatrix} \theta_1^l(1) \ \theta_1^l(2) \dots \ \theta_1^l(u) \\ \vdots \ \ddots \ \vdots \\ \theta_F^l(1) \ \dots \ \theta_F^l(u) \end{bmatrix}. \tag{8}$$

After that, $\boldsymbol{\Theta}^l$ is converted into a vector by flattening process and the resulting vector can be expressed as $\theta^l = [\theta_1^l(1), ..., \theta_1^l(u), ..., \theta_F^l(1), ..., \theta_F^l(u)]^T$. The fine detector's FCNN part executes θ^l with using $\{\mathbf{a}, \mathbf{b}\}$ parameters, where

$\mathbf{a} = [\mathbf{a}_1, \mathbf{a}_2]$ includes weight parameters and $\mathbf{b} = [b_1, b_2]$ contains bias parameters. This part consists of only two layers, first layer and output layer have τ and PT neurons, respectively. Fine detector's output is obtained as

$$\hat{\mathbf{s}}^l = f_{sigmoid}(\mathbf{a}_2(f_{tanh}(\mathbf{a}_1 \theta^l + b_1)) + b_2, \tag{9}$$

where $f_{sigmoid}, f_{tanh}$ are activation functions. Finally, SMX-IM Block Combiner composes the fine detector's output into transmitted information bits. Before using the proposed Deep-SMX-IM detector, it has to be trained offline with the data to be generated at training signal-to-noise ratio (SNR) value by simulations. While the training step uses fixed SNR value, the testing step uses a range of SNR values. Deciding training SNR value has a key role against overfitting. The training executes on total trainable parameters, which consist of $\mathbf{w}^f, \mathbf{c}^f, \mathbf{a}, \mathbf{b}$, for the purpose of minimizing the loss function. It can be expressed as $loss(\mathbf{s}^l, \hat{\mathbf{s}}^l) = \|\mathbf{s}^l - \hat{\mathbf{s}}^l\|$. In the training stage, total trainable parameters are randomly initialized at first. Throughout the training, stochastic gradient descent algorithm executes on these parameters, it can be expressed as

$$\epsilon_+ = \epsilon - \eta \bigtriangledown loss(\mathbf{s}^l_B, \hat{\mathbf{s}}^l_B), \tag{10}$$

where ϵ, η, B, represent total trainable parameters, learning rate and batch size respectively.

4   Complexity Analysis for Deep-SMX-IM

In this section, we have assessed the computational complexity of ZF, maximum likelihood (ML) and Deep-SMX-IM detectors in terms of number of CM. The results are provided in Table 1. Here, $\Psi_{J \times I}$ and $\Phi_{J \times I}$ stand for $J \times I$ matrices, $\psi_{J \times 1}$ and $\phi_{J \times 1}$ denotes $J \times 1$ vectors. Note that, Deep-SMX-IM has only real operation and one CM can be executed with three real multiplications at least. So, the number of multiplications of Deep-SMX-IM is divided by three for the purpose of expressing them as CM. As seen in Table 2, while ML detector has the highest complexity, Deep-SMX-IM and ZF have approximately the same complexity in terms of order of magnitude of CMs.

5   Simulation Results and Discussion

In this section, the performance of Deep-SMX-IM detector has been evaluated for Rayleigh fading with Extended Pedestrian A (EPA) channel model [1] employing BPSK and 4-QAM modulation. The raised cosine filter is used as a GFDM prototype filter with a roll-off factor of 0.5. The following GFDM-IM parameters are assumed: $K = 32, M = 3, N_{Ch} = 8, u = 4, v = 2$. In order to select the active subcarrier indices, the lookup Table 5 is used. Note that $K = 32, M = 1$ is assumed for OFDM parameters. Training data including 12×10^5 IM groups, is

Table 1. Comparison of computational complexities

Receiver scheme	Process	Operation	Execution count	CMs
ML	Forming	$\tilde{\mathbf{H}}\Phi_{N\times N}\Psi_{N\times N}{}^{\dagger}$	1	$N_{Ch}N^2RT$
	Decision	$\min\left(\|\phi_{N\times 1}-(\Phi_{N\times N}\psi_{N\times 1})\|^2\right)^{\dagger\dagger}$	$(\alpha Q^v)^{TN/u}$	$(\alpha Q^v)^{TN/u}\left((N^2TRv)/u+NT\right)$
ZF	Forming	$\tilde{\mathbf{H}}\Phi_{N\times N}\Psi_{N\times N}{}^{\dagger}$	RT	$N_{Ch}N^2RT$
	JDD	$\left(\Phi_{NR\times NT}{}^H\Phi_{NR\times NT}\right)^{-1}\Phi_{NR\times NT}{}^H\phi_{NR\times 1}$	1	$2N^3T^2R+N^3T^3+N^2RT$
	Decision	$\min\left(\|\phi_{u\times 1}-\psi_{u\times 1}\|^2\right)$	$(N/u)\alpha Q^v T$	$N\alpha Q^v T$
Deep-SMX-IM	Forming	$\tilde{\mathbf{H}}\Phi_{N\times N}\Psi_{N\times N}{}^{\dagger}$	RT	$N_{Ch}N^2RT$
	JDD	$\left(\Phi_{NR\times NT}{}^H\Phi_{NR\times NT}\right)^{-1}\Phi_{NR\times NT}{}^H\phi_{NR\times 1}$	1	$2N^3T^2R+N^3T^3+N^2RT$
	CNN	$(2FT+T\lambda)/3^{\dagger\dagger\dagger}$	N	$(2FT+F\lambda))N/3$
	FCNN	$(uF\tau+\tau\lambda+\tau pT+pT\delta)/3^{\dagger\dagger\dagger}$	L	$(uT\tau+\tau\lambda+\tau pT+p\delta T)L/3$

† In every row of $\mathbf{H}$, only N_{Ch} out of N elements are non-zero.
†† In ψ, only vML complex elements are nonzero.
††† λ and δ denote to number of real multiplications needed for f_{tanh} and $f_{sigmoid}$, respectively.

Table 2. The total number of CMs

Config. (T,R)	ML	ZF	Deep-SMX-IM
BPSK $(2,2)$	3.29067×10^{76}	4.53561×10^8	4.53282×10^8
4-QAM $(2,2)$	3.08764×10^{134}	4.53840×10^8	4.53328×10^8
BPSK $(4,4)$	1.83301×10^{178}	2.31930×10^{11}	2.31929×10^{11}
4-QAM $(4,4)$	1.08149×10^{294}	2.31931×10^{11}	2.31929×10^{11}

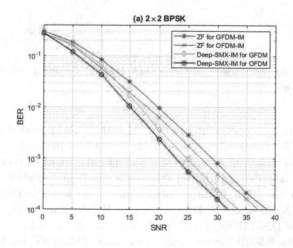

Fig. 2. Performance comparison of Deep-SMX-IM and ZF for 2×2 SMX-IM using BPSK transmission

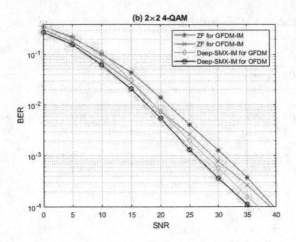

Fig. 3. Performance comparison of Deep-SMX-IM and ZF for 2×2 SMX-IM using 4-QAM transmission.

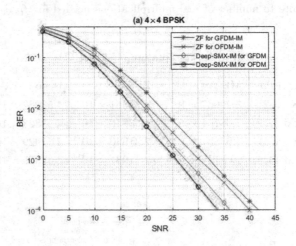

Fig. 4. Performance comparison of Deep-SMX-IM and ZF for 4×4 SMX-IM using BPSK transmission.

generated at SNR 15 dB according to GFDM and OFDM parameters. The Deep-SMX-IM model is trained 120 epochs with $B = 1000$. In Table 3 and 4, Deep-SMX-IM fine detector model parameters and summary can be seen respectively. In order to find a global minimum, stochastic gradient based Adam optimizer [6], is used with 8×10^{-4} learning rate. Figure 2 depicts the BER comparison of the Deep-SMX-IM and ZF detectors using BPSK along with SMX-GFDM-IM and SMX-OFDM-IM for 2×2 SMX systems. As seen from Fig. 2, Deep-SMX-IM provides 5.5 dB better BER performance than ZF for BPSK. Figure 3 depicts the BER comparison of the Deep-SMX-IM and ZF detectors using 4-QAM along with SMX-GFDM-IM and SMX-OFDM-IM for 2×2 SMX systems.

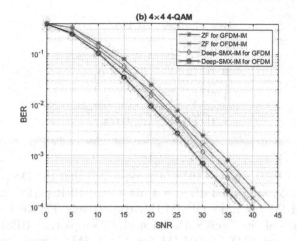

Fig. 5. Performance comparison of Deep-SMX-IM and ZF for 4×4 SMX-IM using 4-QAM transmission.

Table 3. Deep-SMX-IM model parameters

Antenna configuration	Modulation	Parameter	Value
2×2	BPSK	F	64
		τ	128
	4QAM	F	64
		τ	256
4×4	BPSK	F	128
		τ	256
	4QAM	F	128
		τ	512

Table 4. Fine detector model summary

Layer	Output shape	Activation func.
Input ($\mathbf{\Psi}^l$)	$(B, u, 2, T)$	None
CNN	$(B, 1, u, F)$	f_{tanh}
Flattening	(B, uF)	None
1. Layer of FCNN	(B, τ)	f_{tanh}
2. Layer of FCNN	(B, pT)	$f_{sigmoid}$

Table 5. A look-up table example for $u = 4, v = 2$.

Bits	Indices	IM block
[0 0]	$\{1,2\}$	$\begin{bmatrix} s_\chi & s_\zeta & 0 & 0 \end{bmatrix}^T$
[0 1]	$\{2,3\}$	$\begin{bmatrix} 0 & s_\chi & s_\zeta & 0 \end{bmatrix}^T$
[1 0]	$\{3,4\}$	$\begin{bmatrix} 0 & 0 & s_\chi & s_\zeta \end{bmatrix}^T$
[1 1]	$\{1,4\}$	$\begin{bmatrix} s_\chi & 0 & 0 & s_\zeta \end{bmatrix}^T$

From Fig. 3 for a BER value of 10^{-4}, it is observed that the Deep-SMX-IM for GFDM and OFDM achieves 3 dB and 4 dB better BER performance than ZF for SMX-GFDM-IM and SMX-OFDM-IM, respectively. Figure 4 compares the BER performance of the Deep-SMX-IM and ZF employing BPSK along with SMX-GFDM-IM and SMX-OFDM-IM for 4×4 SMX transmission. As seen from Fig. 4, the Deep-SMX-IM for GFDM and OFDM achieves 5.5 better BER performance than ZF for BPSK at a BER value of 10^{-4} for SMX-GFDM-IM and SMX-OFDM-IM, respectively. Figure 5 compares the BER performance of the Deep-SMX-IM and ZF employing 4-QAM along with SMX-GFDM-IM and SMX-OFDM-IM for 4×4 SMX transmission. As seen from Fig. 4, Deep-SMX-IM for GFDM and OFDM achieves 3 dB better BER performance than ZF for 4-QAM at BER value 10^{-4}. As seen from Fig. 2 and 4, as spectral efficiency and the modulation order increases, model's learning capacity decreases. However, Deep-SMX-IM continues to retain its advantage over the classical linear detector in all conceivable system parameters. In Table 2, the number of CMs required for Fig. 2 and 4 are provided. Notice that, while DeepConvIM in [20] has an intermediate solution for GFDM-IM, Deep-SMX-IM can be assessed as an efficient solution in terms of computational complexity.

6   Conclusion

A novel DL-aided detector, called Deep-SMX-IM, has been proposed for SMX transmission with IM. It has been shown that the proposed deep learning-aided detector provides important BER performance improvement compare to ZF detector with the approximately same complexity in terms of order of magnitude of CMs. Our results highlight that the significant advantages of deep learning techniques should be engineered to overcome the challenges of wireless communications arising from the distinct characteristics of time, frequency and spatial domains. As future work, performance analysis of Deep-SMX-IM can be examined for the different coarse detectors, e.g. minimum mean-squared error (MMSE), ML. Furthermore, it may be considered to apply the proposed methods in different index modulation concepts.

References

1. 3GPP: Base station (BS) radio transmission and reception. Technical spec. 36.104 V14.4.0, June 2017
2. 3GPP: NR, physical layer, general description. tech. spec. 38.201, December 2017
3. 3GPP: Study on new radio (NR) access technologies. Technical report 38.912 V14.1.0, June 2017
4. Basar, E., Aygölü, Ü., Panayırcı, E., Poor, H.V.: Orthogonal frequency division multiplexing with index modulation. IEEE Trans. Signal Process. **61**(22), 5536–5549 (2013)
5. Corlay, V., et al.: Multilevel MIMO detection with deep learning. arXiv:1812.01571 (2018)
6. Kingma, D.P., Ba, J.: Adam: a method for stochastic optimization. arXiv preprint arXiv:1412.6980 (2014)
7. Goodfellow, I., Bengio, Y., Courville, A.: Deep Learning. MIT Press (2016). http://www.deeplearningbook.org
8. He, H., et al.: A model-driven deep learning network for MIMO detection. arXiv:1809.09336 (2018)
9. Huang, H., et al.: Deeplearning for physical-layer 5G wireless techniques: opportunities, challenges and solutions. IEEE Wireless Commun. **27**(1), 214–222 (2020). https://doi.org/10.1109/mwc.2019.1900027
10. Luong, T.V., Ko, Y., Vien, N.A., Nguyen, D.H.N., Matthaiou, M.: Deep learning-based detector for OFDM-IM. IEEE Wireless Commun. Lett. **8**(4), 1159–1162 (2019). https://doi.org/10.1109/LWC.2019.2909893
11. Michailow, N., et al.: Generalized frequency division multiplexing for 5th generation cellular networks. IEEE Trans. Commun. **62**(9), 3045–3061 (2014)
12. Ozturk, E., Basar, E., Cirpan, H.: Generalized frequency division multiplexing with index modulation. In: Proceeding IEEE GLOBECOM Workshops. Washington DC, USA, December 2016
13. Ozturk, E., Basar, E., Cirpan, H.: Spatial modulation GFDM: a low complexity MIMO-GFDM system for 5G wireless networks. In: Proceeding 4th IEEE International Black Sea Conference Communication Networking. Varna, Bulgaria, June 2016
14. Ozturk, E., Basar, E., Cirpan, H.: Generalized frequency division multiplexing with flexible index modulation. IEEE Access **5**, 24727–24746 (2017)
15. Ozturk, E., Basar, E., Cirpan, H.: Generalized frequency division multiplexing with space and frequency index modulations. In: Proceeding 5th IEEE International Black Sea Conference Communication Networking. Istanbul, Turkey, June 2017
16. Ozturk, E., Basar, E., Cirpan, H.: Generalized frequency division multiplexing with flexible index modulation numerology. IEEE Signal Process. Lett. **25**(10), 1480–1484 (2018)
17. Ozturk, E., Basar, E., Cirpan, H.: Multiple-input multiple-output generalized frequency division multiplexing with index modulation numerology. Phys. Commun. **34**, 27–37 (2019)
18. O'Shea, T., Hoydis, J.: An introduction to deep learning for the physical layer. IEEE Trans. Cogn. Commun. Netw. **3**(4), 563–575 (2017). https://doi.org/10.1109/TCCN.2017.2758370
19. Samuel, N., Diskin, N., Wiesel, A.: Deep MIMO detection. arXiv preprint arXiv:1706.0115 (2018)

20. Turhan, M., Ozturk, E., Cirpan, H.: Deep convolutional learning-aided detector for generalized frequency division multiplexing with index modulation. In: IEEE International Symposium on Personal, Indoor and Mobile Radio Communications. Istanbul, Turkey (2019)
21. Turhan, M., Ozturk, E., Cirpan, H.: Deep learning aided generalized frequency division multiplexing. In: Proceeding 3rd International Balkan Conference Communication Networking. Skopje, North Macedonia (2019)
22. Wunder, G., et al.: 5GNOW: non-orthogonal, asynchronous waveforms for future mobile applications. IEEE Commun. Mag. **52**(2), 97–105 (2014)

A Self-gated Activation Function SINSIG Based on the Sine Trigonometric for Neural Network Models

Khalid Douge[1](✉), Aissam Berrahou[2](✉), Youssef Talibi Alaoui[3](✉),
and Mohammed Talibi Alaoui[1](✉)

[1] Sidi Mohamed Ben Abdellah University, Fez, Morocco
khaliddouge@gmail.com, mohammed.talibialaoui@usmba.ac.ma
[2] Mohamed V University, Rabat, Morocco
aissam.berrahou@gmail.com
[3] Mohamed I University, Oujda, Morocco
y.talibi.alaoui@gmail.com

Abstract. Deep learning models are based on a succession of multiple layers of artificial neural networks, which allows us to approach the resolution of several mathematical transformations and feed the next layer. This process is turned by exploiting the principle of non-linearity of the activation function that determine the output of neural network layer in aim to facilitate the learning process during training. Indeed, to improve the performance of these functions, it is essential to understand their non-linear behavior, in particular concerning their negative parts. In this context, the enhanced new activation functions which were implemented after ReLU function exploit the negative values to further optimize the gradient descent. In this paper, we propose a new activation function which is based on a trigonometric function and allows to further overcome the gradient problem, with less computation time compared to that of Mish function. The experiments that are performed over multiple datasets challenge show that the proposed activation function gives a high test accuracy than both ReLU and Mish functions in many deep network models.

Keywords: Activation functions · Neural networks · Test accuracy · Deep learning

1 Introduction

Recently, Artificial neural networks (ANN) have been successful in the demonstration of their performance, they opened the way to applications that have contributed to the development of several sectors, including computer vision, image recognition and many more [1]. In fact, thanks to the existence of a large amount of data, several ANN models have been developed and tested. The aim of any model is to increase the accuracy of data recognition. For this, a transformation of characteristics, based on the non-linear separation is required, this is where the role of the activation function becomes clear in neural networks.

© Springer Nature Switzerland AG 2021
É. Renault et al. (Eds.): MLN 2020, LNCS 12629, pp. 237–244, 2021.
https://doi.org/10.1007/978-3-030-70866-5_15

It turns out to be complicated to judge which activation function performs better on each model. Many ways have been taken in the search for new activation functions. In this context, since the beginning of neural networks, several activation functions have been developed such as Sigmoid, Tanh, ReLU and others. ReLU has always been used as the standard activation function according the performance and research that has been done. Afterwards, Ramachandran et al. 2017 [2] developed *Swish* which came to perform more of RELU's achievements. Indeed, the advantage of *swish* compared to ReLU is that it has a non-zero negative part, which allows it to better optimize the gradient descent, however its positive part suffers from some variations and is not perfectly linear. Recently, D. Misra, (2020) [3] developed Mish activation function.

$$Mish(x) = x \tanh(\ln(1 + e^x))$$ (1)

The intuition behind Mish is twofold. First, its positive part is more linear than that of swish but still not perfect as compared to ReLU. Second, Mish has an important negative part values. These characteristics make Mish a powerful activation function that allows neural networks to converge rapidly and is also stable. However, the biggest drawback of Mish is its computation time, which exceeds other usual activation functions.

Our contribution in this work is to propose and test a new activation function for deep learning models which perform the existing activation functions. The rest of this paper is organized as follows,

First, we gave a mathematical definition of the proposed activation function, then we tested it speed convergence and its behavior when changing hyper-parameters, in particular when increasing the number of layers with a custom model on MNIST dataset. We trained it with several network models on CIFAR10 and CIFAR100 datasets then we compare it with Mish and ReLU in term of test accuracy.

2   Proposed Activation Function

In this work, we propose SinSig, a new self-gated non-monotonic activation function, which we can define mathematically as:

$$f(x) = x \sin(\frac{\pi}{2} sigmoid(x))$$ (2)

where $sigmoid(x) = \frac{1}{(1+e^{-x})}$.

SinSig has an important negative value which could push the mean of the activations close to zero, thus accelerates the learning process. The positive part is approximately linear, with a few variations as compared to Msih as plotted in Fig. 1(a).

Sinsig is a smooth function, it minimum value is approximately −0.4291. In the positive part we can observe that is more closest to the linear transformation than Mish it is.

The first order derivative of SinSig is given as follows,

$$f'(x) = \sin(\frac{\pi}{2}\sigma(x)) + x sin'(\frac{\pi}{2}\sigma(x))$$
$$= \sin(\frac{\pi}{2}\sigma(x)) + \frac{\pi}{2}x\sigma'(x)\cos(\frac{\pi}{2}\sigma(x))$$

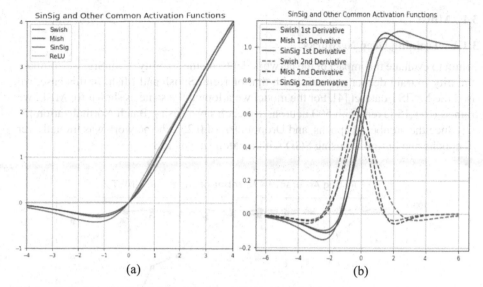

Fig. 1. (a) SinSig, ReLU, Mish, and Swish activation functions. (b) The 1st and 2nd derivatives of SinSig, Mish and Swish activation functions.

$$= \sin(\frac{\pi}{2}\sigma(x)) + \frac{\pi}{2}xe^{-x}\sigma(x)\cos(\frac{\pi}{2}\sigma(x)) \tag{3}$$

Where $\sigma(x) = sigmoid(x)$.

Like both Mish and Swish, the SinSig first order derivative has some values that slightly exceed the unit value and some negative values. The maximum value is near 1.0588 while the minimum value is near -0.1548. As shown in the Fig. 1 we can remark that the SinSig first derivative values are more significant than that of Mish.

The passage of SinSig between positive and negative values is faster. Generally, SinSig will be able to update the weight of neurons better than both Mish and Swish.

Comparing the second derivatives we can say the same thing that was said about the first derivatives.

We can conclude that the power of SinSig compared to Mish, Swish and ReLu, is generally its negative part.

3 Experiment

In this part, we will detail all the tests that have been done to evaluate the performance of our proposed activation function. For this task, we used MNIST, CIFAR10 and CIFAR100 datasets and compare the obtained results for each activation function.

Indeed, we will see to how the activation function impacts the model performances when we adjust several hyper-parameters.

3.1 Mnist

3.1.1 Training Deep Networks

In aim to evaluate the impact of the network depth on the stability of SinSig, we compare it ability to train deeper networks with that of Relu, Swish and Mish. In this case, we used the MNIST dataset [4]. For the model we adopted the same as shown for Mish; we trained over 15 layers with 500 neurons for each layer, using Batch Normalization [5] to reduce the number of epochs, and Dropout [6] of 0.25. The network was trained over 10 epochs and optimized using SGD with a batch size of 128.

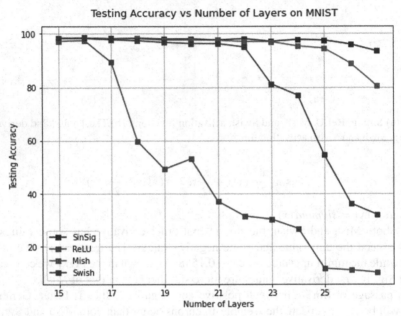

Fig. 2. Testing accuracy while increasing the number of layer on MNIST with SinSig, ReLU, Mish and Swish

We proved as shown in the Fig. 2 that the use of SinSig as activation function for the defined model has avoided the overfitting problem with up to 25 layers. However, the ReLU test accuracy is remarkably decreasing with a model of more than 17 layers. Mish outperform Swish but suffer also from decreasing. SinSig perform almost equally up to 25 layers which make it stable and less sensitive to overfitting compared to other activation functions.

3.1.2 Speed Convergence

As can be seen clearly in Fig. 3, SinSig converges faster than the other benchmarked functions. For only one epoch, it reached a test accuracy of the order of 0.8832 and it remained the best until the last epoch. The same regarding the Test Loss, SinSig recorded just 0.4010 as Testing Loss on the first epoch.

3.2   CIFAR 10

To investigate the performance of our proposed activation function, we used three variants ResNet architecture and one of MobileNet, SE-Net, ShuffleNet and SqueezeNet.

The testing accuracy is reported in Table 1. the training is performed over 200 or 100 epochs according the depth of the used network. The batch size is 128 for all simulation reported in this work. The used learning rate is gradually decreased for ResNet Models but fixed for the others.

We have trained some models on challenging dataset CIFAR10. During the various simulations carried out as we can see, SinSig consistently outperforms both Mish and ReLU with higher Top-1 testing accuracy by replacing the activation function while keeping every other parameter of the model to be constant.

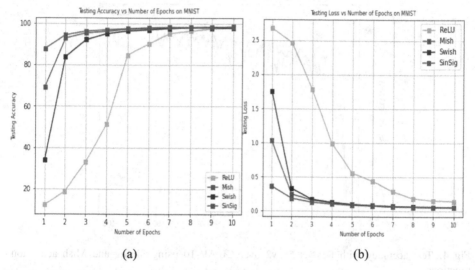

(a)   (b)

Fig. 3. (a) Testing Accuracy while the number of epoch increasing in 15 Layers Model on MNIST with different activations functions. (b) Test Loss while the number of epoch increasing in 15 Layers Model on MNIST with different activations functions.

For illustration, we plotted the graph of test accuracy with ResNet20 v2 on CIFAR10 using SinSig and Mish. we observe that SinSig perform Mish especially in the epochs after 82, where the learning rate is decreased (Fig. 4).

3.3   Cifar 100

This dataset is just like the CIFAR-10, except it has 100 classes containing 600 images each. There are 500 training images and 100 testing images per class. We trained this dataset with the same hyper-parameters as we have trained CIFAR10. We observe that SinSig still perform both Mish and ReLU (Table 2).

Table 1. Testing accuracy on CIFAR10

Model	SinSig	ReLU	Mish
ResNet 20 v2 [7]	**0.9223**	0.9171	0.9202
ResNet 110 v2 [7]	**0.9316**	0.9193	0.9258
MobileNet v2 [8]	**0.8642**	0.8594	0.8605
SE-Net18 [9]	**0.9089**	0.9016	0.9053
SqueezeNet [10]	**0.8848**	0.8785	0.8813
ShuffleNet [11]	**0.8761**	0.8705	0.8731

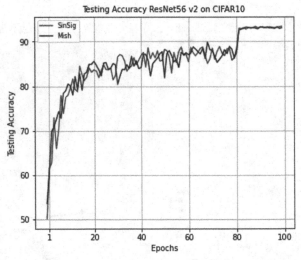

Fig. 4. Test accuracy with ResNet 56 v2 over CIFAR-10 using SinSig and Mish activation functions

Table 2. Testing accuracy on CIFAR100

Model	SinSig	ReLU	Mish
ResNet 20 v2 [7]	**0.7095**	0.7054	0.7086
ResNet 110 v2 [7]	**0.7459**	0.7301	0.7441
MobileNet v2 [8]	**0.5742**	0.5619	0.5706
SE-Net18 [9]	**0.6444**	0.6272	0.6439
SqueezeNet [10]	**0.6336**	0.6093	0.6307
ShuffleNet [11]	**0.5968**	0.5798	0.5919

3.4 Time Computation

In order to strengthen the comparison of our proposed activation function with the other common activation functions, especially with Mish, we have evaluated their computation time as well as that of their first and second derivatives. For this we calculated, on the same machine, the average time computation of 10^5 iterations. The results are given in the table below (Table 3):

Table 3. Time computation (by μs)

Model	SinSig	Mish
Original Function	**52.8065**	96.7686
First order derivative	**196.2436**	249.2705
Second Order Derivative	**696.6381**	699.9982

SinSing function require less time computation than Mish function.

4 Further Work

In our work, SinSig was not considered as a parametric activation function, it was always a single formula that was used during all the tests. However, our proposed function could get more performance if we consider that we can still push the positive part even more towards the unit, by giving up a little of its negative part. In short, SinSig could be parameterized by making a compromise in this way.

So far, the study of the proposed activation function still requires its popularization through other research work such as exploring its ability on the ImageNet dataset with existing models as well as the study of the eventual parameter, his dependency on the hyper-parameters of the network.

5 Conclusion

In this paper, we have presented and tested a new activation function which enhance the performances of Neural Network Models by performing the other common activation functions such as Mish, Swish and ReLU.

SinSig is a smooth non-monotonic self-gated function similar to Mish, but introduce more non-linearity that allows to make the model more accurate and more robust in the face of the adjustment in different hyper-parameters. In this context, we have seen for example that SinSig is less sensitive to the increase of the depth of the network and can better prevent the problem of the overfitting.

References

1. Kumar Roy, S., Manna, S., Ram Dubey, S., Chaudhuri, B.B.: LiSHT: Non-Parametric Linearly Scaled Hyperbolic Tangent Activation Function for Neural Networks. https://arxiv.org/pdf/1901.05894.pdf. Accessed 1 Jan 2019
2. Le, Q.V., Ramachandran, P., Zoph, B.: Swish: a Self-Gated activation function (2017)
3. Misra, D.: Mish: A Self Regularized Non-Monotonic Neural Activation Function. https://arxiv.org/pdf/1908.08681.pdf. Accessed 13 Aug 2020
4. LeCun, Y., Cortes, C., Burges, C.J.: Mnist handwritten digit database. ATT Labs. https://yann.lecun.com/exdb/mnist. Accessed 2 2010
5. Ioffe, S., Szegedy, C.: Batch normalization: Accelerating deep network training by reducing internal covariate shift. arXiv preprint arXiv:1502.03167 (2015)
6. Srivastava, N., Hinton, G., Krizhevsky, A., Sutskever, I., Salakhutdinov, R.: Dropout: a simple way to prevent neural networks from overfitting. J Mach. Learn. Res. **15**(1), 1929–1958 (2014)
7. He, K., Zhang, X., Ren, S., Sun, J.: 'Identity Mappings in Deep Residual Networks
8. Sandler, M., Howard, A., Zhu, M., et al.: MobileNetV2: inverted residuals and linear bottlenecks. In: 2018 IEEE Conference on Computer Vision and Pattern Recognition (CVPR), Salt Lake City, USA, June 2018, pp. 4510–4520 (2018)
9. Hu, J., Shen, L., Sun, G.: Squeeze-and-excitation networks. In: 2018 IEEE Conference on Computer Vision and Pattern Recognition (CVPR), Salt Lake City, USA, June 2018, pp. 7132–7141 (2018)
10. Forrest, N. Iandola, S. Han, M.W., Moskewicz, K. Ashraf, W.J., Dally, K.: Keutzer' SqueezeNet: AlexNet-level accuracy with 50x fewer parameters and <0.5MB model size. https://arxiv.org/abs/1602.07360. Accessed 24 Feb 2016
11. Zhang, X., Zhou, X., Lin, M., et al.: Shufflenet: An extremely efficient convolutional neural network for mobile devices. In: 2018 IEEE Conference on Computer Vision and Pattern Recognition (CVPR), Salt Lake City, USA, June 2018, pp. 6848–6856 (2018)

Spectral Analysis for Automatic Speech Recognition and Enhancement

Jane Oruh and Serestina Viriri[✉]

School of Mathematics, Statistics and Computer Science,
University of KwaZulu-Natal, Durban 4000, South Africa
viriris@ukzn.ac.za

Abstract. Accurate recognition of noisy speech signal is still an obstacle for wider application of speech recognition technology. The robustness of a speech recognition system is heavily influenced by the ability to handle the presence of background noise. In this work, a Short Time Fourier Transform (STFT) filtering technique for the enhancement and recognition of the speech signal is presented. Conventionally, STFT filtering has been applied in speech analysis. However, in this study the combination of modified STFT with Adaptive window width based on the Chirp Rate, termed ASTFT, in conjunction with Spectrogram Features is proposed for optimal speech recognition and enhancement. LibriSpeech ASR Corpus is the benchmark dataset for this experiment. The spectrum from the enhanced Speech signal is estimated using several spectrogram features to obtain a unit peak amplitude. Priori Signal-to-Noise Ratio (SNR) estimation is performed on the modified STFT speech signal, and it achieved an SNR of 31.86 dB which is considered to be an effectively clean speech signal.

Keywords: Noise reduction · STFT filtering · Spectrum estimation · Automatic speech recognition · Speech enhancement · Signal-to-Noise-Ratio

1 Introduction

Speech is used to communicate information from a speaker to one or more listeners. The speaker produces a speech signal in the form of pressure waves that travels from speaker's mouth to listener's ears. The signal consists of variations in pressure as a function of time and is measured directly in front of the mouth, the primary sound source. The conversion of speech signal into a useful message (its corresponding text) is called automatic speech recognition [1].

Enhancement and recognition of speech signals in the presence of reverberation and noise, remains a challenging problem in many applications. Many past methods are prone to generating artifacts in the enhanced speech, and must trade off noise reduction against speech distortion. Recent approaches have started to address this issue, demonstrating improvements in both objective speech quality and automatic speech recognition [2].

© Springer Nature Switzerland AG 2021
E. Renault et al. (Eds.): MLN 2020, LNCS 12629, pp. 245–254, 2021.
https://doi.org/10.1007/978-3-030-70866-5_16

Basically, the recorded speech signal in a real-world application may be corrupted by various noise types, interferences, echoes and reverberation resulting from the acoustic environment and enclosure. These degradation can significantly reduce the intelligibility of the speech signal by human listeners and also deteriorate the performance of speech coding and recognition systems [3]. Improving intelligibility and/or overall perceptual quality of degraded speech signals using signal processing techniques, is the objective of speech enhancement [4]. Hence, a spectral filtering method based on the Short Time Fourier Transform (STFT) filtering are necessary for high performance speech enhancement system, as demonstrated in this work.

The STFT guarantees positivity and is computationally efficient and very robust against noise. However, it suffers poor time-frequency resolution [5]. So there is need to adapt the short-time Fourier transform to the characteristics of the signal at each time point in order to enhance time-frequency resolution [6]. This has led to the idea of deploying an adaptive short-time Fourier transform (ASTFT) based on Chirp Rate for the effective performance of the proposed model.

2   Related Work

Many recent studies on speech enhancement have focused on spectral analysis of the speech signal. The analysis of the speech signal typically starts with a time-dependent Fourier transform, also called Short-Time Fourier Transform (STFT), which is a method to analyze signals whose Fourier transform (i.e. spectrum) changes over time, as it is the case of the speech signal [7].

Speech analysis for Automatic Speech Recognition (ASR) systems typically starts with a Short-Time Fourier Transform (STFT), that implies selecting a fixed point in the time-frequency resolution trade-off. In state-of-the-art ASR systems, the most widely used acoustic features; Mel Frequency Cepstral Coefficients (MFCC), Perceptual Linear Prediction Coefficient (PLP) are based on the Short-Time Fourier Transform (STFT) [8].

The short-time spectrum of the signal is the magnitude of a Fourier Transform of the waveform after it has been multiplied by a time window function of appropriate duration. To obtain a more useful representation of the speech signal in terms of parameters that contains relevant information in an efficient format, Short time Fourier transform amongst the other features, is best used to obtain the energy spectrum of the speech signal [1].

For STFT application, Parachami [9] stated that in the frequency domain speech enhancement, the spectrum of a clean speech signal is estimated through the modification of its noisy speech spectrum and then it is used to obtain the enhanced speech signal in the time domain, this can be compared to the approach deployed for the audio speech enhancement in this paper.

The STFT is an advanced method for ASR enhancement. It extracts the core information from waveforms in the form of short-time spectra of the speech signal as a function of time known as Short Time Fourier Transform (STFT).

Simply, in the continuous-time case, the function to be transformed is multiplied by a window function which is non-zero for only a short period of time.

The Basic concept for Short-Time Fourier Transform is to break up the signal in time domain to a number of signals of shorter duration, then transform each signal to frequency domain. A window function of finite length is chosen, and the signal using the window is truncated. The Fourier Transform of the truncated window is computed and the result saved. Formulating the process of a Continuous STFT given by

$$STFTx(t)(\tau, \omega) \equiv X(\tau, \omega) = \int_{\infty}^{-\infty} x(t)w(t - \tau)e^{-j\omega t}dt \qquad (1)$$

where $x(t)$ is the time-domain signal to be transformed. τ is the time (slow time; lower resolution than t). ω is the frequency. $w(t)$ is the window function, commonly a Hann window or Gaussian window bell centered around zero, and $X(\tau, \omega)$ a complex function representing the phase and magnitude of the signal over time and frequency (this is essentially the Fourier Transform of $x(t)\omega(t-\tau)$ a complex function representing the phase and magnitude of the signal over time and frequency [10].

One of the classic method for joint time-frequency analysis is the short-time Fourier transform (STFT). STFT is a mathematical transformation associated with FT to determine the frequency and phase of a sine wave in a local region of the time-varying signal. The concept of STFT is to first choose a window function with time-frequency localization. Then assume that the analysis window function $\omega(t)$ was stationary over a short time, which ensures $f(t)\omega(t)$ is a stationary signal within different time widths. STFT uses fixed window functions, the most commonly used include the Hanning window, the Hamming window, and the Blackman-Haris window [11]. The Hamming window, a generalized cosine window, is used in this article. It is usually represented as

$$\omega(t) = a_0 + (1 - a_0)cos(\frac{2\pi t}{T}), 0 \leq t \leq 0 \qquad (2)$$

where $a_0 = 0.53836$. This function is a member of both the cosine-sum and power-of-sine families. The Hamming window can efficiently reflect the attenuation relationship between energy and time at a certain moment [12].

Although speech is a non-stationary signal, it is generally assumed to be quasi-stationary i.e., one with approximately constant statistics over short periods of time and, therefore, can be processed through a short-time Fourier analysis. Note that the modifier 'short-time', implies a finite-time window over which the properties of speech may be assumed to be stationary; it does not refer to the actual duration of the window. In speech processing, the Hamming window function is typically used and its width T_w is normally 20–40 ms [13].

3 Methods and Techniques

3.1 Defining the Dataset

Train-clean-360.tar.gz audio file from LibriSpeech ASR corpus was specifically chosen as the benchmark dataset for this work [14]. LibriSpeech is an ASR corpus of approximately 1000 h of 16 kHz read English speech based on public domain audio books. All the speech samples contained in the dataset are in .wav format, and has been carefully segmented and aligned. The corpus is freely available under the very permissive CC BY 4.0 license [15], and there are example scripts in the open source Kaldi ASR toolkit that demonstrate how high-quality acoustic models can be trained on this data [16].

3.2 Dataset Sampling and Pre-processing

Speech is an analog signal which should be converted to digital form. Sampling and quantization are performed to convert the continuous signal into a series of discrete values. The train-clean-360 is read as an audio file and sampled as a speech waveform, as shown in Fig. 2.

3.3 Processing of the Speech Signal

Once the speech signal has been sampled, the next phase is signal processing to separate the speech signals from background noise [17]. Here, the audio speech signal is filtered using STFT technique, and converted to a mono signal waveform as shown in Fig. 3.

The standard STFT was used to perform a filtering operation on the speech waveform as demonstrated in Fig. 2 to enhance the speech signal. In this case the analysis window $w[n]$ plays the role of the filter impulse response. To illustrate the view, we fix the value of ω at ω_0, and rewrite

$$X(n, \omega_0) = \sum_{m-\infty}^{\infty} (x[m]e^{-j\omega om})w[n-m] \tag{3}$$

Which can be interpreted as the convolution of the signal $(x[n]e^{-j\omega on})$ with the sequence $w[n]$:

$$X(n, \omega_0) = (x[n]e^{-j\omega on}) * w[n] \tag{4}$$

and the product $x[n]e^{-j\omega on}$ can be interpreted as the modulation of $x[n]$ up to frequency ω_0 (i.e., per the frequency shift property of the FT) [18].

4   Adapting the Short-Time Fourier Transform to Signal Based on Chirp Rate

A new time-frequency representation (TFR)-based approach to Chirp Rate based on adaptive short-time Fourier transform (ASTFT) is presented. This basically involves a low-complexity ASTFT based on the chirp modulation [19]. This type of modulation employs sinusoidal waveforms whose instantaneous frequency increases or decreases linearly over time. The chirp technique uses a frequency and amplitude modulated signal as the source wave form (Fig. 3).

The ASTFT analyses the time-domain signal $x(t)$ in (1) to allow further control over the window width, and the form in (1) changed as (5)

$$ASTFT_{tf}(t,f) = \int\limits_{\infty}^{-\infty} x(\tau)\frac{1}{\sqrt{2\pi\sigma(t,f)}}e^{-\frac{t-\tau}{2[\sigma(t,f)]^2}}e^{-j2\pi f\tau}dt \qquad (5)$$

The proposed framework known as Adaptive Short-Time Fourier Transform (ASTFT) ASR is as shown in Fig. 1.

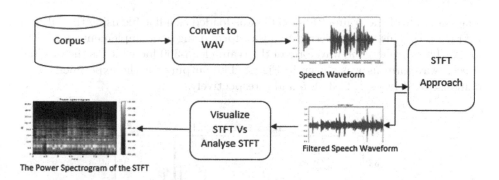

Fig. 1. The ASTFT ASR framework.

5   Evaluation of the Model

For evaluation and the testing of our model, the result of the single parameter STFT will be extended to the Spectrogram. For general purpose time-frequency analysis, the spectrogram remains the dominant technique in use because it is easily computed as the magnitude-squared of the short-time Fourier transform (STFT).

$$STFT(t,f) = \int\limits_{\infty}^{-\infty} x(t-v)w(v)e^{-j2\pi fv}dv \qquad (6)$$

where $u(v)$ is a window function, often taken to be Gaussian in shape [20]. The STFT is adaptively optimizing the performance of the spectrogram representations.

The spectrogram logarithmic of the STFT values with a window size of 20 ms and step of 10 ms known as the power spectrogram is being computed in Fig. 4. In addition, we convert the power spectrogram (amplitude squared) of the enhanced speech signal to decibel (dB) units. A Mel spectrogram for the enhanced speech signal was computed as shown in Fig. 5. This was achieved through the construction of a mel filter bank, while using the pre-computed power spectrogram, and then a normalized Mel Spectrogram was obtained to achieve a unit peak amplitude as shown in Fig. 6.

The power spectrogram of the model yielded an amplitude of 20 dB, which was used in estimating the Priori SNR necessary for the performance evaluation of the model.

6 Results and Discussion

6.1 Experimental Results

The audio file of the train-clean-360 file was loaded as a flac file using the Librosa Library [21]. This was converted to a waveform using code implementation technique. This displays the waveform of the train-clean-360 file which is the original speech waveform as was shown in Fig. 2. The outputs for the experiment is as demonstrated in Figs. 2, 3, 4, 5 and 6 respectively.

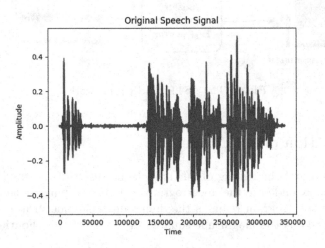

Fig. 2. The waveform of the original speech signal

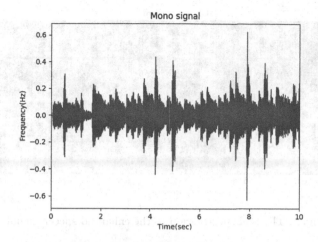

Fig. 3. The enhanced speech signal

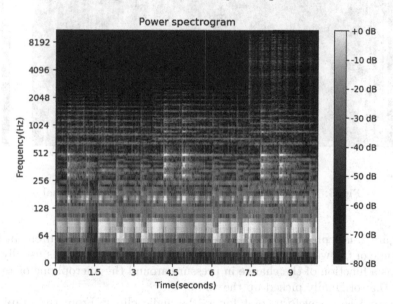

Fig. 4. Spectrogram of the enhanced speech signal

6.2   Discussion

The Proposed approach aims to improve the speech quality along side the intelligibility of the train-clean-360 audio file by time-frequency analysis, filtering, and noise reduction of the speech signal using the STFT techniques and Spectrogram analysis. During the experiment, the enhanced speech signal obtained in Fig. 3 was further analyzed by using spectrogram. Basic patterns of the speech signal can be discovered and analyzed with the help of spectrogram [22]. The spectrogram of the enhanced speech signal is shown in Fig. 4.

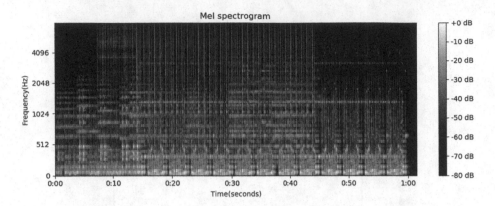

Fig. 5. The Mel spectrogram of the enhanced speech signal

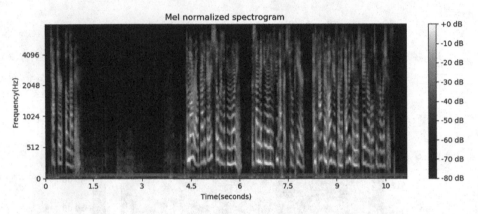

Fig. 6. The Mel spectrogram of the enhanced speech signal

Figure 2, is represented as time series, where the y-axis measurement is the amplitude of the waveform over time in seconds. The amplitude is usually measured as a function of the change in pressure around the microphone or receiver device that originally picked up the audio.

Figure 3 is a waveform plot for a 10-s audio clip y. From the graph, y is monophonic with a filled curve drawn between $[-abs(y), abs(y)]$ on y-axis against time (t) on x-axis. Figure 3 shows the signal processing of the audio speech signal obtained in Fig. 2. Furthermore, it shows the process of noise reduction and the result of noise reduced signal using STFT.

Figure 4 is a power spectrogram of the enhanced speech signal, which is for visualizing the spectrum of frequencies in our filtered speech sample. It is a graph of frequency in Hertz (Hz) against time (t) in decibel (dB). A third-dimension in the graph indicates the amplitude of a frequency at a time which is represented by the intensity or colour of each point in the graph.

Figure 5 is a mel-scaled spectrogram of the STFT. This is suprisingly a spectrogram with the Mel Scale as its y axis. The Mel spectrogram transform the frequency bins into the mel scale. Figure 6 is a Mel Normalized Spectrogram for normalizing the Mel spectrogram to a unit peak amplitude.

7 Performance Parameters

The performance parameter for testing the model is the SNR Metric of the ICSI Speech FAQ which states that "a local SNR of up to 30 dB is effectively a clean signal" [23], given by $SNR\_db = 10log10(S\_e/N\_e)$, where S_e and N_e are the total energy of the speech i.e. $sum(s[n]^2)$ and $sum(n[n]^2)$ respectively.

The performance parameter can be summarized in Table 1.

Table 1. Summary of the performance parameters.

Features	SNR Pre (db)	SNR Post (db)
Simulated noise variance [24]	20.00	26.35
The proposed model	20.00	31.86

8 Conclusion

The proposed model has investigated the performance of STFT for modification of the noisy speech signal and estimated the spectrum using several spectrogram features to achieve a unit peak amplitude. This has shown that combination of processing techniques is necessary to achieve a good speech recognition and enhancement performance. SNR Metric of the ICSI Speech FAQ was used to estimate the Priori SNR of the modified STFT speech signal and compared with that of simulated noise variance for the same amplitude. Results have shown that our model outperforms the simulated noise variance in attaining the set standard of a SNR > 30 dB being considered as a clean speech. The model could be deployed for real time voice processing in the telecommunication industry, to ensure highly accurate, robust, and reliable system in a sound noisy environment.

References

1. Nasreen, P.N., Kumar, A.C., Nabeel, P.A.: Speech analysis for automatic speech recognition. In: Proceedings of International Conference on Computing, Communication and Science (2016)
2. Delcroix, M., et al.: Linear prediction-based dereverberation with advanced speech enhancement and recognition technologies for the REVERB challenge. In: Reverb workshop (2014)
3. Cohen, I., Benesty, J., Gannot, S.: Speech Processing in Modern Communication: Challenges and Perspectives, vol. 3. Springer Science & Business Media, Berlin (2009)

4. Parchami, M., Zhu, W.-P., Champagne, B., Plourde, E.: Recent developments in speech enhancement in the short-time Fourier transform domain. IEEE Circ. Syst. Mag. **16**(3), 45–77 (2016)
5. Kwok, H.K., Jones, D.L.: Improved instantaneous frequency estimation using an adaptive short-time Fourier transform. IEEE Trans. Sig. Process. **48**(10), 2964–2972 (2000)
6. Zhong, J., Huang, Y.: Time-frequency representation based on an adaptive short-time Fourier transform. IEEE Trans. Sig. Process. **58**, 5118–5128 (2010)
7. Toledano, D.T., Fernández-Gallego, M.P., Lozano-Diez, A.: Multi-resolution speech analysis for automatic speech recognition using deep neural networks: experiments on TIMIT. PloS one **13**(10), e0205355 (2018)
8. Tüske, Z., Golik, P., Schlüter, R., Drepper, F.R.: Non-stationary feature extraction for automatic speech recognition. In: 2011 IEEE International Conference on Acoustics, Speech and Signal Processing (ICASSP), pp. 5204–5207. IEEE (2011)
9. Parchami, M.: New Approaches for Speech Enhancement in the Short-Time Fourier Transform Domain. PhD thesis, Concordia University (2016)
10. Ahmadizadeh, M.: An Introduction to Short-Time Fourier Transform (STFT). Advanced Structural Dynamics, April 2014
11. Jurafsky, D., Martin, J.H.: Speech and Language Processing, vol. 3 (2014)
12. Solovyev, R.A., et al.: Deep learning approaches for understanding simple speech commands. In: 2020 IEEE 40th International Conference on Electronics and Nanotechnology (ELNANO), pp. 688–693. IEEE (2020)
13. Paliwal, K.K., Alsteris, L.D.: On the usefulness of STFT phase spectrum in human listening tests. Speech Communi. **45**(2), 153–170 (2005)
14. Dutta, A., Valiveti, G.R.S.: Enhancing the performance of audio visual speech recognition using deep learning techniques. Int. J. Comput. Sci. Commun. **7**(2), 126–135 (2016)
15. Creative Commons. Creative Commons Attribution 4.0 International (CC BY 4.0) License. https://creativecommons.org/licenses/by/4.0/. Accessed 07 Nov 2017
16. Panayotov, V., Chen, G., Povey, D., Khudanpur, S.: Librispeech: an ASR corpus based on public domain audio books. In: 2015 IEEE International Conference on Acoustics, Speech and Signal Processing (ICASSP), pp. 5206–5210. IEEE (2015)
17. Sarma, P., Sarmah, S., Bhuyan, M.P., Hore, K., Das, P.P.: Automatic spoken digit recognition using artificial neural network. Int. J. Sci. Technol. Res. **8**(12), 1400–1404 (2019)
18. Gutierrez-Osuna, R.: Introduction to speech processing. CSE@ TAMU (2016)
19. Pei, S.-C., Huang, S.-G.: STFT with adaptive window width based on the chirp rate. IEEE Trans. Sig. Process. **60**, 4065–4080 (2012)
20. Czerwinski, R.N., Jones, D.L.: Adaptive short-time Fourier analysis. IEEE Sig. Process. Lett. **4**(2), 42–45 (1997)
21. McFee, B., et al.: Librosa: v0.4.0.Zenodo. In: Proceedings of the 14th Python in Science Conference (SCIPY 2015) (2015)
22. Singh, J., Kaur, K.: Speech enhancement for Punjabi language using deep neural network. In: 2019 International Conference on Signal Processing and Communication (ICSC), pp. 202–204. IEEE (2019)
23. F. A. Q. International Computer Science Institute (ICSI) Speech. https://www1.icsi.berkeley.edu/Speech/faq/speechSNR.html. Accessed 17 Sep 2019
24. Athaley, P.D.A.: Audio signal denoising algorithm by adaptive block thresholding using STFT. Int. J. Trend Sci. Res. Dev. **1**(6), 289–300 (2017)

Road Sign Identification with Convolutional Neural Network Using TensorFlow

Mohammed Kherarba[1], Mounir Tahar Abbes[1], Selma Boumerdassi[2(✉)],
Mohammed Meddah[1], Abdelhak Benhamada[1], and Mohammed Senouci[3]

[1] LME, Hassiba Ben Bouali University, Chlef, Algeria
mtaharabbes@yahoo.fr
[2] CNAM/CEDRIC, Paris, France
selma.boumerdassi@inria.fr
[3] University of Oran1, Oran, Algeria

Abstract. With the use and continuous development of deep learning methods, the recognition of images and scenes captured from the real environment has also undergone a major transformation in the techniques and parameters used. In most of the methods, we notice that recognition is based on extraction. This paper proposes a classification technique based on convolutional features in the context of Traffic Sign Detection and Recognition (TSDR) which uses an enriched dataset of traffic signs. This solution offers an additional level of assistance to the driver, allowing better safety of passengers, road users, and cars. An experimental evaluation on publicly available scene image datasets with convolutional features presents results with an accuracy of 94.7% of our classification model.

Keywords: Traffic Sign Detection · TSDR · CNN · Tensorflow

1 Introduction

Traffic signs play an essential role in regulating traffic and provide drivers information regarding the availability of information on a road (restrictions, warnings, possible directions, etc.). Most of the sign plates are standardized by international laws, which facilitate the task of detection by computer systems. We notice in the last decade, the frequent and exponential use of personal means of transport and even public transport; and this will obviously increase the number of accidents on the roads and mainly because of human errors such as exceeding speed, fatigue... All of these parameters make traffic sign recognition systems a very important element in accident prevention. The precise identification of traffic signs depends on internal and external parameters such as weather and factors which affect driver attention. The ability of drivers to identify the traffic lights is strongly affected by their physical and mental alertness; the main risk factors are fatigue, emotional stress, or side effects caused by drugs [1]. Therefore, the development of *automatic traffic sign recognition system* (TSR) is very suitable, as these systems play an essential task in preventing car collisions, saving human lives, and growing driver performance, particularly in cases where traffic signs are damaged or

© Springer Nature Switzerland AG 2021
É. Renault et al. (Eds.): MLN 2020, LNCS 12629, pp. 255–264, 2021.
https://doi.org/10.1007/978-3-030-70866-5_17

hidden. In our study, we use the same classification and strategies used by [2] to identify which method to use in detection.

Detection and recognition are accomplished through the application of ROIs detection and limitation of areas of interest methods to extract the candidate areas from a frame.

In this paper, the learning model built from a convolutional neural network which then makes it possible to compare the candidate route signs by the classification model and this by exploiting the GTSRB sign base. The detection program, implemented in python, is intended to operate in real time. The system has been successfully tested on signage images obtained under varying weather conditions at different resolutions.

The rest of this paper is organized as follows. Section 2 describes fundamental concepts of convolutional neural networks (CNN). Section 3 discusses the related work, and then Sect. 4 presents the proposed solution. Section 5 details and analyzes the results obtained. Finally, Sect. 6 concludes the paper.

2 Fundamental Concepts

This section presents fundamental elements of convolutional neural networks (CNN), the use of CNN models in deep learning has developed the domain of computer vision. There are many types of CNN architectures in the field of artificial intelligence. On the other hand, their basic modules are very comparable. In general, we find three types of layers.

2.1 Convolutional Layer

It is composed of neurons that connect segments of an input image or the outputs of the preceding layer. This layer uses filters and strides to generate feature maps. This action is repeated many times during the process of mapping. Figure 1 shows an example of the operation.

2.2 Pooling Layer

The pooling layer aims to reduce the number of parameters by reducing the resolution of the feature maps. It is usually placed between two convolutional layers, Spatial pooling also called subsampling or down sampling which reduces the dimensionality of each map but retains important information. Figure 1 shows the position of the pooling layer.

2.3 Fully-Connected

A fully connected layer is a group of neurons fully connected to each other. Each neuron has many inputs and one output. Inside each neuron, there is a function applied to all inputs of the neuron and gives one output. The neurons are organized in layers. The first layer is called the input layer, and the last layer is called the output layer. The other layers are called the hidden layer as illustrated in Fig. 2.

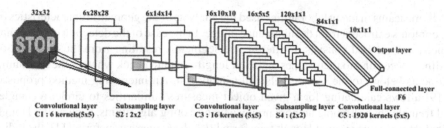

Fig. 1. CNN layers, adopted from [3].

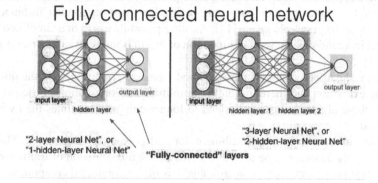

Fig. 2. Fully connected neural network, adopted from [3].

3 Related Work

Traffic signs are manufactured and standardized objects. As such, they therefore exhibit several of geometric, colorimetric, and pictographic characteristics that distinguish them from their environment. These panels are mainly characterized by bright colors, mainly red and blue, simple geometric shapes (polygons, disks).

The difficulties inherent in the analysis of vertical signage, but also the advantages presented by the particular character of the objects considered, explain the great variety of methods proposed in the literature over the past two decades. Their evolution has accompanied the development of modern algorithms for detecting and recognizing patterns as well as the progress of available computer resources. The first study [4] investigated how to recognize small object signs traffic in real-time mode. Authors generate more data using a generic function to increase the accuracy rate.

In many cases, the first detection step is to identify the pixels of a color image corresponding to the characteristic colors of the panel. It is then necessary to set one or more thresholds beyond which the pixel is classified as belonging to the panel or not. Many authors work in the HSV color space and segmentation is most often done by thresholding of the hue component (Hue), which is insensitive to changes in luminance. This approach is used by [5] for the selection of red panels. In real-time applications, the YUV color space, obtained from linear transformations of the RGB space, is used by [6].

Some teams in many laboratories used the analysis of the geometric characteristics of the contours extracted from the luminance image. In the case of the detection of triangular panels, one method consists in filtering the segments, resulting from the detection of outlines, according to their slopes and their lengths, and to check whether the remaining segments belong to the same equilateral triangle [8]. An interesting method proposed in [9] allows calculating for a given object measures similarities to circle, rectangle, and triangle shapes. Some authors propose using voting algorithms, such as the Hough transform (34), the circular Hough transform [10], the Chinese transform [11], the radial symmetry transform [12] and its version extended to detect polygons [13].

Another approach is based on finding a simplified model of the panel in a candidate image. Template matching techniques can be implemented on contour distance maps calculated from the grayscale image [14]. Another possibility is to fit a simplified model of the panel in a candidate image. Optimization of this fit is then carried out using genetic algorithms [15].

However, the effectiveness of these methods remains conditioned by the quality of the outline extraction and therefore on the amplitude and orientation of the gradients. Moreover, these algorithms generally lead to longer computation times than when the color is used.

The most used strategy is to combine color and geometric information. The color segmentation algorithms can be used as a first step for the detection of traffic signs. At the end of this pre-detection, a geometric analysis of the connected components makes it possible to refine the selection of the objects. It is thus possible to quickly eliminate objects according to their sizes [16] or the value of the compactness, defined by the ratio between the area of object and its perimeter squared. However, more elaborate descriptors are often necessary to make a more detailed selection. The choice is most often oriented towards the characterization of the shape outline using, for example, the Fourier descriptors obtained by the normalized Fourier transform of the signature of the object [17]. The circularity of red objects can also be controlled using the approach defined in [18]. The distance from the edges of the outline of the object to the edges of its bounding box of the object is also an interesting characteristic which can make it possible to discriminate circles, triangles and rectangles [5]. The sensitivity of these approaches to the variability of situations (occultation, change of appearance…) represents a major difficulty. From the image obtained from the colorimetric filter, shape-fitting models can also be applied using genetic algorithms [19], simulated annealing algorithms, or clonally or particle swarm selection algorithms.

The efficiency of these methods is linked to the performance of colorimetric segmentation. The grouping of pixels into connected components can in particular be problematic since it conditions the shape of the object to be analyzed subsequently. Over-segmentation results in a division of the object of interest into different parts.

The rest of this section presents the Learning-Based Recognition Techniques:

Some detection work is based on classification methods that consist of comparing the vector of characteristics of observation with a reference vector. To obtain these references, training is conventionally carried out on a set of representative images. Different learning techniques can be used, the best known being Adaboost and SVM (Support Vector Machine). Less classic, a non-supervised method, based on the analysis of the temporal statistical changes of the objects of interest in sequences (appearance, magnification, disappearance) is proposed in [20]. The multi-dimensional histograms (of color or shape measurements) between two successive images are thus compared and the objects are located by histogram rear projection. Another example of a detector based on SVM learning and using characteristics of shape and colors is described for example in [21]. The performance of these methods remains conditioned by the quality of the learning carried out beforehand.

Other methods are based on ANN (Artificial Neural Network). ANNs have exceeded human performance on the classification of road signs [22, 23]. However, their architectures differ considerably from others. Although the classification of traffic signs has been considered in the last ten years, the results could not be evaluated with other datasets until the GTSRB and the GTSDB [24] have been presented.

A concept has been presented in [25] that performed a categorization on ten Malaysian signs through ANN with two layers of feedback and any softmax. Every module was made up of one hundred instances separating them: 70%-15% for sets, validation, and testing correspondingly. Authors in [26] have also used the same dataset to compare two classification techniques, a radial basis function neural network and a convolutional neural network.

Authors in [27] have illustrated a CNN (Convolutional Neural Network) to to recognize and sort speed signs in the USA highway. The proposed solution is based on an adapted version of R-CNN [28] in recognition and Cudaconvnet for classification.

4 Proposed Solution

The proposed solution of detecting and recognizing road signs was based on two parts:

1) Classify the input frames into two classes. The first one contains non-interesting frames and the second groups the frames likely to contain signs. This is done thanks to the Support Vector Machine (SVM) method which allows to detect and limit ROIs area of interest to extract a candidate area from a sign after color segmentation by applying the Gaussian filter, as shown in Fig. 3.
2) Exploit the characteristics of Convolutional Neural Networks (as discussed in the previous paragraph) which subsequently allow the comparison of the candidate panels results from the first step with the model built from the extraction of the characteristics of the road signs collected in the German sign base GTSRB. Figure 4 gives an overview of this part.

The German Benchmark for Traffic Sign Recognition has separated traffic signs into three principal's classes:

1) Danger.
2) Prohibition.
3) Mandatory and contain others as an irrelevant module.

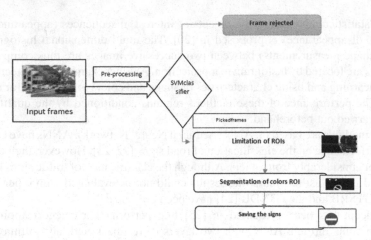

Fig. 3. ROI detection.

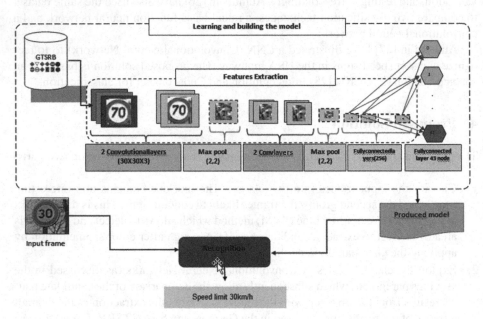

Fig. 4. Representation of the CNN learning phase and the recognition of candidate panels.

Classes two and three belong to the regulatory group defined at the Vienna Convention. Most publicly available datasets only include danger or mandatory signs. This data set has been divided into three subsets for training, validation, and testing by preserving the proportions of the images for each class. Furthermore, the normalization of the images was performed by dividing them by 255 and subtracting the average image, which, in turn, was calculated from the training data set. As a result, the dataset containing 3-channel RGB images was prepared. The training subset contains over 50,000 images, the validation subset contains 4,000 images, and the test subset contains 12,000 images.

5 Results and Discussion

The detection and recognition program, implemented in python and the TensorFlow API from KERAS, are proposed to operate in real time. We tested several configuration settings of the classification model. Through the results obtained in the section on learning the model, we noticed their change according to the different parameters entered in the model, in particular the size of the image, the resolution of the image, and the number of epochs performed in a single task. This led us to conduct three different experiments in the same environment (device and system) to obtain the best possible model after comparing the results of these experiments as shown in Table 1 (Fig. 5).

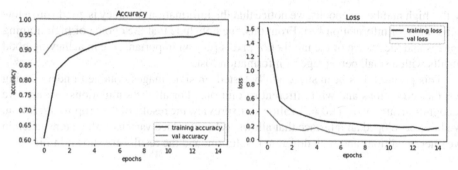

Fig. 5. Accuracy and loss vs epochs.

At this point, we put 15 epochs and set the pixel count to 45 × 45 × 3 pixels. We noticed that the average precision increases to 89.99% or the precision value equals to 96.54% and the average of information loss decreases to 0.7080 or the average value of information lost is 0.4131. The average execution time of each step (image processing) is 48 ms (Fig. 6).

Table 1. Experience parameters

Epochs	Resolution	Time (s)	Time for every state (ms)	Loss	Accuracy
15	45 × 45 pixels	22511 (6 h 15 min)	48	0.4131	96.54%
15	30 × 30 pixels	8294 (2 h 18 min)	18	0.2698	92.25%
50	30 × 30 pixels	30293 (8 h 24 min)	19	0.1383	96.31%

We observe that the higher the number of pixels, the higher the percentage of information lost and the accuracy of the information. When choosing an average pixel count

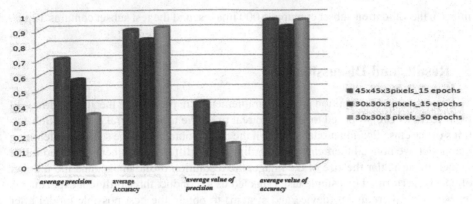

Fig. 6. Overall accuracy classification results.

with a high number of epochs, we notice that the information accuracy is good with a low percentage of information lost. From this, we conclude that the choice of high training epochs and the choice of the number of pixels play an important role in achieving good results with a small percentage of information lost.

This program has been successfully tested on sign images obtained under varying weather conditions and with different resolutions. For all the simulations, we achieve recognition rates of over 90%. To allow us to preview the results of the proposed solution, we have developed an interface that allows us to introduce various traffic signs taken in weather conditions and with different resolution, and the results are shown in Fig. 7.

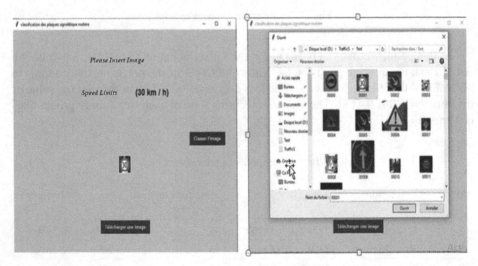

Fig. 7. Sample images from the dataset.

6   Conclusion

This paper presents a high-level representation method based on convolutional features. The application of various deep learning techniques using convolutional neural networks allowed us to apply artificial intelligence in its modern version. There are still several possible things to do on this project:

- Use the edge detector to detect missed panels, and possibly combine color and edge-oriented techniques.
- Correct detection problems in urban areas.
- Combine this algorithm with a program capable of taking photos at well-chosen times, or even of monitoring panels.

References

1. Bui-Minh, T., Ghita, O., Whelan, P.F., Hoang, T.: A robust algorithm for detection and classification of traffic signs in video data. In: 2012 International Conference on Control, Automation and Information Sciences (ICCAIS), Ho Chi Minh City, pp. 108–113 (2012)
2. Saadna, Y., Behloul, A.: An overview of traffic sign detection and classification methods. Int. J. Multimed. Inf. Retrieval **6**(3), 193–210 (2017). https://doi.org/10.1007/s13735-017-0129-8
3. Gu, J., et al.: Recent advances in convolutional neural networks. Pattern Recogn. **77**, 354–377 (2018)
4. Wu, Y., Li, Z., Chen, Y., Nai, K., Yuan, J.: Real-time traffic sign detection and classification towards real traffic scene. Multimed. Tools Appl. **79**, 18201–18219 (2020)
5. De La Escalera, A., Armingol, J.M., Mata, M.: Traffic sign recognition and analysis for intelligent vehicles. Image Vis. Comput. **21**(3), 247–258 (2003)
6. Broggi, A., Cerri, P., Medici, P., Porta, P.P., Ghisio, G.: Real time road signs recognition. In: IEEE Intelligent Vehicles Symposium, Istanbul, 2007, pp. 981-986 (2007)
7. Piccioli, G., De Micheli, E., Parodi, P., Campani, M.: A robust method for road sign detection and recognition. Image Vis. Comput. J. **14**(3), 209–223 (1996)
8. Rosin, P.L.: Measuring shape: ellipticity, rectangularity, and triangularity. Mach. Vis. Appl. MVA. **1**, 952–955 (2003)
9. Habib, A., Ha, M.: Hypothesis generation of instances of road signs in color imagery captured by mobile mapping systems. Int. Arch. Photogramm. Remote Sens. Spatial Inf. Sci. **36**, 159–165 (2007)
10. Garrido, M., Sotelo, M., Martm-Gorostiza, E.: Fast traffic sign detection and recognition under changing lighting conditions. In: IEEE Intelligent Transportation Systems Conference (ITSC 2006), pp. 811–816 (2006)
11. Belaroussi, R., Tarel, J.-P.: A real-time road sign detection using bilateral Chinese transform. In: Bebis, G. (ed.) ISVC 2009. LNCS, vol. 5876, pp. 1161–1170. Springer, Heidelberg (2009). https://doi.org/10.1007/978-3-642-10520-3_111
12. Loy, G., Zelinsky, A.: Fast radial symmetry for detecting points of interest. IEEE Trans. Pattern Anal. Mach. Intell. **25**(8), 959–973 (2003)
13. Loy, G., Barnes, N.: Fast shape-based road sign detection for a driver assistance system. In: IEEE/RSJ International Conference on Intelligent Robots and Systems (IROS), Sendai, 2004, vol. 1, pp. 70–75 (2004)

14. Gavrila, D.M., Philomin, V.: Real-time object detection for "smart" vehicles. In: The Seventh IEEE International Conference on Computer Vision, Kerkyra, Greece, vol. 1, pp. 87–93 (1999)

15. Aoyagi, Y., Asakura, T.: A study on traffic sign recognition in scene image using genetic algorithms and neural networks. In: 22 nd IEEE international Conference on Industrial Electronics, Control and Instrumentation (IECON), Taiwan, vol. 3, pp. 1838–1843 (1996)

16. Foucher, P., Charbonnier, P., Kebbous, H.: Evaluation of a road sign pre-detection system by image analysis. In: The fourth International conference on computer vision theory and applications (VISAPP 2009), Lisbonne, pp. 362–367 (2009)

17. Lafuente-Arroyo, S., Gil-Jimenez, P., Maldonado-Bascon, R., Lopez-Ferreras, F., Maldonado-Bascon, S.: Traffic sign shape classification evaluation I: SVM using distance to borders. In: IEEE Intelligent Vehicles Symposium (IV 2005), Las Vegas, pp. 557–562 (2005)

18. Ishizuka, Y., Hirai, Y.: Segmentation of road sign symbols using opponent-color filters. In: 11th World Congress on Intelligent Transport Systems and Services (ITSWC), Japan, pp. 18–22 (2004)

19. Serna, C.G., Ruichek, Y.: Classification of traffic signs: the European dataset. IEEE Access. **6**, 78, 136–78, 148 (2018)

20. Simon, L., Tarel, J.P., Brémond, R.: Towards the estimation of conspicuity with visual priors. In: Third International Conference on Computer Vision Theory and Applications (VISAPP 2008), Portugal, pp. 323–328 (2008)

21. Stallkamp, J., Schlipsing, M., Salmen, J., Igel, C.: Man vs. computer: benchmarking machine learning algorithms for traffic sign recognition. Neural Netw. **32**, 323–332 (2012)

22. Aghdam, H., Heravi, E.J., Puig, D.: A practical and highly optimized convolutional neural network for classifying traffic signs in real-time. Int. J. Comput. Vision **122**, 246–269 (2017)

23. Ruta, A., Li, Y., Liu, X.: Robust class similarity measure for traffic sign recognition. IEEE Trans. Intell. Transp. Syst. **11**, 846–855 (2010)

24. Raj, K.T.I., Raj, R.G.: Real-time (vision-based) road sign recognition using an artificial neural network. Sensors **17**, 853 (2017). Mdpi

25. Lau, M.M., Lim, K.H., Gopalai, A.A.: Malaysia traffic sign recognition with convolutional neural network. In: IEEE International Conference on Digital Signal Processing (DSP), Singapore, pp. 1006–1010 (2015)

26. Li, Y., Møgelmose, A., Trivedi, M.M.: Pushing the 'speed limit': high-accuracy US traffic sign recognition with convolutional neural networks. IEEE Trans. Intell. Veh. **1**, 167–176 (2016)

27. Girshick, R., Donahue, J., Darrell, T., Malik, J.: Rich feature hierarchies for accurate object detection and semantic segmentation. In: IEEE Conference on Computer Vision and Pattern Recognition, Columbus, pp. 580–587 (2014)

28. Li, J., Wang, Z.: Real-time traffic sign recognition based on efficient CNNs in the wild. IEEE Trans. Intell. Transp. Syst. **20**, 975–984 (2019)

A Semi-automated Approach for Identification of Trends in Android Ransomware Literature

Tanya Gera[1], Jaiteg Singh[1]($\boxtimes$), Deepak Thakur[1], and Parvez Faruki[2]

[1] Chitkara University Institute of Engineering and Technology, Chitkara University, Punjab, India
{tanya.gera,deepak.thakur}@chitkara.edu.in,
jaitegkhaira@gmail.com
[2] AV Parekh Technical Institute, Rajkot, India
parvezfaruki.kg@gmail.com

Abstract. Android ransomware is seen in the highlights of cyber security world reports. Ransomware is considered to be the most popular as well as threatening mobile malware. These are specsial malware used to extort money in return of access and data without user's consent. The exponential growth in mobile trans-actions from 9.47 crore in 2013–14 to 72 crores in 2016–17 could be a potential motivation for numerous ransomware attacks in the recent past. Attackers are consistently working on producing advanced methods to deceit the victim and generate revenue. Therefore, study of Android stealth malware, its detection and analysis gained a substantial interest among researchers, thereby producing suffi-ciently large body of literature in a very short period. Manual reviews do provide insight but they are prone to be biased, time consuming and pose a great challenge on number of articles that needs investigation. This study uses Latent Seman-tic Analysis (LSA), an information modelling technique to deduce core research areas, research trends and widely investigated areas within corpus. This work takes a large corpus of 487 research articles (published during 2009–2019) as input and produce three core research areas and thirty emerging research trends in field of stealth malwares as primary goal. LSA, a semi-automated approach is helpful in achieving a significant innovation over traditional methods of literature review and had shown great performance in many other research fields like medical, sup-ply chain management, open street map etc. The secondary aim of this study is to investigate popular latent topics by mapping core research trends with core research areas. This study also provides prospective future directions for heading researchers.

Keywords: Stealth malware · Ransomware · Latent semantic analysis · Research trends · Topic solution

1 Introduction

In parallel with increasing features and benefits of Smartphone, we cannot neglect the fact that malicious software is also raising at an alarming rate. Android being an open

© Springer Nature Switzerland AG 2021
É. Renault et al. (Eds.): MLN 2020, LNCS 12629, pp. 265–283, 2021.
https://doi.org/10.1007/978-3-030-70866-5_18

source platform; is also susceptible to malware infections. Android users are connected to the internet for almost all the time to leverage the features of Android applications. Taking advantage of this fact, malware authors compromise the users' security by developing malicious applications. The malicious applications can harm the user by stealing their sensitive information like contacts, reading personal messages, call recording, send messages to premium rate numbers, financial loss, gain access of gallery and access to user's location etc., Popularity among users as well as consistency in malware attacks has made Android security as an interesting area in research community. Consistent on-going expansion in research dimensions of android financial malware has brought about humongous content in short span. It is generally challenging to deduce overall insight of trends in cited literature. Manual Reviews or semi-automated review processes [1–3] are two methods to make out decisions about core research areas and trends. Manual reviews can be biased at times and further it does not guarantee the inter and intra document comparison [4]. Moreover, it poses a challenge on field specific expertise to provide accurate inferences. In contrast to past works, this investigation is certifiably not a conventional synthesis on taxonomy of financial malwares or survey of existing [5–8]. This paper expects to recognize latent research trends from sufficiently large corpus of android financial malware. This work uses systematic approach called latent semantic analysis for extracting three core research areas and thirty research trends from corpus of 487 research articles. LSA is popular and proficient in enquiring important related structure.

Rest of the paper is organized as follows: second section presents related work; the third section describes the materials and research methodology adopted for collecting the research literature. The fourth section presents stepwise implementation procedure. The fifth section discusses the experimental results. The sixth section presents discussion over results obtained. The last section concludes the study with future scope for other researchers.

2 Related Work

In 2020, authors [9] performed decision making to analyze trends in blockchain technology using Word2vec-based Latent Semantic Analysis (W2V-LSA). The experimentals results confimed its usefulness and better topic modelling than tradition bibliometric methods. This approach used only 231 abstracts of blockchain related literature instead of full text documents. Systematic information retrieval using automated and semi-automated approaches in any field of research has itself become a trend. Initially, general trend analysis was perfomed first time for one-dimensional data of time [10]. However, recent advancements in text mining, information retrieval and topic modelling techniques has attracted researcher's attention towards trend analysis [11]. Text mining has also leverage machine learning algothims to enhance its capability to mine latent information [12] from user review on websites, newspaper articles, social media information analysis etc. Use of LDA is also seen in literature; but it is difficult to interpret its results without feeding a few parameters and prior knowledge of topics [13]. Other techniques like probabilistic latent semantic analysis approaches uses uni-gram format conversion of a word, which generally get fail to capture the specific context in the document [14]. In the other hand, n-gram format leads to decrease in efficiency of model due to wide dimensionality [15].

3 Material and Methods

3.1 Search Strategy and Data Gathering

With an aim to perform a systematic and detailed literature study, quality articles published in reputed journals like IEEE Transactions on Mobile Computing, IEEE Transactions on Information Forensics and Security, Computers & Security, Future Generation Computer Systems, Digital Investigation and ACM Computing Surveys have been considered. All the tasks involved are organised into research process flow. The overall outline is depicted in the Fig. 1 and discussed here below:

a.) *Article Search Strategy*

First, articles were searched, reviewed and refine on the basis of article search strategy. The keyword set used were like "malware" OR "stealth" OR "advanced" OR "persistent" OR "threats" OR "ransomware" OR "security" OR "privacy" OR "monitoring" OR "application" OR "kernel" OR "android" OR "hybrid" OR "static" OR "dynamic". Further, A fine-tuned inclusion and exclusion criteria used is explained here below in Table 1. It is an iterative process to refine the collection.

Table 1. Inclusion and exclusion criteria

Inclusion criteria	Exclusion criteria
• Articles published between 2009–2018 • Focus on Android stealth malware analysis, detection and prevention	• Articles focus on general malware • Articles focus on other operating systems like iOS, windows, Symbian, Blackberry etc.

• *Initial Paper inferences*

Under initial paper inferences, all the collected papers were thoroughly reviewed and analysed. This process majorly focusses frequency of the articles and detection approach used to augment and refine the collection. Figure 2 shows the resulting statistics as below:

• *Conduct Detailed Article inferences*

Under this task, a comprehensive manual study of research article was performed. After performing the exhaustive and extensive study of articles; a few affecting parameters came out in frame. Further, a comparison among prominent studies of Android stealth malware detection and analysis frameworks to identify objective or technique used for the study (i.e. either analysis or detection) and approach used for detection (i.e. static, dynamic or hybrid).

• *Consolidated Manual Inferences*

After compiling results of both initial and detailed level process, a few manual inferences were also made and lately will be compared with the experimental results produced by this study.

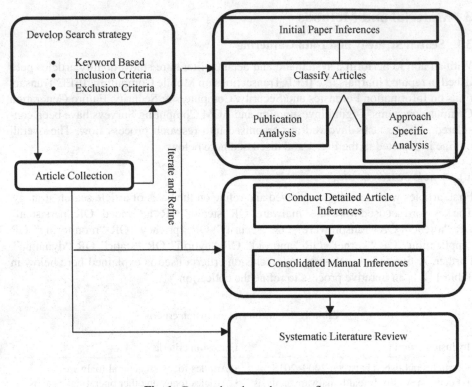

Fig. 1. Research tasks and process flow

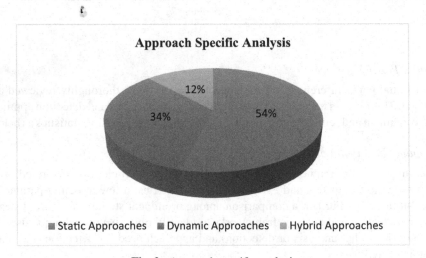

Fig. 2. Approach specific analysis

3.2   Proposed Method

The section presents the overall process flow involved in performing systematic literature study and the expected outcomes at every step are presented in Fig. 3. The whole process of systematic literature study is divided into three steps: first step is data gathering and second is to implement latent semantic analysis to extract latent topics from large gather data repository and last step is to visualize the results to present core research areas and research trends. The results of LSA will definitely help in identifying widely investigated sub-areas of financial malware attack studies.

Step 1: Data Gathering
A large corpus of research articles were gathered as discussed in detail under Sect. 3.1.
Step 2: Implementing LSA
Latent Semantic Analysis (LSA), utilizes a well-established algorithm to convert unstructured raw textual data into organized information objects, and further analyze these objects to recognize patterns for learning [3]. It employs a systematic and comprehensive approach to uncover the research trends in huge literature dataset [2]. Multiple contexts of terms belonging to different documents were collected followed by deriving the associations between related concepts that represent a latent class. A latent class can be defined as a topic solution representing multiple entities retaining similar semantics and associated patterns. The analytical tasks, which were performed in context of LSA are segregated into three modules namely Pre-LSA module, Core LSA module and Post LSA module.

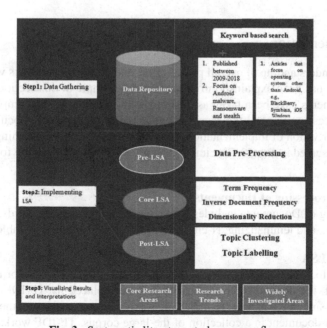

Fig. 3. Systematic literature study: process flow

(i)   *Pre-LSA:* This step involves data pre-processing and generating document-term matrix of the corpus.
  - Data Pre-processing: It involves tokenization, removing stop words, normalization, stemming & lemmatizing and character filtering.
  - Document-Term Matrix: LSA uses bag of word (BoW) model, which results in a document-term matrix (containing term frequency in a document). LSA model typically replace raw counts in the document-term matrix with a Term Frequency-Inverse Document Frequency (TF-IDF) score. TF-IDF assigns a weight for term t in document d.

(ii) *Core-LSA:* As LSA learns latent topics by performing matrix decomposition on the document-term matrix, therefore Singular Value Decomposition (SVD) technique is used. The dimensionality reduction over matrix is implemented through an application of truncated SVD.

(iii) *Post-LSA:* Finally, terms and documents, represented in the space defined by the SVD dimensions, are analysed by implementing a specific analytic method such as queries (doscument or term comparisons), clustering, or factor analysis.

Step 3: Visualizing Results
This work aims to achieve strong mapping between dominant core research areas and research trends among the corpus of 487 abstracts of research articles related to Android financial malware attack and their studies. Latent Semantic Analysis (LSA) had been applied, as discussed in Sect. 4. As a result of which three core research areas followed by ten, twenty and thirty topic solutions were presented as research trends as described in Sect. 5.

4   Implementation

For a better understanding of overall process flow and all the related tasks while implementing LSA has been explained through a sample example in this section. Here, five sample statements have been taken as documents as in Table 2. The complete process from pre-processing to resulting scores is explained through sample documents so as to have a clear vision of implementation. However, actual results containing 487 documents are presented in Sect. 5. The implementation of LSA involves the following key steps:

Step 1: **Pre-processing of Documents**
Pre-processing of Documents includes tokenization, removing stop words, normalization, stemming and lemmatizing, N-character filtering as explained in Table 3.

Step 2: **Core LSA**
As the task is to mine the relevant terms that provide useful or quality information about the document, the document-term matrix has to be replaced by TF-IDF weights for further processing of the matrix. TF-IDF weight is a measure to interpret the importance of a term to a document in a collection of the large corpus. TF-IDF works on the fact that relevant words are not necessarily frequent words. The TF-IDF weight is build-up of two terms that need to be calculated beforehand:

- **TF (Term Frequency)**: It provides the Term Frequency (TF) with normalized value, which can be computed as the division of the number of occurrences of a term in a document and the total number of terms in that document, refer Eq. 1. Sample TF scores in context with an example taken are shown in Table 4.

$$TF(t, d) = \frac{Number\ of\ occurences\ of\ term\ t\ appears\ in\ document\ d}{Total\ number\ of\ terms\ in\ the\ document} \qquad (1)$$

- **IDF (Inverse Document Frequency)**: IDF takes care of the importance of each term within the document. IDF is calculated as a log of the division of count of documents in the corpus divided by the count of documents where the specific term appears. However, it is quite obvious that some terms, like "is", "the", "of", and "that" or certain domain-specific words, will appear many a time but may not have much importance. Thus, arises a need to weigh down the importance of most occurred terms while scaling up the rare ones. Therefore, IDF scores are important and can be computed as given in Eq. 2. IDF scores for sample documents are presented in Table 5.

$$IDF(t, d) = log \frac{Total\ number\ of\ documents}{Number\ of\ documents\ with\ term\ t\ in\ it} \qquad (2)$$

Hence, after the calculation of TF and IDF scores, the final document-term matrix with TF-IDF scores is calculated with the following Eq. 3.

$$w_{t,d} = TF_{t,d} \times Log \frac{N}{df_t} \qquad (3)$$

Table 2. Sample statements

Documents	Sample statements
Document 1	Crypto ransomware encrypts the personal files/folders
Document 2	Lock screen ransomware locks the screen and demands payment
Document 3	Attackers use tactics like spam emails to spread the ransomware
Document 4	Ransomware authors blackmail the victims of the ransomware attack
Document 5	Ransomware can easily evade from signature-based mechanisms

(where t denotes the terms; d denotes each document; N denotes the total number of documents).

Consider Table 6 above which represents the document-term matrix with TF-IDF scores for previously stated example, a term will have more TF-IDF value when its occurrences across the *document* are more but less across the *corpus*. Let's take an example of a domain-specific word i.e. "malware" which will be fairly a very common word in the whole corpus but may appear often in a document hence, it will not have a high TF-IDF score. However, the word "permissions" may appear frequently in a document, and appears less in the rest of the corpus, it will have a higher TF-IDF score.

The corpus for the TF-IDF calculation will be the collection of all pre-processed documents. TF-IDF's implementation was used that is included in the sklearn's package feature_extraction.text of python library and can be easily called using the class TfidfVectorizer.

As shown in the example given above, a large corpus of the Android security dataset resulted in high dimensional TF-IDF matrix. High dimensional matrix is generally expected to be redundant and noisy. Therefore, to uncover the relationship among the words and documents and capture the latent topics within the corpus, dimensionality reduction was performed, as detailed in the following section.

Table 3. Sample outcomes after Pre-processing

S. no	Pre-processing steps	After pre-processing
a.)	**Tokenization** • Converts Large Chunks of text to sentences • Sentences to Words	['Malware', 'application', 'reads', 'the', 'unique', 'device', 'identifier', 'to', 'track', 'the', 'user', 's', 'device']
b.)	**Removing Stop Words** • Remove stop words • Remove Common words	['Malware', 'application', 'reads', 'unique', 'device', 'identifier', 'track', 'user', 'device']
c.)	**Normalization** • Standard Formatting • Upper case to lower case • Numbers to word equivalents	Malware application reads the unique device identifier to track the user s device
d.)	**Stemming and Lemmatizing** • Reduces total number of unique words • Converts the words to their word stem • Past and Future tenses to present • Third form to first form	['malwar', 'applic', 'read', 'uniqu', 'devic', 'identifi', 'track', 'user', 'devic']
e.)	**N-Character Filtering** • Words less than the length 4 were omitted	['malwar', 'applic', 'read', 'uniqu', 'devic', 'identifi', 'track', 'user', 'devic']

Table 4. TF scores

Documents	TF scores
Doc 1	{'crypto': 0.083, 'ransomwar': 0.167, 'encrypt': 0.083, 'person': 0.083, 'file': 0.083, 'folder': 0.083, 'applic': 0.083, 'similar': 0.083, 'function': 0.083, 'call': 0.083, 'packag': 0.083}
Doc 2	{'lock': 0.286, 'screen': 0.286, 'ransomwar': 0.143, 'demand': 0.143, 'payment': 0.143}
Doc 3	{'attack': 0.143, 'tactic': 0.143, 'like': 0.143, 'spam': 0.143, 'email': 0.143, 'spread': 0.143, 'ransomwar': 0.143}
Doc 4	{'ransomwar': 0.333, 'author': 0.167, 'blackmail': 0.167, 'victim': 0.167, 'attack': 0.167}
Doc 5	{'ransomwar': 0.167, 'easili': 0.167, 'evad': 0.167, 'signatur': 0.167, 'base': 0.167, 'mechan': 0.167}

Table 5. IDF scores

Terms	IDF score	Terms	IDF score	Terms	IDF score
Applic	2.098612	evad	2.098612	ransomwar	1.000000
Attack	1.693147	file	2.098612	Screen	2.098612
Author	2.098612	folder	2.098612	signatur	2.098612
Base	2.098612	function	2.098612	Similar	2.098612
blackmail	2.098612	like	2.098612	Spam	2.098612
call	2.098612	lock	2.098612	Spread	2.098612
crypto	2.098612	mechan	2.098612	Tactic	2.098612
demand	2.098612	packag	2.098612	Victim	2.098612
easili	2.098612	payment	2.098612		
email	2.098612	person	2.098612		
encrypt	2.098612				

Step 3: **Dimensionality Reduction Using Singular Vector Decomposition**

TF-IDF matrices produce in the previous step clearly has high dimensional data which is difficult to interpret. So, to convert it into low dimensional vector space, singular vector decomposition is applied. This step converts the tf-idf matrix into two matrices term loading matrices and document loading matrices as shown in Fig. 4. Term loading matrix represents association between terms and topics whereas, document loading matrix represents association between documents and topics. The corresponding scores for both matrices are shown in Table 7 and 8 respectively. Each topic in the matrices signals towards a research theme in the whole corpus. It is totally in hands of the researcher to

Table 6. TF-IDF scores

Terms	Doc 1	Doc 2	Doc 3	Doc 4	Doc 5
Applic	0.302777	0.000000	0.000000	0.000000	0.000000
Attack	0.000000	0.000000	0.332773	0.377851	0.000000
Author	0.000000	0.000000	0.000000	0.468337	0.000000
Base	0.000000	0.000000	0.000000	0.000000	0.437393
blackmail	0.000000	0.000000	0.000000	0.468337	0.000000
Call	0.302777	0.000000	0.000000	0.000000	0.000000
Crypto	0.302777	0.000000	0.000000	0.000000	0.000000
Demand	0.000000	0.312698	0.000000	0.000000	0.000000
Easily	0.000000	0.000000	0.000000	0.000000	0.437393
Email	0.000000	0.000000	0.412464	0.000000	0.000000
Encrypt	0.302777	0.000000	0.000000	0.000000	0.000000
Evad	0.000000	0.000000	0.000000	0.000000	0.437393
File	0.302777	0.000000	0.000000	0.000000	0.000000
Folder	0.302777	0.000000	0.000000	0.000000	0.000000
function	0.302777	0.000000	0.000000	0.000000	0.000000
Like	0.000000	0.000000	0.412464	0.000000	0.000000
Lock	0.000000	0.625395	0.000000	0.000000	0.000000
mechan	0.000000	0.000000	0.000000	0.000000	0.437393
Package	0.302777	0.000000	0.000000	0.000000	0.000000
payment	0.000000	0.312698	0.000000	0.000000	0.000000
Person	0.302777	0.000000	0.000000	0.000000	0.000000
ransomwar	0.288550	0.149002	0.196541	0.446330	0.208420
Screen	0.000000	0.625395	0.000000	0.000000	0.000000
signatur	0.000000	0.000000	0.000000	0.000000	0.437393
Similar	0.302777	0.000000	0.000000	0.000000	0.000000
Spam	0.000000	0.000000	0.412464	0.000000	0.000000
Spread	0.000000	0.000000	0.412464	0.000000	0.000000
Tactic	0.000000	0.000000	0.412464	0.000000	0.000000
Victim	0.000000	0.000000	0.000000	0.468337	0.000000

alter the value of number of topic solutions they want at a particular time. Researchers also face a significant issue at times during implementing this step i.e. multiple documents can be mapped into same topic solutions. So, to avoid an iterative process of k-mean clustering is further applied to visualise the results clearer and easy to interpret.

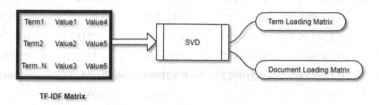

Fig. 4. Applying SVD over TF-IDF matrix

Table 7. Term loading matrix scores

Terms	Topic 0	Topic 1	Topic 2	Topic 3	Topic 4
Applic	0.107961	0.068837	0.068550	0.258226	−0.076493
Attack	0.353090	−0.247212	−0.110429	−0.088438	0.094649
Author	0.251933	−0.095740	−0.035791	−0.016441	0.401476
Base	0.124356	0.158389	0.316281	−0.221523	−0.062988
blackmail	0.251933	−0.095740	−0.035791	−0.016441	0.401476
Call	0.107961	0.068837	0.068550	0.258226	−0.076493
Crypto	0.107961	0.068837	0.068550	0.258226	−0.076493
demand	0.067287	0.226574	−0.195981	−0.057421	−0.028939
Easily	0.124356	0.158389	0.316281	−0.221523	−0.062988
Email	0.185713	−0.210672	−0.101084	−0.093175	−0.284161
Encrypt	0.107961	0.068837	0.068550	0.258226	−0.076493
Evad	0.124356	0.158389	0.316281	−0.221523	−0.062988
File	0.107961	0.068837	0.068550	0.258226	−0.076493
Folder	0.107961	0.068837	0.068550	0.258226	−0.076493
function	0.107961	0.068837	0.068550	0.258226	−0.076493
Like	0.185713	−0.210672	−0.101084	−0.093175	−0.284161
Lock	0.134575	0.453147	−0.391961	−0.114842	−0.057878
mechan	0.124356	0.158389	0.316281	−0.221523	−0.062988
Package	0.107961	0.068837	0.068550	0.258226	−0.076493
payment	0.067287	0.226574	−0.195981	−0.057421	−0.028939

(continued)

Table 7. (*continued*)

Terms	Topic 0	Topic 1	Topic 2	Topic 3	Topic 4
Person	0.107961	0.068837	0.068550	0.258226	−0.076493
ransomwar	0.522796	0.057412	0.040377	0.053107	0.130504
Screen	0.134575	0.453147	−0.391961	−0.114842	−0.057878
signatur	0.124356	0.158389	0.316281	−0.221523	−0.062988
Similar	0.107961	0.068837	0.068550	0.258226	−0.076493
Spam	0.185713	−0.210672	−0.101084	−0.093175	−0.284161
Spread	0.185713	−0.210672	−0.101084	−0.093175	−0.284161
Tactic	0.185713	−0.210672	−0.101084	−0.093175	−0.284161

Table 8. Document loading matrix scores

	Topic 0	Topic 1	Topic 2	Topic 3	Topic 4
Doc 1	0.477735	0.224989	0.219206	0.797172	−0.193947
Doc 2	0.288304	0.717045	−0.606811	−0.171642	−0.071047
Doc 3	0.603249	−0.505456	−0.237279	−0.211149	−0.528886
Doc 4	0.720724	−0.202300	−0.073990	−0.032813	0.658088
Doc 5	0.380924	0.358358	0.700109	−0.473394	−0.110552

Step 4: **Topic Clustering and Topic Labelling**

Researchers can vary the number of topic solutions, but cannot go beyond number of documents at maximum. Here, a problem may arise in which multiple documents can be mapped into same topic solutions. So, to avoid an iterative process of k-mean clustering is further applied to visualise the results clearer and easy to interpret. K-Means clustering is used to form clusters of semantically related terms. Here, in this study five, ten, twenty and thirty topic solutions are successfully discovered and presented in subsequent section. Loading weights of the terms for every topic solution are sorted in descending order so as to give suitable labels for high loading values with help of subject field experts. This is being done for all five, ten, twenty and thirty topic solutions. Though we tried to provide results for fifty topic solution but it resulted in overlapping and repetitions of terms. However, this could also be due the dataset chosen.

5   Experimental Results

This study identifies three core research areas and research themes trends across the wide literature set of 487 articles focussing on Android stealth malware. Semantic analysis results in three dominant core research areas. However, iterative process of varying the number of topic solutions results in ten, twenty and thirty topic solutions. Each topic solution is considered as $TS_{x.y}$ where x denotes solution number whereas y is yth factor of the xth number topic solution. For example, TS10.2 should be considered as 2nd factor of ten solutions. The Detained description along with their publication counts as discussed in subsequent section.

The graphs plotted for all topic solutions also provides information about the publication count for each topic label during three different time periods within 2009–2019, Figs. 4 and 5. The publication count associated with each topic solution represents the importance of the corresponding research area within that topic solution. Further, to uncover the research trends and future directions in the field of Android security, thirty topic solutions were also discovered as shown in Table 12. The semantic mapping between fifty topic solutions and five core research areas helps to identify research trends within each core research area, as presented in Table 10.

5.1 Core Research Areas

Systematic literature Analysis over Android stealth malware corpus results in identifying three core areas as Table 9.

Table 9. Core research areas

Topic no.	Topic label	Top loading terms
TS3.1	App Structure Monitoring	signature, bytecode, graph, context, dalvik, flow, permission, component, control, library, program, service, method, object, entry, event, field, code, data, path
TS3.2	App·Behaviour Monitoring	kernel, privilege, escalation, policy, control, enforcement, security, exploit, memory, vulnerability, library, native, context, component, linux, mechanism, access, resource, sandbox, virtual
TS3.3	Hybrid Level Monitoring	dynamic, analysis, static, cloud, taint, application, instruction, execution, component, sensitive, bytecode, android, library, program, native, object, string, dalvik, class, event

Core Research Areas (TS3.1) is found to be the most trending area, whereas (TS3.2) behavioural analysis of applications seems to be utilised and became in trend from 2014 onwards. TS3.3 produces promising results in analysing and detecting of smart malwares. Likewise, Table 10 and 11 shows ten topic solution and twenty topic solutions, their sensible labels and their count values as follows:

Table 10. Ten topic solutions

Topic no	Topic label	2009–2013	2014–2019	2009–2019
TS10.1	Emulator Based Analysis	23	26	49
TS10.2	Dynamic Code Loading	19	32	51
TS10.3	High Battery Consumption	22	21	43
TS10.4	Context Monitoring	21	30	51
TS10.5	API Call Monitoring	18	31	49
TS10.6	Dalvik Byte Code Analysis	21	17	38
TS10.7	Permission Based Analysis	22	31	53
TS10.8	Classification Based on App Behavior	19	38	57
TS10.9	Graph Based Analysis	13	29	42
TS10.10	Feature Based Analysis	23	31	54

Table 11. Twenty topic solutions

Topic no	Topic label	2009–2013	2014–2019	2009–2019
TS20.1	Obfuscated Code Analysis	5	6	11
TS20.2	Privacy Leakage Monitoring	19	7	26
TS20.3	Hybrid Analysis	5	24	29
TS20.4	Pattern Assessment	9	12	21
TS20.5	Permission Based Analysis	22	20	42
TS20.6	Kernel Level Check	12	18	30
TS20.7	Signature Based Analysis	16	19	35
TS20.8	Classification Based on App Behavior	11	21	32
TS20.9	Dynamic Code Loading	8	21	29
TS20.10	Emulator Based Analysis	7	14	21
TS20.11	Taint Analysis	4	12	16
TS20.12	Graph Based Analysis	12	13	25
TS20.13	Flow Monitoring	3	9	12
TS20.14	API Call Monitoring	6	12	18
TS20.15	User Interactions	3	13	16
TS20.16	Context Monitoring	12	22	34
TS20.17	Feature Based Analysis	4	7	11
TS20.18	Dalvik Byte Code Analysis	8	15	23
TS20.19	High Battery Consumption	4	19	23
TS20.20	Text Based Analysis	4	29	33

The results show that under ten topic solutions (TS10.8) Classification Based on App Behavior is most widely used method for analysis and detection of Android malware. Specifically, TS10.8 became in trend between the year 2014–2018. However, (TS10.7) i.e. Permission based analysis remained trending during 2009–2014. Under twenty topic solutions, results showed in Table 11 clearly depicts (TS20.5), (TS20.6), (TS20.7), (TS20.8) and (TS20.16) became most trending factors that pose a significantly impact on malware detection process.

5.2   Research Themes/Trends

After an extensive iterative process, thirty topic solutions considered to be the major core research trends/themes across wide corpus of Android ransomware. Table 12 shows list of topic solutions along with their count of articles falling under the same solution.

Table 12. Research trends

Topic no	Label	2009–2013	2014–2019	2009–2019
TS30.1	Kernel Level Check	6	15	21
TS30.2	Dynamic Code Loading	4	19	23
TS30.3	Classification Based on App Behavior	9	8	17
TS30.4	Obfuscated Code Analysis	2	7	9
TS30.5	Ransomware detection	5	10	15
TS30.6	User Interaction	15	11	26
TS30.7	Runtime Analysis	3	6	9
TS30.8	Emulator Based Analysis	4	4	8
TS30.9	Privacy Leakage Monitoring	2	7	9
TS30.10	Flow Monitoring	4	8	12
TS30.11	Sandboxing Techniques	1	15	16
TS30.12	Text-based analysis	2	7	9
TS30.13	Fuzz Testing	8	4	12
TS30.14	Pattern Assessment	6	13	19
TS30.15	Graph Based Analysis	3	13	16

(*continued*)

Table 12. (*continued*)

Topic no	Label	2009–2013	2014–2019	2009–2019
TS30.16	Function Call Monitoring	3	6	9
TS30.17	Hybrid Analysis	4	12	16
TS30.18	Permission Based Analysis	10	11	21
TS30.19	Dalvik Byte Code Analysis	5	24	29
TS30.20	High Battery Consumption	5	18	23
TS30.21	File Operation	1	12	13
TS30.22	Context Monitoring	7	12	19
TS30.23	API Call Monitoring	7	22	29
TS30.24	VM Monitoring	2	10	12
TS30.25	Taint Analysis	6	15	21
TS30.26	Signature based Similarity	2	10	12
TS30.27	Context Monitoring	2	14	16
TS30.28	Privilege Escalation	9	11	20
TS30.29	Feature Based Analysis	6	9	15
TS30.30	System Events	1	10	11

Among thirty core research area, research trends like Dynamic Code Loading (TS30.2), Dalvik Byte Code Analysis (TS30.19), API Call Monitoring (TS30.23). Results also shows that these topic solutions are found to be most widely investigated areas. The areas remained in trend throughout the year 2014–2019.

5.3   Widely Investigated Areas

Another aim of this study was to identify topmost widely investigated sub-areas inside each core research areas. The results of experimentation helped to interpret significant analysis patterns. Though static analysis is more popular than dynamic but it may be due to corpus selected for analysis. That dominance of behaviour-based study is more than structure-based studies. Here, permission based, graph based, context based and signature based turned out to be most impactful sub-areas under App structural monitoring (TS3.1). As depicted in Fig. 5; out of 145 articles, 6% of them focussed on permission-based analysis, likewise 9% were of signatures-based similarity. Likewise, other widely investigated sub-areas can be observed from Fig. 5 below:

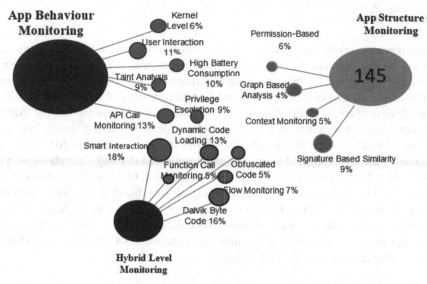

Fig. 5. Widely investigated sub-areas

6   Discussion

To prove the significance of this study, results framed out after complete LSA experimentation is compared with a few inferences made manually from literature review:

- Need of robust hybrid solutions:
  Manual reviews states, there exists comparatively less hybrid approaches for malware detection when compared to Static and Dynamic approaches. Our study also shows hybrid solutions are found to be only 21%, comparatively very less than app structure monitoring and app behaviour monitoring.
- Stealthy Behaviour:
  Recent reported ransomware silently installed its packages in mobile phones. They first gained administrative rights to exploit the user's privacy and subsequently encrypts device data. To detect such malware kernel level scanning is required (Ahmadian et al. 2015; Tam et al. 2017; Rashidi and Fung 2015). Our results state out of 244 articles, kernel level check was 9%.
- Intense hardware requirement for real time analysis:
  Analysing infections on the mobile phone itself requires high consumption of CPU power, battery and memory, while analysis on the server might suffer from high communication overhead and delayed response (Al-rimy et al. 2018; Alazab et al. 2012; Xu et al. 2016). These clearly sign towards the requirement of designing a hybrid framework for effective detection and predictions.

7 Conclusion

Manual literature review may give bias and incomplete interpretations. This study aims to review a large literature set of 487 articles systematically to identify latent research themes. Implementing LSA helped to three core research areas and thirty research trends. Results suggested that overall analysis and detection process of ransomware revolves around three core areas i.e. app structure, app behaviour and hybrid monitoring. Among thirty core research area, research trends like dynamic code loading, dalvik byte code analysis, API call monitoring. Results also shows that these topic solutions are found to be most widely investigated areas. The areas remained in trend throughout the year 2014–2019 and are expected to be in remain in future also. Our study also shows hybrid solutions are found to be only 21%, comparatively very less than app structure monitoring and app behaviour monitoring. For future researchers, analysis at kernel level is need of the hour to detect financial attacks. Researchers can apply same methodology in one or more other research fields by little or almost no changes. LSA can help researchers to identify research trends in other fields also.

References

1. Lee, S., Song, J., Kim, Y.: An empirical comparison of four text mining methods. J. Comput. Inf. Syst. **51**(1), 1–10 (2010)
2. Evangelopoulos, N., Zhang, X., Prybutok, V.R.: Latent semantic analysis: five methodological recommendations. Eur. J. Inf. Syst. **21**(1), 70–86 (2012)
3. Delen, D., Crossland, M.D.: Seeding the survey and analysis of research literature with text mining. Expert Syst. Appl. **34**(3), 1707–1720 (2008)
4. Yalcinkaya, M., Singh, V.: Patterns and trends in building information modeling (BIM) research: a latent semantic analysis. Autom. Constr. **59**, 68–80 (2015)
5. Becher, M., Freiling, F.C., Hoffmann, J., Holz, T., Uellenbeck, S., Wolf, C.: Mobile security catching up? Revealing the nuts and bolts of the security of mobile devices. In: 2011 IEEE Symposium on Security and Privacy, pp. 96–111 (2011)
6. Enck, W.: Defending users against smartphone apps: techniques and future directions. In: Jajodia, S., Mazumdar, C. (eds.) ICISS 2011. LNCS, vol. 7093, pp. 49–70. Springer, Heidelberg (2011). https://doi.org/10.1007/978-3-642-25560-1_3
7. Suarez-Tangil, G., Tapiador, J.E., Peris-Lopez, P., Ribagorda, A.: Evolution, detection and analysis of malware for smart devices. IEEE Commun. Surv. Tutorials **16**(2), 961–987 (2013)
8. Faruki, P., Bharmal, A., Laxmi, V., Gaur, M.S., Conti, M., Rajarajan, M.: Evaluation of android anti-malware techniques against dalvik bytecode obfuscation. In: 2014 IEEE 13th International Conference on Trust, Security and Privacy in Computing and Communications, pp. 414–421 (2014)
9. Kim, S., Park, H., Lee, J.: Word2vec-based latent semantic analysis (W2V-LSA) for topic modeling: a study on blockchain technology trend analysis. Expert Syst. Appl. **152**, 113401 (2020)
10. Kivikunnas, S.: Overview of process trend analysis methods and applications. In: ERUDIT Workshop on Applications in Pulp and Paper Industry, pp. 395–408 (1998)
11. Kang, H.J., Kim, C., Kang, K.: Analysis of the trends in biochemical research using latent dirichlet allocation (LDA). Processes **7**(6), 379 (2019)
12. Kim, Y.M., Delen, D.: Medical informatics research trend analysis: a text mining approach. Health Inform. J. **24**(4), 432–452 (2018)

13. Alghamdi, R., Alfalqi, K.: A survey of topic modeling in text mining. Int. J. Adv. Comput. Sci. Appl. (IJACSA) **6**(1), 147–153 (2015)
14. Lu, Y., Zhai, C.: Opinion integration through semi-supervised topic modeling. In: Proceedings of the 17th International Conference on World Wide Web, pp. 121–130 (2008)
15. Bengio, Y., Ducharme, R., Vincent, P., Jauvin, C.: A neural probabilistic language model. J. Mach. Learn. Res. **3**, 1137–1155 (2003)

Towards Machine Learning in Distributed Array DBMS: Networking Considerations

Ramon Antonio Rodriges Zalipynis[(✉)]

HSE University, Moscow, Russia
rodriges@gis.land

Abstract. Computer networks are veins of modern distributed systems. Array DBMS (Data Base Management Systems) operate on big data which is naturally modeled as arrays, e.g. Earth remote sensing data and numerical simulation. Big data makes array DBMS to be distributed and highly utilize computer networks. The R&D area of array DBMS is relatively young and machine learning is just paving its way to array DBMS. Hence, existing work is this area is rather sparse and is just emerging. This paper considers distributed, large matrix multiplication (LMM) executed directly inside array DBMS. LMM is the core operation for many machine learning techniques on big data. LMM directly inside array DBMS is not well studied and optimized. We present novel LMM approaches for array DBMS and analyze the intricacies of LMM in array DBMS including execution plan construction and network utilization. We carry out performance evaluation in Microsoft Azure Cloud on a network cluster of virtual machines, report insights derived from the experiments, and present our vision for the future machine learning R&D directions based on LMM directly inside array DBMS.

Keywords: Distributed machine learning · Experimental evaluation · Performance analysis · Cloud computing · Matrices · Vision

1 Introduction

A distributed array DBMS handles big multi-dimensional or N-d arrays. It is obvious that computer networks are veins of this class of systems and many related research directions are open-ended. The area of array DBMS is currently experiencing an R&D surge due to the rapid growth of array volumes.

Machine learning (ML) is just paving its way to array DBMS and only a handful of papers exist on this important topic [17,28]. It is very beneficial to have machine learning capabilities inside a DBMS, so many vendors integrate ML techniques directly into their systems [15]. Large matrix multiplication (LMM) is the core operation for a vast variety of ML algorithms on big data. However, as soon as LMM is performed inside an array DBMS, it should be designed using the unique facilities inherent exclusively to array DBMS. Hence, running LMM inside an array DBMS requires novel, dedicated approaches specific to

© Springer Nature Switzerland AG 2021
E. Renault et al. (Eds.): MLN 2020, LNCS 12629, pp. 284–304, 2021.
https://doi.org/10.1007/978-3-030-70866-5_19

array DBMS. Moreover, array DBMS should perform the multiplication on large matrices in a distributed fashion utilizing computer networks.

An N-d array is a native model for many important data types including time series, satellite imagery, LiDAR data, weather & climate forecasts, and other physical measurements and numerical simulations. N-d arrays are crucial for vital daily tasks like urban planning, agriculture monitoring, forestry control, and rapid-response in disaster management [1].

First array DBMS and add-ons, e.g. RASDAMAN [3], POSTGIS [19], ORACLE SPATIAL [16], appeared long ago. However, only the last decade flourished with a significant number of groups carrying out R&D on geospatial array management: CHRONOSDB (2018) [21,23], SCIDB (2008) [6] (array DBMS); DataCube (2017) [13], EarthServer (2016) [4] (national initiatives); TILEDB (2016) [18], SAGA (2014) [29] (array stores); GOOGLE EARTH ENGINE (2012) [9], DASK (2018) [7] (array engines), and others [21,28]. Full-stack array DBMS stand out by providing high-level, end-to-end array management, e.g. query languages, execution techniques, and interoperability.

The aforementioned array DBMS R&D surge is caused by the rapid growth of big geospatial array data. For example, Maxar, a commercial company, alone acquires about 80 TB/day and has already accumulated over 100 PB of satellite imagery in the Amazon Cloud [14].

Sophisticated array processing techniques are just beginning to emerge. Only recently, top-k queries [5] and similarity array joins [32] were fist introduced. To date, only one work is devoted to distributed caching in array DBMS [33]. Data compression has been recently evaluated in array DBMS [24] while a handful of new formats enable direct querying of compressed arrays [10]. Novel indexing techniques appear which are specifically designed for array DBMS [25]. A group of multidimensional aggregation types was studied in [22]. Array DBMS are beginning to back up interactive visualization [2].

Hence, the array DBMS R&D area can be viewed as young by right: no commonly accepted standards have yet been established, architectures and implementations are to be improved and mature, and many R&D opportunities are attractive and unexplored.

In this paper we take this opportunity to advance the integration of the core machine learning operation into array DBMS and study its behavior within the computer network in the Cloud. We describe the details in the Related Work and the sections that follow.

2   Related Work

As array DBMS is a relatively young R&D area, the related work is rather sparse. In the context of array DBMS, matrix multiplication was studied only by the group of Ordonez (see [17] and references therein to the earlier works by the same group). However, all previous work has several limitations:

1. **Generality limitation.** All previous work in the context of array DBMS was focused on optimizing the multiplication of ZZ^T, where Z is $n \times m$ matrix. Clearly, this is only a special case of matrix multiplication.
2. **Vendor dependency.** Only SciDB was used for integrating novel matrix multiplication techniques.
3. **Array DBMS underutilization.** Previous work treat array DBMS as a parallel engine and does not show how to utilize the peculiarities and potential of array DBMS: how to build execution plans to perform matrix multiplication, how matrix multiplication can be interleaved with other array operations, and what is the network behavior during this operation.
4. **Architectural limitations.** Given Z of shape $n \times m$, all SciDB chunks must have the shape of $1 \times m$. Many real-world matrices do not have large n values, e.g. the KDDnet dataset used in [17] has $n = 34$. Hence, authors artificially replicated KDDnet data to make $n = 100$ for their experiments. In addition, this may hinder other array operations or require changing chunk shape each time a matrix multiplication is required.
5. **Scalability limitations.** Authors of previous work require that any $1 \times m$ SciDB chunk must fit into main memory. Very large matrices may require secondary storage I/O which was not taken into account. Moreover, the coordinator node must be involved into the computation and aggregate intermediate multiplication results. This may appear to be a bottleneck in a multi-tenancy setting.

Currently only 3 array DBMS exist: RasDaMan [3], ChronosDB [21], and SciDB [6]. Only the Community Edition of RasDaMan is freely available which is not distributed (does not use computer network to perform array processing). SciDB delegates linear algebra operations to external libraries such as BLAS or LAPACK [26]. This incurs redundant data conversion between internal SciDB storage format and the library input data format.

In this paper we present novel matrix multiplication techniques specially designed for a distributed array DBMS. Unlike existing approaches, our techniques have the following properties.

1. **Generality.** Our techniques are able to multiply arbitrary matrices.
2. **Array DBMS utilization.** We rely on array DBMS capabilities such as execution plan construction and its automatic parallel execution by array DBMS in a computer network.
3. **Architectural flexibility.** Input matrices can possess any internal structure to be eligible for multiplication.
4. **Scalability limitations.** The computation is performed in a massively parallel fashion and the coordinator node only schedules tasks for execution and does not participate in computations.

3 Array DBMS Background

In this section we provide the information on array DBMS data model and architecture peculiarities which are required to design the algorithms and approaches presented in the rest of the paper. Currently array DBMS do not have commonly accepted standards on their data models. Hence, we follow the data model and architecture of CHRONOSDB [22], one of the state-of-the-art array DBMS.

3.1 Multidimensional Array Model

In this paper, an N-dimensional array (N-d array) is the mapping $A : D_1 \times D_2 \times \cdots \times D_N \mapsto \mathbb{T}$, where $N > 0$, $D_i = [0, l_i) \subset \mathbb{Z}$, $0 < l_i$ is a finite integer which is said to be the *size* or *length* of ith dimension, and $\mathbb{T}$ is a standard numeric type. In this paper, $i \in [1, N] \subset \mathbb{Z}$. Let us denote the N-d array A by

$$A\langle l_1, l_2, \ldots, l_N \rangle : \mathbb{T} \tag{1}$$

By $l_1 \times l_2 \times \cdots \times l_N$ denote the *shape* of A, by $|A|$ denote the *size* of A such that $|A| = \prod_i l_i$. A *cell* value of A with index $(x_1, x_2, \ldots, x_N)$ is referred to as $A[x_1, x_2, \ldots, x_N]$, where $x_i \in D_i$. Each cell value of A is of type $\mathbb{T}$. An array may be initialized after its definition by listing its cell values: $A\langle 2, 2 \rangle$: int = $\{\{1, 2\}, \{3, 4\}\}$, where A is 2-d array of integers, $A[1, 0] = 3$, $|A| = 4$, and the shape of A is 2×2. A graphical array illustration is in Fig. 1.

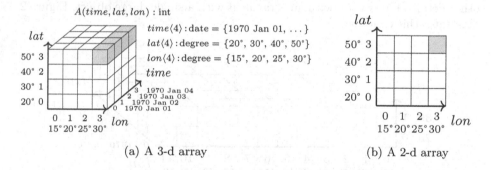

(a) A 3-d array   (b) A 2-d array

Fig. 1. Array examples.

Indexes x_i are optionally mapped to specific values of ith dimension by *coordinate* arrays $A.d_i\langle l_i \rangle : \mathbb{T}_i$, where $\mathbb{T}_i$ is a totally ordered set, and $d_i[j] < d_i[j+1]$ for all $j \in D_i$. In this case, A is defined as

$$A(d_1, d_2, \ldots, d_N) : \mathbb{T} \tag{2}$$

A *hyperslab* $A' \sqsubseteq A$ is an N-d subarray of A defined by the notation

$$A[b_1 : e_1, \ldots, b_N : e_N] = A'(d'_1, \ldots, d'_N) \tag{3}$$

where $b_i, e_i \in \mathbb{Z}$, $0 \leqslant b_i \leqslant e_i < l_i$, $d'_i = d_i[b_i : e_i]$, $|d'_i| = e_i - b_i + 1$, and for all $y_i \in [0, e_i - b_i]$ the following holds

$$A'[y_1, \ldots, y_N] = A[y_1 + b_1, \ldots, y_N + b_N] \tag{4a}$$

$$d'_i[y_i] = d_i[y_i + b_i] \tag{4b}$$

Equations (4a) and (4b) state that A and A' have a common coordinate subspace over which cell values of A and A' coincide. Note that the original dimensionality is preserved even if some $b_i = e_i$ (in this case, "$: e_i$" may be omitted in (3)). Also, "$b_i : e_i$" may be omitted in (3) if $b_i = 0$ and $e_i = |d_i| - 1$.

3.2 Datasets

A *dataset* $\mathbb{D} = (A, H, M, P)$ contains a *user-level* array $A(d_1, \ldots, d_N) : \mathbb{T}$ and the set of *system-level* arrays $P = \{(A_{key}, key, M_{key}, node_{key})\}$, where $A_{key} \sqsubseteq A$, $key = \langle N \rangle : \text{int}$ is the key for A_{key}, $node_{key}$ is the ID of the cluster node storing A_{key}, M_{key} is metadata for A_{key}, $H \langle t_1, \ldots, t_N \rangle : \text{int}$ such that $A_{key} = A[h_1 : h'_1, \ldots, h_N : h'_N]$, where $h_i = H[k_1, \ldots, k_i, \ldots, k_N]$, and $h'_i = H[k_1, \ldots, k_i + 1, \ldots, k_N]$. This means that array A is divided by $(N - 1)$-d hyperplanes on subarrays to enable more manageable processing of arrays. Let us call a user-level array and a system-level array an array and a subarray respectively for short. Let us refer to subarray A_{key} by key as $\mathbb{D}\langle\langle key \rangle\rangle$ or $\mathbb{D}\langle\langle k_1, k_2, \ldots, k_N \rangle\rangle$.

A user-level array is never stored explicitly: operations with A are mapped to a sequence of operations with respective subarrays from P. Hence, users are not aware of the quantity of subarrays, their properties, network location, and other details. Users work with an array as if with a single, large object. Figure 2 illustrates this concept.

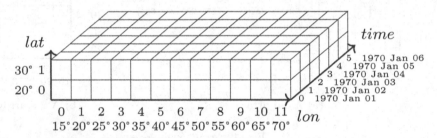

(a) User-level array (perceived by users as a single entity).

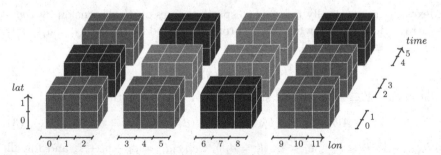

(b) Respective system-level arrays split by hyperplanes (subarrays of the same color reside on the same cluster node in a computer cluster network, best viewed in color).

Fig. 2. User-level array and subarrays.

Dataset metadata $M = \{(key, val)\}$ includes general dataset properties (name, description, contacts, etc.) and metadata valid for all $p \in P$ (data type $\mathbb{T}$, storage format, etc.). For example, $M = \{(name = \text{"Matrix 1"}), (type = \text{int}16), (format = \text{NetCDF})\}$. Let us refer to an element in a tuple $p = (A_{key}, key, \dots) \in P$ as $p.A$ for A_{key}, $p.key$ for key, etc.

3.3   Network Architecture of Distributed Array DBMS

Let us take CHRONOSDB as an example. It runs in a network on a computer cluster of commodity hardware. Subarrays of diverse file formats are distributed among cluster nodes without changing their formats. A subarray is always stored entirely on a node in contrast to parallel or distributed file systems. However, many array operations require exchanging subarrays over the network. Workers are launched at each node and are responsible for data processing. A single coordinator at a dedicated node receives client queries and coordinates workers.

The coordinator stores metadata for all datasets and subarrays. Consider a dataset $\mathbb{D} = (A, H, M, P)$. Arrays $A.d_i$ and elements of $\forall p \in P$ except $p.A$ are stored on the coordinator. In practice, array axes usually have coordinates such that $A.d_i[j] = start + j \times step$, where $j \in [0, |A.d_i|) \subset \mathbb{N}$, $start, step \in \mathbb{R}$. Only $|A.d_i|$, $start$ and $step$ values have to be usually stored. The array model merit is that it has been designed to be generic as much as possible but allowing the establishment of 1:1 mapping of a $p \in P$ to a real dataset file at the same time.

When workers receive the `discover` command, they connect to the coordinator and receive the dataset metadata and its file naming rules. Workers scan their local file systems to discover $p \in P$ by parsing file names or reading file metadata. Workers transmit to the coordinator the discovered information.

3.4   Array DBMS Execution Plans

Again, let us take CHRONOSDB as an example. Its execution plan is a directed acyclic graph $G = (V, E)$, where $E = \{((\mathbb{D}_i, key_i^j), (\mathbb{D}_k', key_m)) : \mathbb{D}_i\langle\langle key_i^j \rangle\rangle$ is required to compute $\mathbb{D}'\langle\langle key_m \rangle\rangle\}$. For an array operation, $\mathbb{D}_i$ is an input dataset and $\mathbb{D}_k'$ is the output dataset. To enable query execution in an array DBMS, we must develop an algorithm that builds an execution plan G.

Input subarrays may reside on different cluster nodes. An output subarray is produced locally by a worker node W. Hence, other worker nodes stream the required subarrays to worker W over the network to enable the generation of output subarrays by worker W. This is where the major network I/O takes place.

4   Novel Distributed Matrix Multiplication Techniques

This section presents novel distributed matrix multiplication techniques in the context of array DBMS. Unlike previous matrix multiplication approaches, our techniques rely on the peculiarities of array DBMS. We show how to enable massively parallel matrix multiplication directly inside array DBMS using subarrays.

We perform the multiplication in several steps generating interim subarrays. Furthermore, we show how to construct array DBMS execution plans for matrix multiplication. Input matrices may have arbitrary chunk/subarray shapes.

4.1 Computing Z^T

Matrix transpose operation is frequently used in expressions that involve matrix multiplication. Hence, we discuss it before matrix multiplications in terms of the data model presented before.

Reshaping operation $\Psi : A, \pi \mapsto A'$ takes as input an N-d array $A(d_1, \ldots, d_N) : \mathbb{T}$ and the permutation mapping $\pi : i \mapsto j$, where $i, j \in [1, N] \subset \mathbb{N}$, $\pi(i) \neq \pi(j)$ for $i \neq j$, and $\bigcup_i \{\pi(i)\} = [1, N]$. The reshaping operation outputs the N-d array $A'(d_{\pi(1)}, \ldots, d_{\pi(N)}) : \mathbb{T}$ such that $A[x_1, \ldots, x_N] = A'[x_{\pi(1)}, \ldots, x_{\pi(N)}]$, where $x_i \in [0, |d_i|) \subset \mathbb{N}$ for all i.

Transpose is a special case of the reshaping. For a 2-d array, π can be defined as $0 \mapsto 1$ and $1 \mapsto 0$. Transposing and reshaping of an array are reduced to independent transposing/reshaping of its subarrays. Figure 3 depicts array Z with subarrays shaped 3×4 and its reshaped version Z^T with subarrays shaped 4×3, the dimensions of Z and Z^T are swapped. Note that a subarray with key (k_1, k_2) of Z^T is a transposed version of the subarray with key (k_2, k_1) of Z.

This nice property of the transposition will help us to design a massively parallel algorithm to compute the multiplication of ZZ^T.

4.2 Computing ZZ^T

We first start from the topic to which the previous work was devoted: computing ZZ^T. We show how to compute this expression for arbitrary chunk/subarray shapes for Z and demonstrate the pattern of the respective execution plan.

Figure 3 illustrates the idea of computing ZZ^T on a tiny Z matrix shaped 6×8. Figure 4 gives the respective execution plan for the example in Fig. 3.

We use letters and numbers for array axes to facilitate the visual tracking of transformations applied to matrices. Matrix Z has 6×8 cells and is split into 4 subarrays shaped 3×4. Thick lines distinguish subarray borders. A subarray is referred to by key. Keys are drawn on the opposite side of array axes. For example, $Z.(1, 2)$ refers to the subarray $Z[a : c, 5 : 8]$. It is clear that $Z.(1, 1)^T = Z^T.(1, 1)$. The computation of ZZ^T happens in 3 stages.

First, we compute the transpose of Z which is Z^T. Since a subarray can be transposed independently of other subarrays, this happens in a massively parallel fashion. Figure 4 shows this step, its is a 1:1 operation. Our idea is to reverse the source subarray keys to get the subarray keys for Z^T.

Next, we perform matrix multiplications of subarray by subarray. For example, we have to multiply $Z.(1, 1)$ on two subarrays: $Z^T.(1, 1)$ and $Z^T.(1, 2)$. However, this will generate only partial results: to compute $ZZ^T.(1, 1)$ we need both $Z.(1, 1) \times Z^T.(1, 1)$ and $Z.(1, 2) \times Z^T.(2, 1)$. Hence, each subarray by subarray multiplication generates an interim subarray with partial results.

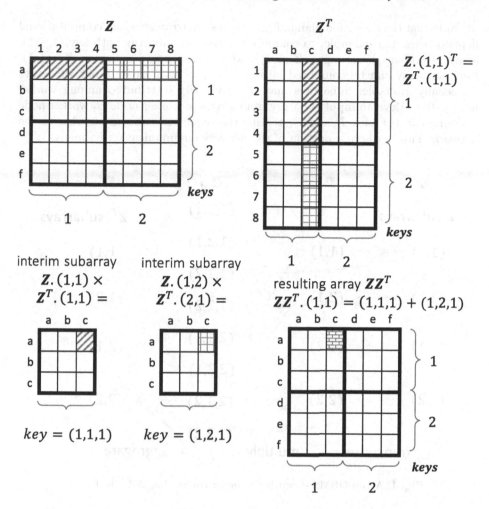

Fig. 3. Computing ZZ^T when Z has arbitrary chunk/subarray shapes.

Figure 3 shows two such interim subarrays with keys $(1,1,1)$ and $(1,2,1)$. Our another idea is operate with keys of length 3 for these interim subarrays: interim subarrays with keys $(1,1,1)$ and $(1,2,1)$ both contribute to the resulting subarray with key $(1,1)$ and we need a way to distinguish such interim subarrays, Fig. 4.

Finally, the final step is to aggregate interim subarrays into the subarray which will contain the final results, Fig. 3 and Fig. 4. This is an element-wise summation of cells of interim subarrays and setting the subarray keys accordingly. Note that we keep Z^T and interim subarrays in separate datasets, so the keys are independent of Z. In this way, distributed array DBMS can perform the multiplication of ZZ^T in a massively parallel fashion.

Note that the execution plan in Fig. 4 shows subarray keys and computational dependencies. For example, to compute the interim subarray with key $(1,1,1)$, array DBMS needs as input $Z.(1,1)$ and the transposed subarray $Z^T.(1,1)$. Each subarray can be computed locally by a worker if all its input subarrays are locally available. Subarrays may be arbitrarily distributed among worker nodes. To retrieve an input subarray from a remote worker node B, worker node A communicates with B and B streams the requested subarray to A over the network. This is where a portion of the network communication happens.

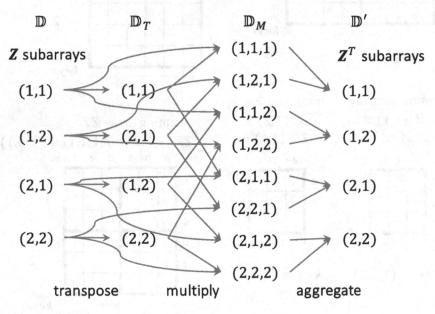

Fig. 4. Array DBMS execution plan for computing ZZ^T in Fig. 3

Now we present a novel algorithm for computing ZZ^T, algorithm 1. The novelty of Algorithm 1 is that it builds an array DBMS execution plan for computing ZZ^T, Fig. 4. The algorithm itself does not parallelize computations or performs multiplications, but yields an execution plan suitable for massively parallel execution by a distributed array DBMS. The algorithm accepts Z with subarrays of arbitrary shapes, not with a fixed shape of $1 \times m$ as in previous work.

To build the execution plan for computing ZZ^T, we need to call AGGREGATE as it uses all previous functions. Two intermediate datasets will be created: $\mathbb{D}_T$ and $\mathbb{D}_M$ which store the transposed version of Z and partial aggregates respectively, Fig. 4.

Algorithm 1. Distributed computation of ZZ^T in array DBMS

Input: $\mathbb{D}, \mathbb{D}.Z$, where Z has arbitrary chunk/subarray shapes $\quad \triangleright\ Z$ is a 2-d array
Output: $\mathbb{D}', \mathbb{D}'.P = ZZ^T$, multiplication result
1: **function** TRANSPOSE $\hfill \triangleright$ Step 1
2: $\quad \mathbb{D}_T.P \leftarrow \{\}$ $\hfill \triangleright$ subarrays of Z^T, fig. 4
3: $\quad$ **for each** $p \in \mathbb{D}.P$ **do** $\hfill \triangleright$ it is a $1:1$ operation
4: $\qquad (k_1, k_2) \leftarrow p.key$
5: $\qquad key^T \leftarrow (k_2, k_1)$ $\hfill \triangleright$ the key of a transposed subarray
6: $\qquad p^T \leftarrow p.A^T$ $\hfill \triangleright$ transposing a source subarray
7: $\qquad p^T.key \leftarrow key^T$
8: $\qquad \mathbb{D}_T.P \leftarrow \mathbb{D}_T.P \cup \{p^T\}$ $\hfill \triangleright$ add subarray p^T to the dataset
9: $\quad$ **return** $\mathbb{D}_T$
10: **function** PARTIAL-MUL $\hfill \triangleright$ Step 2
11: $\quad \mathbb{D}_M.P \leftarrow \{\}$ $\hfill \triangleright$ interim subarrays, partial aggregates, fig. 4
12: $\quad K_R \leftarrow \{\}$ $\hfill \triangleright$ the set with resulting keys of subarrays ZZ^T
13: $\quad \mathbb{D}_T \leftarrow$ TRANSPOSE ()
14: $\quad$ **for each** $p \in \mathbb{D}.P$ **do**
15: $\qquad (k_1, k_2) \leftarrow p.key$
16: $\qquad P_M \leftarrow \{p : p \in \mathbb{D}_T.P \wedge p.key[0] = k_2\}$ $\hfill \triangleright\ Z^T$ subarrays to multiply on
17: $\qquad$ **for each** $p_M \in P_M$ **do**
18: $\qquad\quad A_M \leftarrow p.A \times p_M.A$ $\hfill \triangleright$ matrix multiplication
19: $\qquad\quad k_3 \leftarrow p_M.key[1]$
20: $\qquad\quad key \leftarrow (k_1, k_2, k_3)$ $\hfill \triangleright$ constructing a unique key of length 3
21: $\qquad\quad \mathbb{D}_M.P \leftarrow \mathbb{D}_M.P \cup \{(A_M, key)\}$
22: $\qquad\quad K_R \leftarrow K_R \cup (k_1, k_3)$ $\triangleright$ save resulting keys to avoid re-computing them
23: $\quad$ **return** $(\mathbb{D}_M, K_R)$
24: **function** AGGREGATE $\hfill \triangleright$ Step 3
25: $\quad \mathbb{D}'.P \leftarrow \{\}$ $\hfill \triangleright$ resulting subarrays, fig. 4
26: $\quad (\mathbb{D}_M, K_R) \leftarrow$ PARTIAL-MUL()
27: $\quad$ **for each** $key \in K_R$ **do** $\hfill \triangleright$ link input interim subarrays
28: $\qquad P_R \leftarrow \{p : p \in \mathbb{D}_M.P \wedge p.key[0] = key[0] \wedge p.key[2] = key[1]\}$
29: $\qquad p_R \leftarrow \sum P_R$ $\hfill \triangleright$ element-wise sum
30: $\qquad \mathbb{D}'.P \leftarrow \mathbb{D}'.P \cup \{(p_R, key)\}$
31: $\quad$ **return** $\mathbb{D}'$

4.3 Computing ZY

In this section we consider the multiplication of arbitrary matrices Z and Y such that the shape of Z is $m \times n$ and the shape of Y is $n \times m$. In addition, the shapes of Z and Y subarrays may differ.

The main idea is to *retile* subarrays of Y to match the shape of Z^T subarrays. Let us call this new array as Y', Fig. 5. After that, we can compute $ZY' = ZY$ using the same algorithm as for ZZ^T, Algorithm 1.

Thick lines in Fig. 5 separate distinct subarrays. We can define retiling as an operation that transforms subarrays of an N-d array to a given shape $s_1 \times s_2 \times \cdots \times s_N$. The basic idea is to cut each $p \in P$ into smaller pieces $P' = \{p' : p' \sqsubseteq p\}$, assign each piece a key, and merge all pieces with the same key into a single, new subarray.

For example, we can cut $p_1 = Y[1 : 2, a : b]$ from $Y.(1,1)$, $p_2 = Y[3 : 4, a : b]$ from $Y.(2,1)$, (which are complete subarrays) and the strips $p_3 = Y[1 : 2, c : c]$ from $Y.(1,2)$ and $p_4 = Y[3 : 4, c : c]$ from $Y.(2,2)$. After that, we can merge p_1, p_2, p_3, p_4 to get $Y'[1 : 4, a : c] = Y'.(1,1)$.

The retiling of Y is just an additional step just as computing Z^T before computing ZY.

We must replace TRANSPOSE with RETILE in Algorithm 1 and again call AGGREGATE to get ZY.

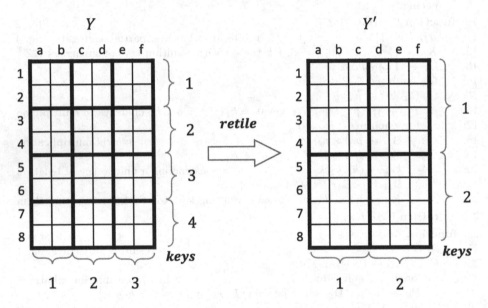

Fig. 5. Retiling $Y \mapsto Y'$ to match the subarray shapes of Z, Fig. 3.

5   Experimental Evaluation and Networking Statistics

Microsoft Azure Cloud was taken for the experiments. Azure cluster creation, scaling up and down with given network parameters, number of virtual machines was fully automated using Java Azure SDK [21]. We used Ubuntu Linux 16.04 LTS. We rented standard D4S V3 machines with 4 CPU cores, 16 GB RAM, and 512 GB local SSD drive. We deployed 4 virtual machines for worker nodes and 1 for the coordinator node. We also used a standard Azure virtual network.

CHRONOSDB is fully written in Java, ran one worker per node on the Oracle-JDK version 14 (64 bit). We used NetCDF file format to store subarrays. For our experiments, without loss of generality, we created several arrays Z and X with varying sizes to run experiments on. Arrays were filled by random numbers. The data type we used was a 16-bit signed integer. A worker node contained

$|P|/4$ subarrays which are evenly distributed after creation, where P is the set of subarrays and 4 is the number of cluster nodes. Resulting subarrays are also evenly assigned to the cluster nodes. When computing $ZX = Y$ (and during any other query execution), the DBMS measures the execution time and collects rich statistics about the network activity which we further analyze.

We evaluated cold query runs (a query is executed for the first time). Respective OS commands were issued to free `pagecache`, `dentries` and `inodes` each time before executing a cold query to prevent data caching at various OS levels. CHRONOSDB benefits from native OS caching and is much faster during hot runs when the same query is executed for the second time on the same data.

Let us show several visualization examples of real execution plans constructed automatically by CHRONOSDB to compute $ZX = Y$. Subarrays of Z, X, and Y are colored in lavender, chocolate, and lime green respectively. Intermediate subarrays used for partial aggregation are colored in black ($\mathbb{D}_M$, Fig. 4). Visualizations are created in Gephi [8]. Figure 6 shows a tiny example where all input and resulting arrays have 4 subarrays.

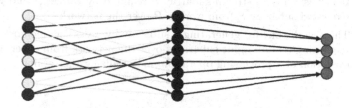

Fig. 6. Generated array DBMS execution plan for $ZX = Y$ (4 subarrays).

Figure 7 shows a portion of the execution plan for large $Z = 8192 \times 4096$ and $X = 4096 \times 8192$ arrays with 64 subarrays each. It is possible to see that Y subarrays (in lime green) are aggregated from 8 intermediate subarrays (in black).

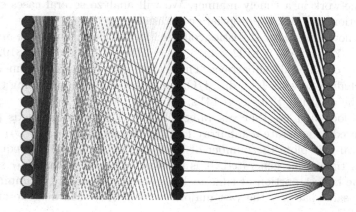

Fig. 7. A portion of the execution plan for $ZX = Y$ (8×8 subarrays).

After the query was executed, our array DBMS provides rich statistics for each subarray, Table 1. Trunning and Tsuccess are general indicators of the array DBMS performance. The smaller the values, the faster array DBMS answers the queries (performs matrix multiplication in our case).

Table 1. Array DBMS execution statistics: runtime and networking

Parameter	Description
Trunning	The time spent for running an operation on a subarray (multiply, aggregate, or transpose in the case of matrix multiplication)
Twait	The time for a subarray waiting before it is scheduled for execution: Some of its input subarrays must arrive from other workers over the network
Tarrived	The time elapsed since worker A received a remote subarray located at worker B over the network after Worker A requested this subarray from worker B
Trequest	The time elapsed since worker A received a message from worker B such that it is ready to stream subarray S required by worker A and worker A requested subarray S from worker B over the network
Tsuccess	The overall time for generating a subarray by a worker (includes requesting remote input subarrays over the network from other workers, execution of a processing operation, materializing the result)

The networking statistics collected by array DBMS makes it possible to answer many important questions and spot interesting insights. For example, large Twait values may indicate load imbalance when some workers already need input subarrays for some of their tasks but remote workers are not ready. Large Tarrived values may indicate network bottlenecks, while small Tarrived values suggest that array DBMS efficiently uses the network. Trequest values suggest whether a worker is overloaded or not. Small Trequest values indicate that a worker is not too busy and can respond to subarray streaming queries over the network in a timely manner. We will analyze several cases of matrix multiplication and make appropriate conclusions.

Let us now explore the statistics collected for the execution of the multiplication $\boldsymbol{ZX} = \boldsymbol{Y}$, where $\boldsymbol{Z} = 8192 \times 4096$, $\boldsymbol{X} = 4096 \times 8192$, and $\boldsymbol{Y} = 8192 \times 8192$. Each array contains 64 subarrays. The overall execution time from issuing a query, including network I/O, disk I/O, computations, to committing the metadata for the resulting $\boldsymbol{Y}$ array is 141 s.

Let us look first on Tsuccess values, Fig. 8. The horizontal axis plots the time required to generate a subarray for $\boldsymbol{Y}$ or an interim dataset $\mathbb{D}_M$ (Fig. 4). The vertical axis plots the number of $\boldsymbol{Y}$ or $\mathbb{D}_M$ subarrays which required this amount of time to be generated. It is possible to conclude that $\mathbb{D}_M$ subarrays require the largest portion of time to be generated. Similarly, Trunning values reveal the same pattern: the vast majority of subarrays can be generated quite fast, Fig. 9.

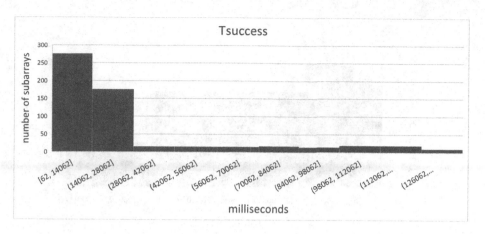

Fig. 8. The distribution of Tsuccess values for $Z = 8192 \times 4096$, $X = 4096 \times 8192$.

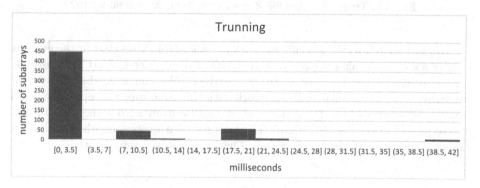

Fig. 9. The distribution of Trunning values for $Z = 8192 \times 4096$, $X = 4096 \times 8192$.

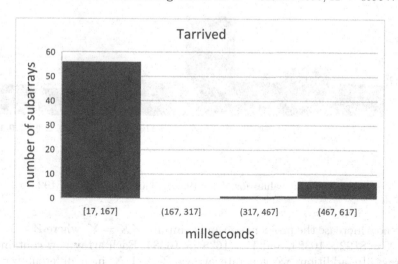

Fig. 10. Tarrived values for $Z = 8192 \times 4096$, $X = 4096 \times 8192$.

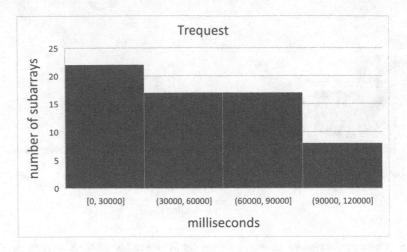

Fig. 11. Trequest values for $Z = 8192 \times 4096$, $X = 4096 \times 8192$.

We now switch to the specific networking statistics. Note that in the current setting a subarray takes approximately 2 MB of disk space in NetCDF3 format. Figure 10 shows the network statistics Tarrived. We conclude that almost all subarrays were streamed over the network with the same speed except some small subset. All workers have the same load as can be drawn from the distribution of Trequest values, Fig. 11. Figure 12 also reveals a quite smooth distribution except some peak at the graph center. This is possibly a set of outliers.

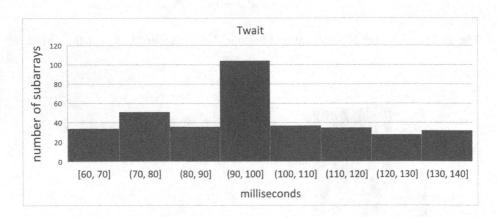

Fig. 12. Twait values for $Z = 8192 \times 4096$, $X = 4096 \times 8192$.

We now increase the problem size and compute $ZX = Y$, where $Z = 16384 \times 8192$, $X = 8192 \times 16384$, and $Y = 16384 \times 16384$. Each array now contains 256 subarrays. In addition, we generate arrays Z and X in a different way. We distribute subarrays of Z among the cluster nodes in a round-robin fashion as previously, but we distribute subarrays of X in a round-robin fashion in the reverse order. Hence, array DBMS will have to perform much more network I/O to generate intermediate and resulting subarrays.

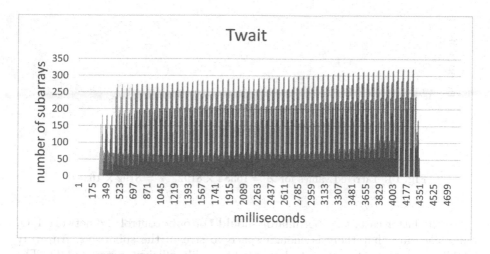

Fig. 13. Twait values for $Z = 16384 \times 8192$, $X = 8192 \times 16384$.

Figure 13 shows the distribution of wait times until all remote input subarrays arrive for a certain subarray. The chart is rather flat since in total the array DBMS has to produce 256 subarrays. The cluster consists of 16 virtual CPUs and we cannot launch the production of all output subarrays in parallel. Hence, many subarrays just wait until they can be produced. We can conclude that the load is evenly distributed since the chart is flat: approximately the same number of subarrays at each time step have to wait, there are no noticeable bottlenecks.

It is interesting to explore the distribution of Tarrived values as they are directly related to the network I/O, Fig. 14. We can see that matrix multiplication is a rather CPU-heavy operation. Under a heavy CPU load, the Tarrived distribution becomes significantly skewed. We assume that this happens due to the following reason. Network I/O is controlled by CPU, so when a busy worker receives a streaming request, it is simply not able to respond in a timely fashion.

We can take into account this lesson in future work. For example, it is possible to stream subarrays beforehand as the execution plan is known in advance. In this case, we could save several seconds of runtime: Tarrived reaches 250 milliseconds for hundreds of subarrays. It might be possible to reduce this value by transferring such subarrays beforehand.

Another interesting possibility, but which is harder to implement, would be organizing some time "windows" when no computations take place. During such windows, CPU will be free to perform only subarray streaming. When there are many subarrays ready for streaming just before such a window, all CPU resources could be invested into the organization of an efficient network I/O.

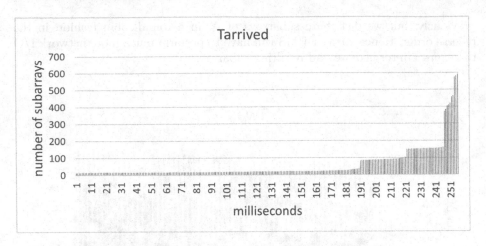

Fig. 14. Tarrived values for $Z = 16384 \times 8192$, $X = 8192 \times 16384$.

In the latter case, the coordinator should not only control the network I/O, but make sure that there is enough space to stream the subarrays. An array DBMS may support several techniques to enable efficient network I/O. The choice between the techniques may be driven by a machine learning approach.

6 Machine Learning Directly Inside Array DBMS

Now we present our vision for the future machine learning R&D directions based on large matrix multiplication directly inside a distributed array DBMS. Dozens of machine learning techniques rely on matrix multiplication. It is crucial for an array DBMS to provide a robust and fast matrix multiplication operator.

Value Prediction. Linear regression serves as a baseline approach to build predictive models in machine learning. Linear regression is easy to build and interpret. In addition, it often performs well in practice. Data scientists often build linear regression models first and then decide whether it is worth trying to build more complicated models [27].

To find the optimal fit for a least squares linear regression model, we can compute

$$(A^T A)^{-1} A^T b \tag{5}$$

where we represent n feature vectors as an $n \times m$ matrix A and n target values as an $n \times 1$ vector b [27]. Note that in order to compute the expression in eq. (5), we must perform several operations of matrix multiplication.

Dimensionality reduction is a fundamental data preprocessing step in machine learning. Singular Value Decomposition (SVD) and Principal Component Analysis (PCA) are popular techniques to reduce the number of input variables by finding dependencies between them. For example, SVD and PCA are often used for analyzing Earth remote sensing data [20].

Let X be an $n \times m$ matrix with column-wise zero empirical mean. Each row of X represents a feature vector. The principal components decomposition P of X can be computed as

$$P = XW \tag{6}$$

where W is an $m \times m$ matrix whose columns are the eigenvectors of X^TX.

To find the singular value decomposition of a real $n \times m$ matrix, we can compute

$$U\Sigma V^T \tag{7}$$

where U is an $n \times n$ matrix, Σ is an $n \times m$ diagonal matrix, and V is an $m \times m$ matrix [27].

Matrix multiplication serves as a basic operation to compute both SVD and PCA, Eqs. (6) and (7). Moreover, SVD and PCA are usually computed in several steps which also use matrix multiplication. Hence, an array DBMS must be able to quickly multiply large matrices.

Previous work proposed to precompute statistics to accelerate PCA. They also identified that matrix multiplication inside array DBMS is useful for linear regression and PCA [17]. We now further complement the list of machine learning approaches that can benefit from having matrix multiplication in array DBMS.

Document clustering, topic modeling & computer vision have tasks that can be solved by a popular machine learning algorithm called Gaussian Non-Negative Matrix Factorization (GNMF) [11]. GNMF is iterative. Given the maximum number of iterations, at each iteration GNMF updates matrices H and W such that

$$H = H * (W^TV/W^TWH) \tag{8a}$$

$$W = W * (VH^T/WHH^T) \tag{8b}$$

where $*$ and/denote cell-wise multiplication and devision respectively. GNMF represents a class of algorithms where the execution plan can be built in advance for several steps ahead. Given the large number of matrix multiplications, an array DBMS must be able to ensure load balance, correct choice of subarray shapes at each iteration, and high-performance network I/O.

Deep Learning is one of the most popular modern machine learning trends. Convolutional neural networks frequently find their way in many real-world applications [12,31]. The output of a fully-connected layer can be generated using matrix multiplication [30]. Each input of such a layer has k elements, the layer has n neurons each with k weights. If m is a sample size, the layer output is the multiplication of an $m \times k$ matrix by a $k \times n$ matrix. Convolution operation can also be reduced to matrix multiplication [30]. Therefore, the computation in the perceptron network is expressed as

$$Y = W^TX \tag{9}$$

where $W \in \mathbb{R}^{(k,m)}$, $X \in \mathbb{R}^{(k,n)}$, and $Y \in \mathbb{R}^{(m,n)}$ [30]. It is obvious that matrix multiplication plays a key role in state-of-the-art neural network architectures.

Networking considerations for future R&D in the area of machine learning in array DBMS include the following. First of all, novel techniques are required to optimize subarray streaming over the network during query execution. Subarray sizes must be dynamically varied to balance computations and network I/O. Finally, novel networking R&D areas such as in-networking computing [34] can be exploited to accelerate matrix multiplication and related operations, e.g. aggregating partial results within a network, rather than by CPU nodes.

7   Conclusions

Computer networks are vital for distributed array DBMS and their massively parallel operations. Machine learning in a distributed array DBMS highly utilizes computer networks. Array DBMS, as other DBMS types, need machine learning algorithms directly inside the system to be able to provide fast analytical services out-of-the-box within a single system. Matrix multiplication is a core component for a vast majority of machine learning algorithms.

We presented novel distributed matrix multiplication approaches that utilize computer networks and array DBMS capabilities: execution plan and massively parallel computations. In addition, we carried out experimental evaluation of our techniques on a computer cluster network in Microsoft Azure Cloud. We collected rich statistics on the running time and network I/O. We analyzed the statistics and reported insights that we found.

Our work may inspire a broad corpus of future research directions. Array DBMS may redesign approaches from high performance computing traditionally held on supercomputers to improve matrix multiplication algorithms on commodity clusters and networks. In addition, array DBMS may enable just-in-time compilation to make computations faster on individual cluster nodes. Furthermore, it would be interesting to devise optimizations possible for larger execution plans including machine learning algorithms for big data on top of distributed matrix multiplication.

References

1. ArcGIS book (2020). https://learn.arcgis.com/en/arcgis-imagery-book/
2. Battle, L., Chang, R., Stonebraker, M.: Dynamic prefetching of data tiles for interactive visualization. In: SIGMOD, pp. 1363–1375 (2016)
3. Baumann, P., et al.: The multidimensional database system RasDaMan. In: ACM SIGMOD, pp. 575–577 (1998)
4. Baumann, P., et al.: Big data analytics for Earth sciences: the EarthServer approach. Int. J. Digit. Earth **9**(1), 3–29 (2016)
5. Choi, D., Park, C.S., Chung, Y.D.: Progressive top-k subarray query processing in array databases. Proc. VLDB Endow. **12**(9), 989–1001 (2019)
6. Cudre-Mauroux, P., et al.: A demonstration of SciDB: a science-oriented DBMS. PVLDB **2**(2), 1534–1537 (2009)
7. Dask (2020). https://dask.org/

8. Gephi (2020). https://gephi.org/
9. Gorelick, N., et al.: Google Earth Engine: planetary-scale geospatial analysis for everyone. Remote Sens. Environ. **202**, 18–27 (2017)
10. Ladra, S., Paramá, J.R., Silva-Coira, F.: Scalable and queryable compressed storage structure for raster data. Inf. Syst. **72**, 179–204 (2017)
11. Lee, D.D., Seung, H.S.: Algorithms for non-negative matrix factorization. In: Advances in Neural Information Processing Systems, pp. 556–562 (2001)
12. Lee, S., et al.: DeepRoof: a data-driven approach for solar potential estimation using rooftop imagery. In: SIGKDD, pp. 2105–2113 (2019)
13. Lewis, A., et al.: The Australian geoscience data cube-foundations and lessons learned. Remote Sens. Environ. **202**, 276–292 (2017)
14. Maxar: 80 TB/day (2017). https://youtu.be/mkKkSRIxU8M
15. Machine learning inside DBMS (2020). https://analyticsindiamag.com/top-databases-used-in-machine-learning-projects/
16. Oracle spatial (2020). http://www.oracle.com/database/technologies/spatialandgraph.html
17. Ordonez, C., Zhang, Y., Johnsson, S.L.: Scalable machine learning computing a data summarization matrix with a parallel array DBMS. Distrib. Parallel Databases **37**(3), 329–350 (2019)
18. Papadopoulos, S., et al.: The TileDB array data storage manager. PVLDB **10**(4), 349–360 (2016)
19. PostGIS (2020). http://postgis.net/
20. Richards, J.A.: Remote Sensing Digital Image Analysis. Springer, Heidelberg (2013). https://doi.org/10.1007/978-3-642-30062-2
21. Rodriges Zalipynis, R.A.: ChronosDB: distributed, file based, geospatial array DBMS. PVLDB **11**(10), 1247–1261 (2018)
22. Rodriges Zalipynis, R.A.: Generic distributed in situ aggregation for earth remote sensing imagery. In: van der Aalst, W.M.P., et al. (eds.) AIST 2018. LNCS, vol. 11179, pp. 331–342. Springer, Cham (2018). https://doi.org/10.1007/978-3-030-11027-7_31
23. Rodriges Zalipynis, R.A.: ChronosDB in action: manage, process, and visualize big geospatial arrays in the Cloud. In: Proceedings of the 2019 International Conference on Management of Data, SIGMOD Conference 2019, Amsterdam, The Netherlands, 30 June-5 July 2019, pp. 1985–1988. ACM (2019). https://doi.org/10.1145/3299869.3320242
24. Rodriges Zalipynis, R.A.: Evaluating array DBMS compression techniques for big environmental datasets. In: 10th IEEE International Conference on Intelligent Data Acquisition and Advanced Computing Systems: Technology and Applications, IDAACS 2019, Metz, France, 18–21 September 2019, pp. 859–863. IEEE (2019). https://doi.org/10.1109/IDAACS.2019.8924326
25. Rodriges Zalipynis, R.A.: BitFun: fast answers to queries with tunable functions in geospatial array DBMS. PVLDB **13**(12), 2909–2912 (2020). http://www.vldb.org/pvldb/vol13/p2909-zalipynis.pdf
26. SciDB GEMM (2020). https://paradigm4.atlassian.net/wiki/spaces/scidb/pages/730268668/gemm
27. Skiena, S.S.: The Data Science Design Manual. Springer, Cham (2017). https://doi.org/10.1007/978-3-319-55444-0
28. Villarroya, S., Baumann, P.: On the integration of machine learning and array databases. In: ICDE, pp. 1786–1789. IEEE (2020)
29. Wang, Y., et al.: SAGA: array storage as a DB with support for structural aggregations. In: SSDBM (2014)

30. Zhang, H., Cheng, X., Zang, H., Park, D.H.: Compiler-level matrix multiplication optimization for deep learning. arXiv preprint arXiv:1909.10616 (2019)
31. Zhang, L., Zhang, L., Du, B.: Deep learning for remote sensing data: a technical tutorial on the state of the art. IEEE Geosci. Remote Sens. Mag. 4(2), 22–40 (2016)
32. Zhao, W., Rusu, F., Dong, B., Wu, K.: Similarity join over array data. In: SIGMOD, pp. 2007–2022 (2016)
33. Zhao, W., et al.: Distributed caching for processing raw arrays. In: SSDBM (2018)
34. Zilberman, N.: In-network computing (2019). https://www.sigarch.org/in-network-computing-draft/

Deep Learning Environment Perception and Self-tracking for Autonomous and Connected Vehicles

Ihab Benamer[1]([⊠]), Arslane Yahiouche[1]([⊠]), and Afifa Ghenai[2]([⊠])

[1] Constantine 2 - Abdelhamid Mehri University, Constantine, Algeria
{ihab.benamer,arslane.yahiouche}@univ-constantine2.dz
[2] LIRE Laboratory, Constantine 2 - Abdelhamid Mehri University, Constantine, Algeria
afifa.ghenai@univ-constantine2.dz

Abstract. Autonomous and Connected Vehicle (CAV) refers to an intelligent vehicle that is capable of moving, making its own decisions without the assistance of a human driver and ensure the communication with its environment. CAVs will not only change the way we travel, their deployment will make an impact on the evolution of society in terms of safety, environment and urban planning. In the automotive industry, researchers and developers are actively pushing approaches based on artificial intelligence, in particular, deep learning to enhance autonomous driving. However, before an autonomous vehicle finds its way into the road, it must first overcome a set of challenges regarding functional safety and driving efficiency.

This paper proposes an autonomous driving approach based on deep learning and computer vision, by guaranteeing the basic driving functions, the communication between the vehicle and its environment, obstacles detection and traffic signs identification. The obtained results show the effectiveness of the environment perception, the lane tracking and the appropriate decisions making.

Keywords: Autonomous and Connected Vehicle (CAV) · Communication · Deep learning · Autonomous driving · Computer vision · Lane tracking · Decision making

1 Introduction

According to statistics estimated in 2019, the World Health Organization deplores 1.35 million loss of human life on road accidents worldwide, a number considered catastrophic in addition to 5.85 gigatons of CO2 emitted by transport road vehicles therefore 74% of all types of transport emissions in the world [1]. The impact of the use of the autonomous and connected vehicles is significant, especially in terms of human and environmental safety. Several efforts have been made to mitigate these risks by automating driving and adding new features to vehicles to ensure the safety of the driver, passengers and the environment surrounding the vehicle and also to make driving more pleasant and easier such as ABS systems, speed regulators, parking assistance, etc. [2].

© Springer Nature Switzerland AG 2021
É. Renault et al. (Eds.): MLN 2020, LNCS 12629, pp. 305–319, 2021.
https://doi.org/10.1007/978-3-030-70866-5_20

However, these efforts are considered insufficient as long as human driver are still taking the control, which has led to thinking of new solutions given the remarkable rise of new technologies in the field of artificial intelligence.

The complete automation of land transport has become the new challenge of the century. The alliance of mechanical, computer and electronic fields has deduced that automating the actions of a driver is the adequate solution to the various problems caused by land vehicles. To imitate the gestures and reflexes of the human driver and therefore make the vehicle autonomous, we must initially extract these gestures and actions which we can call characteristics. The good performance of machine learning algorithms relies heavily on the representation of these characteristics; however, identifying these characteristics remains a very difficult challenge [3].

The goal of this paper is to propose an approach of autonomous driving, which satisfies the three main axes previously mentioned and solves problems related to already existing solutions based on the latest technologies and techniques of deep learning.

The remainder of the paper is structured as follows: the key concepts indispensable to present the proposed environment perception and self-tracking approach are recalled in Sect. 2. Section 3 discusses the related literature on the field of road traffic control and autonomous vehicle control. Section 4 is dedicated to detail our proposed approach. Finally, Sect. 5 concludes the paper and outlines our future work.

2 Basic Concepts

2.1 Autonomous and Connected Vehicle

An autonomous vehicle is a vehicle capable of full detection of the surrounding objects, cars and pedestrians and to fully operate without a human intervention. A human passenger is not necessary taking the control of the vehicle at all times, nor is a human passenger must be present in the vehicle at all. An autonomous vehicle can go anywhere like a traditional vehicle and do everything an experienced human driver can do.

The autonomous vehicle relies on sensors, actuators, complex algorithms, machine learning models and powerful processors. Sensors such video cameras monitor the position of nearby vehicles, detect traffic lights, read traffic signs, track other vehicles and search for pedestrians. A sophisticated algorithm then processes all of these inputs, plan a trajectory, and sends the appropriate commands to the vehicle's actuators, which controls acceleration, braking and steering [4].

A connected vehicle is a vehicle that uses different communication components and technologies to communicate with the driver, other vehicles on the road (vehicle-to-vehicle (V2V)), road infrastructure (vehicle-to-infrastructure (V2I)), pedestrians (vehicle-to-pedestrian (V2P)) and the Cloud (vehicle-to-cloud (V2C)). This components and technologies can be used not only to improve driving safety, but also to improve the efficiency of vehicle, performance and travel times [5].

Therefore, an autonomous and connected vehicle is a vehicle capable of operating automatically without the intervention of a human driver and capable of communicating with the other various driving actors such as vehicles, infrastructures and drivers [6].

3 Related Work

There are several works that address issues related to the controls of autonomous and connected vehicles and the management of road traffic. Each work handles the problem from a specific perspective which gives us the opportunity to study the field in a general way. We present in the following, the most relevant works:

In [7], authors propose an approach that helps optimize road traffic, consisting of both autonomous vehicles and vehicles driven by humans. The traffic management is decentralized and based on controllers with reinforcement learning agents, each responsible for its own control region, but capable of at least partially observing the regions of control of neighboring agents. The inputs of the control system are the following distance and lane placement of the autonomous vehicles and the human-vehicles are uncontrolled. The given driving commands are basic and do not make the vehicle completely autonomous.

In [8], authors propose an algorithm to collect training data from a video game, which has an environment quite similar to the real world. The proposed algorithm collects both the image frame of the game screen and the value of the control key (keyboard key, mouse, joystick...) in order to generate a new dataset for potential machine learning model training. This method can be used to enlarge the actual navigation dataset, but not to fully replace it.

In [9], authors propose an approach to achieve high-level driving automation, which makes the vehicle capable of making decisions and controlling its movements in complex scenarios. The authors propose an architecture based on hierarchical reinforcement learning model for decision making and the control of lane changing.

In [10], authors propose a dynamically reconfigurable system for autonomous driving vehicles which is capable of detecting vehicles and pedestrians in real time. Their approach uses different deep learning-based methods under different lighting conditions to achieve better results.

In [11], authors present a new method of controlling traffic signals using the combination of IoT and video processing techniques. In the models offered, the programming of the traffic lights is determined according to the density and the number of passing vehicles. This method is implemented by a Raspberry-Pi board and the OpenCV tool. This method can be used only in intersections with reduced traffic.

Table 1 summarizes the comparison between the above-mentioned articles.

Table 1. Related work comparison.

	Machine learning	Vehicle connection	Vehicle control	Overall performance
Harshal Maske, Tianshu Chu Uroš, Kalabić [7]	High support	Fully connected	Low control	Medium
Juntae Kim, Geun Young Lim, Youngi Kim [8]	High support	Not connected	Partially controlled	High
Tianyu Shi, Pin Wang, Xuxin Cheng, Ching-Yao Chan, Ding Huang [9]	High support	Not connected	Fully controlled	Medium
Maryam Hemmati, Morteza Biglari-Abhari, Smail Niar [10]	Medium support	Fully connected	Partially controlled	High
Meisam Razavi, Mehdi Hamidkhani, Rasool Sadeghi [11]	Medium support	Not connected	Partially controlled	Medium

After having gathered and analyzed a bench of existing studies in the field of autonomous and connected vehicles, which are the previous five studies, each study deals with the problem from a given perspective that was not addressed in the other studies.

So after having inspired by the solutions proposed in these studies and after doing our own research, we have now a global vision on the problems relating to autonomous driving.

4 The Proposed Approach

4.1 Main Idea

We aim with our work to provide an autonomous driving approach based on modern machine learning technologies and more specifically deep learning.

The proposed approach must address most problems related to the existing works in order to accomplish a system capable of:

– Identify and avoid obstacles (vehicles, animals, pedestrians and other objects).
– Identify and track the lane of the road.
– Identify and classify road signs and make the appropriate decisions.
– Ensure the communication between vehicles, the cloud and infrastructure.

The main idea of our approach is to develop an IoT (Internet of Things) system based on deep learning models to solve the problems related to the autonomous driving of autonomous and connected vehicles [12].

Our driving system make it possible for vehicles to communicate with each other, with the cloud and with road infrastructure. This communication helps to share information related to the state of the road, weather condition, road traffic condition, abnormal driving conditions (e.g. road accidents, closed road ...) and improve performance in order to reduce decision-making time.

The cloud server uses Spring Cloud's technologies to ensure better cyber-security, a light communication based on Rest micro-services which helps to log all the system states and all the different driving commands, the server has a discovery service to ensure system scalability and the ease of integration for new vehicles [13].

Our system uses a computer vision algorithm and two deep learning models based on the YOLOv4 algorithm to operate:

• The first is a pre-trained model used to detect the various objects spotted by the vehicle sensors, then to classify these objects in categories (pedestrians, vehicles, obstacles ... etc.).
• The second model is a traffic sign detection and classification model.
• In addition, we have proposed an algorithm for detecting and tracking the lane of the road [14].

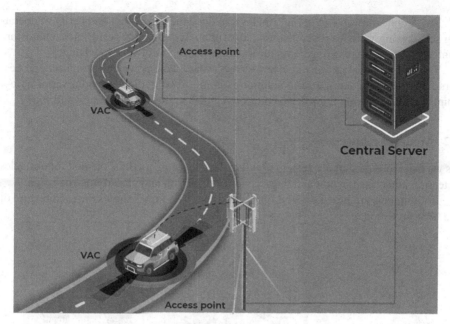

Fig. 1. The concept of the proposed approach.

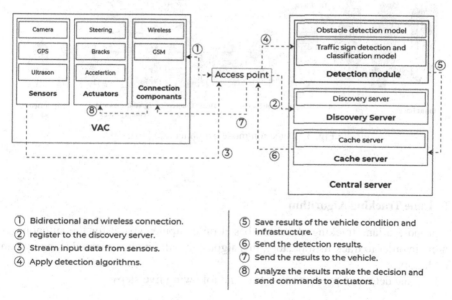

① Bidirectional and wireless connection.
② register to the discovery server.
③ Stream input data from sensors.
④ Apply detection algorithms.

⑤ Save results of the vehicle condition and infrastructure.
⑥ Send the detection results.
⑦ Send the results to the vehicle.
⑧ Analyze the results make the decision and send commands to actuators.

Fig. 2. The system architecture.

The autonomous vehicle can make its own decision based on the results of Deep learning models, computer vision algorithm results and sensors inputs to control its direction, acceleration and braking as shown in Fig. 1 and Fig. 2.

The architecture shown in Fig. 1 helps to reduce the hardware complexity of operating self-driving vehicles (deep learning algorithms require high performance hardware configurations) and real-time mass information sharing with very low latency.

The system uses centralized cloud servers which have powerful computing processors where most of our software solution is going to be deployed, and a set of access points installed around the road to connect vehicles to the server as shown in Fig. 2.

4.2 Models Architecture

As we described earlier, our system uses two deep learning models: the first for obstacles detection and classification and the second for detecting and classifying road sign.

Both models are based on YOLOv4 algorithm. The Fig. 3 shows the architecture of the YOLOv4 model:

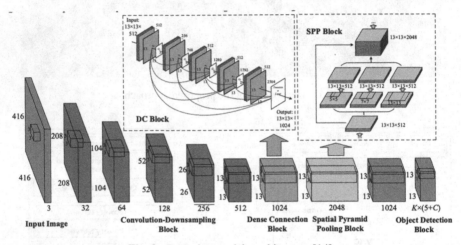

Fig. 3. Detection models architecture [14].

4.3 Lane Tracking Algorithm

We propose a lane tracking algorithm based on computer vision for the autonomous vehicle in order to track its trajectory, this algorithm will be implemented using python CV2 library.

The lane detection process consists of the following five steps:

Camera Calibration

Most cameras use lenses for illuminating the image to help focus and for other processing, but these lenses as a result will bend the edges of the image due to refraction. This phenomenon called radial distortion leads to an effect that distorts the edges of the images.

Fig. 4. Example of camera calibration.

So we need to solve the radial distortion problem before continuing with lane detection. The Fig. 4 shows an example of radial distortion calibration.

Perspective Transformation

The second step is to align our visual systems to visualize the road ahead in a way that is like a bird view, this will help calculate the curvature of the road, therefore, will help us predict the steering angle for the next few tens of meters.

The Fig. 5 represents the post-transformation input and output image respectively.

Fig. 5. Bird view perspective transformation.

Limitation

The next step is to segment the input image. The input image in most cases composed by three channels: Red, Green and Blue (RGB), the pixels colors that interests us are either white or yellow, based on this assumption, the input image can be converted to a gray image at one single channel (grayscale), thus eliminate channels that we do not need. Next, we apply Canny edge detection algorithm. The Fig. 6 shows the result:

Lane Pixel Scanning

Once the input image is preprocessed, the next step is to locate and map the lanes in the image space. The approach would be to plot a histogram of pixels that are not equal to zero (the whitest pixels) in the lower half of the binary image.

Then the concept of the sliding window approach will be applied here, which is essentially a window with a margin placed around the center of the row (one for each i.e. left and right lanes) as shown in Fig. 7.

Fig. 6. Gray scaling & Canny edge detection result. (Color figure online)

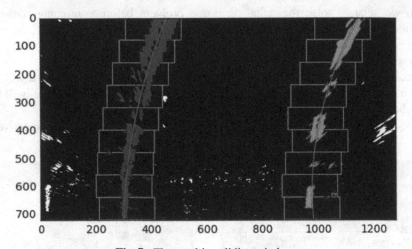

Fig. 7. The resulting sliding windows.

Recognition of the Lane Area

The last step would be to highlight the lane area to track by fitting a polygon between the edge points of the limited area and projecting it back onto the original image, as shown in Fig. 8. The lane tracking area and radius of curvature were calculated in image space based on pixel values which are not the same as real world space, therefore, they need to be converted to real world values, and this involves measuring the length and width of the section of track onto which we are projecting our distorted image.

4.4 Detection Model

In general, there are three main steps to create a deep learning model: data preparation, model generation, and model deployment.

- Data preparation focuses on preparing data for model training and testing, covering topics such as finding datasets, recording data, labeling images, data augmentation etc.
- Model generation involves the development of the network architectures, the training of network and the evaluation of trained models.

Fig. 8. The final result.

– The trained model is then pruned and optimized for specific target hardware during the deployment phase.

Data Preparation

In this section we present the data sets used and the preprocessing required for the implementation of the models, in our approach we used two models and therefore we have to present the two datasets used.

Microsoft COCO (Common Objects in Context): is a dataset with the aim of advancing the state of the art of object recognition by matching object recognition to scene understanding. This is achieved by putting together images of complex everyday scenes containing common objects in their natural context. Objects are labeled using instance segmentations to facilitate precise location of objects.

The dataset contains photos of 91 types of objects (see Table 4) that would be easily recognized by a 4-year-old.

With a total of 2.5 million instances tagged in 328k images, the creation of the dataset relied on the extensive involvement of workers with new user interfaces for category discovery, instance discovery and segmentation of instances [15].

RTSD (Russian Traffic Sign Dataset): is a new public dataset of traffic sign images. The dataset is intended for training and testing of traffic sign recognition algorithms. We describe the structure of the dataset and the guidelines for working with the dataset. The evaluation of modern detection and classification algorithms conducted using the proposed dataset has shown that existing methods of recognizing a large class of traffic signs do not achieve accuracy and completeness required for a number of applications.

The images are obtained from a wide screen digital video camera which captures 5 images per second.

The image resolution is between 1280 × 720 and 1920 × 1080. The images are captured at different seasons (spring, autumn, winter), at different hours of the day (morning, afternoon, evening) and in different weather conditions (rain, snow, bright sun) [16].

The dataset uses coco format labels which use 5 variables for each bounding box: The X position of the frame, the Y position of the frame, the height of the frame, the width of the frame and the traffic sign class.

As we are going to use the algorithm of YOLOv4, the labels must be in the format of YOLO which is: the class identifier, the central point X normalized between 0 and 1, the central point Y normalized between 0 and 1, the height normalized between 0 and 1, the normalized width between 0 and 1.

Apart from the annotations, there are some necessary files related to the creation of our model:

- classes.names: containing the categories of objects, each row contains a category, the dataset comes with non-significant names and therefore for each class we associate the name of the corresponding sign (e.g. stop sign).
- train.txt: we need to divide our dataset into two subsets: the first subset for model training and the second subset for validation. This file contains the names (without extensions) of the images that will be used for model training.
- val.txt: contains the rest of the images that will be used for the model training validation. This file has the same format as train.txt
- obj.data: this file will contain the following information: number of classes, the path of train file, the path of the valid file, the path of the names file and backup path.
- Yolo.cfg: contains the architecture of the model.

Model Training

After preparing the data needed for training and testing process, the next step is to generate the detector model. The following sections present how we generated this model.

Creation of the Darknet Environment

We used the Darknet framework for model building and so we need to build the Darknet environment from its Github repository [17].

CPUs are designed for more general computing workloads. GPUs on the other hand are less flexible, but GPUs are designed to calculate the same instructions in parallel. Deep Neural Network DNNs are structured very uniformly so that at each layer of the network thousands of identical artificial neurons perform the same calculation. Therefore, the structure of a DNN matches quite well the types of computations that a GPU can do efficiently.

So, we must enable graphics card support to build Darknet so that we can then use the Darknet executable to run and train the object detector.

Creating Configuration File

The YOLOv4 algorithm needs some specific files to know how and what to train. We have already presented in the previous section the contents and how to create the obj.data and classes.names files. We still have to create yolo.cfg file.

The configuration file (yolo.cfg) is used for model generation, this file contains the model architecture as well as the information needed for the model training.

Files Verification

After creating the configuration files, we need to verify that each dataset file or configuration file is present and located in the right folder. Verifying configuration files can be done manually, but for verifying the dataset files we need python script to automate the verification.

Model Training

Now we finally get to generate the model. The training process can take a several hours to several days. After every 1000 epochs, the network weights (parameters values) are saved in the backup folder. This gives us the opportunity to start from there in case of any failure.

Model Deployment

Once the traffic sign detector model has been successfully trained, we get an obj_last.weights file which contains the weight parameter values for the neural network. For the obstacle detector model we already have the pretrained weights file.

The file format .weights is a proprietary format for Darknet framework, in order to make deployment easier, we need to convert it to a more user-friendly format, so we used the Tensorflow framework for that and therefore we need to convert the .weight file to a .tf file (Tensorflow format).

Our approach is based on a cloud environment for the deployment of the models and for communication and data exchange between the different elements of the system (vehicle, infrastructure…).

The final step is to create this cloud environment, so we used Spring cloud which make it easy to implementing a discovery service for the various micro-services, centralize the configuration and the processing, establish a light and reliable connections and to persist the different states of the cloud system, infrastructure and autonomous and connected vehicles.

Our System is divided into 4 levels:

– Discovery server (Central server).
– Gateway server (API gateway proxy).
– Detection server.
– Client applications.

Client Applications

Client applications are the applications installed in each autonomous and connected vehicle, its role is to perceive the environment and to pass the driving commands (target speed, steering angle, etc.) to the actuators.

Client applications are a python application that receive the sensor data to apply the lane detection algorithm, and at the same time pass the inputs from the video sensors to the detection server after registering with the discovery server. These applications also communicate with the central server to get the current traffic status, weather condition road condition and other information to make the best decision.

Discovery Server
The detection server is where we will deploy the detector models, this server receives the live video streams from the client applications (vehicles), apply the detection algorithms and then send the results via micro-services to the central server for the persistence and to the vehicles to make the appropriate decisions.

The detection server is a python application that uses the Flask framework to create APIs.

Gateway Server (API Gateway Proxy)
Provide a common interface for micro-services to discover and communicate with each other.

The Gateway Server is a Spring boot application that uses Netflix Eureka Client technologies.

Discovery Server (Central Server)
Contains the global configuration of the entire system, this server acts as a directory that other applications (micro-services) can discover and register through the proxy server (API gateway).

The Discovery Server is a Spring boot application that uses Netflix Eureka technologies.

4.5 Obtained Results

In this section, we discuss the obtained results after training the model and the overall performance of the software solution.

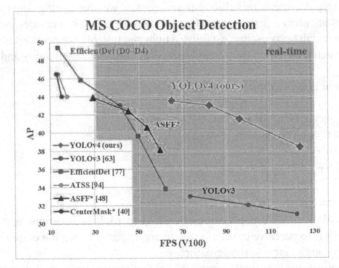

Fig. 9. Comparison between the different detection algorithms [18].

For the obstacle detection model, we could see the model performance from the article by YOLOv4 which shows a comparison between yolov4, its previous version and other popular detection algorithms based on the model performance and the frames per second (FPS).

As shown in Fig. 9, YOLOv4 achieved an AP (Average Accuracy) value of 43.5% (65.7% AP50) on the MS COCO dataset and a real-time speed of 65 FPS using the graphics card Tesla V100 chart, beating the fastest and most accurate detectors in terms of speed and accuracy. YOLOv4 is twice as fast as EfficientDet with comparable performance.

Based on the graph (Fig. 10) obtained after the end of the training cycle of our classifier model for traffic signs, we can see that we get a decent result despite not having done the whole training cycle (we let's train the model for only 10,000 iterations).

The loss value is less than 0.2 which is a good sign, on the other hand the validation graph shows that we have a good value for the validation loss which is up to 89%.

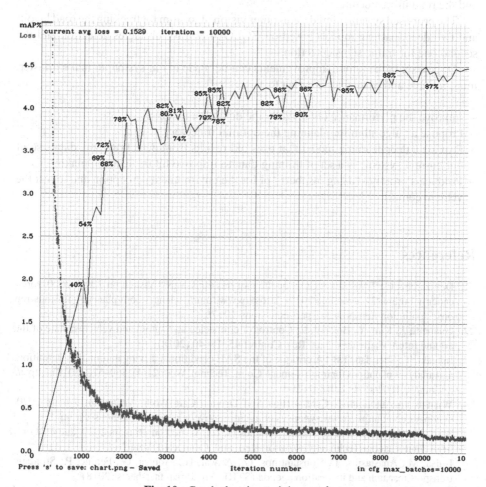

Fig. 10. Graph showing training results.

5 Conclusion

Artificial intelligence plays a major part in the development of the self-driving vehicles. The performance of machine learning and especially deep learning has made the virtual reality. However, as the autonomous and connected vehicles remain a recent field, there still persists a lack of optimization and innovation.

In the literature, many works have been addressing the problems relating to autonomous driving. Each work is focused on a specific perspective some limitations that must be overcome. To work around these issues, we came up with a solution based on deep learning and computer vision in order to take into account three pillars, the interconnection via micro-services between different vehicles, the cloud, pedestrians and the road infrastructure.

The proposed system offered a computer vision algorithm for tracking the road layout and two deep learning models based on YOLOv4 one for obstacles detection and the second one for traffic signs detection.

Of course, in the future, improvements can be made to our system to make it faster, more reliable and adaptable. We plan to:

- Improve the performance obtained by the detection models.
- Propose an algorithm for planning the optimal trajectory.
- Provide a deep learning model that uses old results from detections (the database of our current system) as a set of training data to improve future decision making.
- Produce a prototype using raspberry pi for the environment perception and to test the vehicle control.

References

1. Deluzarche, C.: Transport et CO2: quelle part des émissions ? Visited on 7 août 2020, robo-cademy Futura (27 novembre 2019). https://www.futura-sciences.com/planete/questions-rep onses/pollution-transport-co2-part-emissions-1017/
2. Guanetti, J., Kim, Y., Borrelli, F.: Control of connected and automated vehicles: state of the art and future challenges. Ann. Rev. Control **45**, 18–40 (2018)
3. Inagaki, T., Sheridan, T.B.: A critique of the SAE conditional driving automation definition, and analyses of options for improvement. Cogn. Technol. Work **21**(4), 569–578 (2018). https://doi.org/10.1007/s10111-018-0471-5
4. What is an Autonomous Car? – How Self-Driving Cars Work. Visited on 9 mars 2020, synopsys.com: https://www.synopsys.com/automotive/what-is-autonomous-car.html
5. Staff, C.: Connected and automated vehicles (s. d.). Visited on 16 mars 2020, autocaat.org. https://autocaat.org/Technologies/Automated_and_Connected_Vehicles/
6. Luisa, M., Anastasios, T., Konstantinos, G., Anwar, H., Monica, G., Ferenc, P.: Strategic transport research and innovation agenda (STRIA) roadmap factsheets (2018)
7. Maske, H., Chu, T., Kalabić, U.: Large-scale traffic control using autonomous vehicles and decentralized deep reinforcement learning (2019)
8. Kim, J., Lim, G.Y., Kim, Y.: Deep learning algorithm using virtual environment data for self-driven car (2019)

9. Shi, T., Wang, P., Cheng, X., Chan, C.-Y., Huang, D.: Driving decision and control for automated lane change behavior based on deep reinforcement learning (30 octobre 2019)
10. Hemmati, M., Biglari-Abhari, M., Niar, S.: Adaptive vehicle detection for real-time autonomous driving system (2019)
11. Razavi, M., Hamidkhani, M., Sadeghi, R.: Smart traffic light scheduling in smart city using image and video processing (2019)
12. Kumar, S., Tiwari, P., Zymbler, M.: Internet of Things is a revolutionary approach for future technology enhancement: a review. J. Big Data **6**(1), 1–21 (2019). https://doi.org/10.1186/s40537-019-0268-2
13. Christudas, B.: Spring cloud. In: Practical Microservices Architectural Patterns. Apress, Berkeley (2019)
14. Bochkovskiy, A., Wang, C.-Y., Liao, H.-Y.M.: Yolov4: optimal speed and accuracy of object detection. arXiv preprint arXiv:2004.10934 (2020)
15. Lin, T.-Y., et al.: Microsoft COCO: common objects in context. In: Fleet, D., Pajdla, T., Schiele, B., Tuytelaars, T. (eds.) ECCV 2014. LNCS, vol. 8693, pp. 740–755. Springer, Cham (2014). https://doi.org/10.1007/978-3-319-10602-1_48
16. Shakhuro, V., Konushin, A.: Russian traffic sign images dataset. Comput. Opt. **40**, 294–300 (2016)
17. Joseph, R.: Darknet: open source neural networks in C (2016)
18. Bochkovskiy, A., Wang, C.-Y., Liao, H.-Y.M.: Yolov4: optimal speed and accuracy of object detection. Digital Image (2020)

Remote Sensing Scene Classification Based on Effective Feature Learning by Deep Residual Networks

Ronald Tombe and Serestina Viriri[✉]

School of Mathematics, Statistics and Computer Science,
University of KwaZulu-Natal, Durban 4000, South Africa
viriris@ukzn.ac.za

Abstract. Remote sensing image scene interpretation has many applications on land use land covers; thanks to many satellite technologies innovations that generate high-quality images periodically for analysis and interpretation through computer vision techniques. In recent literature, deep learning techniques have demonstrated to be effective in image feature learning thus aiding several computer vision applications on land use land cover. However, most deep learning techniques suffer from problems such as the vanishing gradients, network over-fitting, among other challenges of which the different literature works have attempted to address from varying perspectives. The goal of machine learning in remote sensing is to learn image feature patterns extracted by computer vision techniques for scene classification tasks. Many applications that utilize data from remote sensing are on the surge, this include, aerial surveillance and security, smart farming, among others. These applications require to process satellite image information effectively and reliably for appropriate responses. This research proposes the deep residual feature learning network that is effective in image feature learning which can be utilizable in a networked environment for appropriate decision making processes. The proposed strategy utilizes short-cut connections and mapping functions for deep feature learning. The proposed technique is evaluated on two publicly available remote sensing datasets and it attains superior classification accuracy results of 96.30% and 92.56% respectively on the Ucmerced and Whu-siri datasets, improving the state-of-the-art significantly.

Keywords: Remote sensing · Inception architecture · Batch normalization · Activation functions · Residual feature learning

1 Introduction

Remote sensing is a topic of great interest owing its capability to power many applications such as, change detection and classification [1,2], scene labeling [5], image recognition [6], scene parsing [13], feature localization [14] street scene

© Springer Nature Switzerland AG 2021
É. Renault et al. (Eds.): MLN 2020, LNCS 12629, pp. 320–336, 2021.
https://doi.org/10.1007/978-3-030-70866-5_21

segmentation [15] among other applications. There are huge volumes of image data that have been collected due to the advancements of satellite and remote sensing sensor technologies. The remotely sensed images contain diverse semantic information on land cover land use. The key problem in remote sensing is to develop algorithms for effective image feature processing to recognize and classify scene images into correct categories independent of scale, illumination, color, clutter, pose positions among the other image features. An interpretation of remote sensing images is a challenging undertaking whose implications are of significance to the many land use applications. The fundamental question is, how can a remote sensing system learn image feature representations effectively given the ever-increasing volumes of image data which are periodically generated with sensor and satellite technology?

To achieve high scene classification accuracy results on remote sensing applications, effective image feature extraction is a key prerequisite [2] before the classification tasks. Most recent literature of remote sensing [1, 2] demonstrate that state-of-the-art classification-results are attained with the deep learning feature extraction techniques. Deep learning [16] provides an architecture for models that are made up of multiple processing layers to learn feature representations of data with many levels of abstractions. Deep learning uncovers complex structures in large datasets by applying effective algorithms to show how a machine should modify its internal parameters that apply to compute the image feature representations, in each layer depending on those in a former layer. The effectiveness of deep learning in feature characterization is apparent in current literature [1, 2] on remote sensing image scene classification tasks. Deep convolutional neural network architectures [23, 24] combine low, mid, and high-level features; extracted by lower, mid, and upper layers of the network respectively. Literature evidence [23, 24] show that the network-depth is of significance in feature-representation thus yields superior classification results [1, 3] on several image datasets. Since deep learning models attain superior classification results, a question arises, is better image-feature learning-dependent entirely by stacking several layers? An interesting finding from literature [2, 6, 24] is that the classification performance with deep learning model seem to depend on their architectural designs. The vanishing/exploding gradient is another problem to deep learning models which hinders network convergence [25, 26]; the layer normalization [28] strategy mitigates [6] this problem. Further, the network degradation problem which is apparent with the increase in depth of the deep learning network models has been proven to be drastically alleviated through adjustments of the network weights by gradually varying the learning rates [29] as the network depth increases. Furthermore, in their work [30, 33], they show that network overfitting is not as a consequence of network-degradation, but rather a network design issue.

The improvements in deep learning network architecture design's quality improve the performance of many computer vision tasks that rely on well-learned visual-features [24]. The VGGNet [44] has an impressive architecture on feature representation, however, evaluating this network architecture entails high computation cost [24]. The inception network [23] has been utilized in big-data applications [31,32] for processing huge volume of data with less computational costs. The inception network is complex to modify on different use-cases, for instance, if it is scaled up without considerations of its computations parts, the computation cost can increase significantly, this scenario is elaborated further in [24]. The residual learning [6] achieves superior accuracy in the ImageNet Challenge [6] with a lower computational cost (that is, quick convergence) hence faster training. Inspired with this finding, this work proposes a deep-residual-feature-learning network that is effective in image feature learning in the context of remote sensing. The operation mechanisms of the proposed deep residual feature learning are grounded on the various literature philosophies, i.e., the inception architecture [21], batch normalization [28], residual connections [6], and network layer weight's adjustments [29]. The subsequent sections of the paper are structured as follows: Sect. 2 provides a concise and comprehensive literature review on works related to this work, Sect. 3.1 presents this paper's proposed deep residual feature learning method while Sect. 3 discusses the materials and methods which apply in this work. Section 4 presents the results, analysis and discussions. Finally, Sect. 5 concludes this paper.

2   Related Work

This section reviews literature works that are related to this work from seven aspects; this entails 1. inception architecture, 2. activation functions, 3. batch-Normalization, 4. weight initialization, 5. residual feature learning, 6. residual mapping functions, 7. Machine learning, remote sensing, computer vision, and networking.

2.1   Inception Architecture

Result of Arora et al. [34] states that for a dataset to be representable enough with the probability distribution of a very sparse deep-network, then the network architecture can be assembled layer after another through the analysis of correlation statistics for the predecessor layer activations and neurons with very correlated outputs. This statement resonates with the Hebbian principle [35]: neurons that fire together, wire together. Achieving this requires to establish how the structure of an optimum local sparse convolutional-vision network may be approximated and represented with existing dense components [23]. An essential need is the determination of an optimum local architecture which repeats its spatiality, following the claim [34]. Correlated cluster groups form high-correlated units. Units from the initial layers are connected to subsequent ones. It is taken that every unit from the initial layer corresponds to a given region of an input image which groups to feature maps. The fundamental idea of the inception architecture [21] is the introduction of sparsity by replacing fully connected layers with sparse ones, even within the convolutions. This forms "Inception-modules"

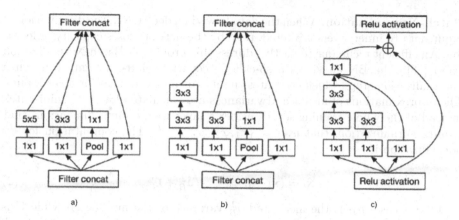

Fig. 1. (a) inception-module-V2 with dimensionality-reduction [23] (b) Enhanced inception module-V3 for computer vision [24] (c) Inception module-V4 of residual connections for learning [21]

and they are stacked layer-wise. Given that their output-correlation statistics can vary since high abstraction of features is captured with higher layers, there is a decrease of spatial-concentration. This necessitates feature embedding in dense compressed form to reduce high dimensionality problem [23]. Szegedy et al. [23] apply 1×1 convolutions for dimensionality reduction prior to utilizing the costly 3×3 and 5×5 convolutions. Additionally, the inception architecture uses the rectified linear unit (ReLU) activation's with the convolutions (see Fig. 1(c)) to improve on the network sparse representation. Figure 1 depicts the various versions of the inception architecture. Inception networks are generally very deep and they do replace a filter integration phase of inception architectures with residual connections. This allows the inception to take advantage of the residual strategy whereas it retains its computation efficiency [21].

2.2   Activation Function: Rectified Linear Unit

The use of an appropriate activation function can greatly improve the CNN performance [21]. The rectified linear unit (ReLU) [42] is a common non-saturated activation function which is defined as:

$$f_{h,i,j} = max(z_{h,i,j}, 0) \tag{1}$$

In this case, $z_{h,i,j}$ is the activation function's input at position (h, i) on the jth channel. The ReLU function retains the positive part and it discards the negative part. The operation max(.) of the ReLU computes faster than tanh or sigmoid functions. Further, it introduces sparsity to the units of the hidden layer thereby allowing the network to achieve sparse representations.

Batch-Normalization. When the data flow via a deep CNN, the distribution of input-data to inner layers is altered, thereby the network loses capacity of learning. An efficient technique to partly address this problem is Batch-Normalization (BN) [28]. The BN alleviates a so-called "covariate-shift-problem" through a "normalization-step" which estimates the means and variances for layer inputs. The approximations of means and variances are calculated for every mini-batch instead of the whole training set. Consider a layer with k input dimensions that require normalization, that is, $x = \{x_1, x_2, \ldots, x_k\}^T$. The computation for l^{th} normalization is:

$$\hat{\mathbf{x}}_l = (\mathbf{x}_l - \mu_B)/ \sqrt[root]{\delta_B^2 + C} \tag{2}$$

In this case, μ_B is the mean and δ_B^2 variance of the mini-batch while C is a constant. The normalized input $\hat{\mathbf{x}}_l$ representation power is improved through transformation to:

$$y_l = \mathbf{BN}_{\beta,\gamma}(\mathbf{x}_l) = \gamma \hat{\mathbf{x}}_l + \beta \tag{3}$$

The parameters β and γ are learned. Batch-normalization is advantageous to global data-normalization in many ways. First, it minimizes the inner layers covariant-shift. Second, Batch-normalization minimizes the reliance of gradients on the ratio of parameters, this yields a beneficial result on the flow of gradient through CNN. Additionally, batch-normalization regularizes the model therefore there is no need for Dropouts. Finally, with BN, the use of saturating nonlinear functions is possible as the BN mitigates risks of getting stuck into a model-saturated-state.

2.3   Weight Initialization

Overfitting is a major problem in deep CNN that can be minimized by network parameter regularization. A Deep convNet contains a massive number of parameters and it is difficult to train it given the non-convex nature of its loss function [17]. To attain quick convergence in training while avoiding the vanishing gradient problem, an appropriate weight initialization of the network is a key prerequisite [18,19]. Appropriate weight initialization can ensure attainment in breaking the symmetry between inner units of the same layer and the bias parameter can be set to zero. Poor network initialization, for instance, if every layer scales their input with a factor k, the resultant output scales the original inputs by k^L, L denotes the number of network layers. It follows that the value for $k > 1$ leading very large values in output layers whereas if $k < 1$ results to the problem of diminishing gradients and output values. [22] uses a Gaussian distribution with a zero-mean and standard deviation of 0.01 for initialization. Xavier [26] apply Gaussian distribution with a variance of $2/(n_{input} + n_{output})$ and a zero-mean. In this case, n_{input} denotes the number of neurons inputs while n_{output} depicts the number of output neurons that feed to another layer. Therefore, the weight initialization [26] automatically establishes the initialization scale from the number of inputs and outputs. Caffe [20] is a variant of the

Xavier method and it applies the n_{input} only, thus its easier to implement. He et al. [29] extends the Xavier model and derives a robust weight initialization method that considers ReLU. Their method [29] facilitates training of very deep models [23] which converges whereas with the Xavier technique [26] it does not converge.

2.4   Residual Feature Learning

In image classification, the integration of local image descriptors form compact encoded feature representations, that is, Vector of Locally Aggregated Descriptors (VLAD) [36] that can be seen as visual vocabularies of a dictionary [38]. Fisher Vectors [39] can also be thought of as probabilistic representations of VLAD. These aggregation strategies are very powerful for shallow image feature representation in image classification [40]. With vector quantization, residual vector encoding [27] demonstrates to be more effective [6] as compared to original vector encoding.

2.5   Residual Mapping Functions

To address the network degradation problem, [6] proposes explicit fitting of residual mappings to layers instead of the implicit stacking of layers which assumes "desired fits of underlying mappings". Formally, the "desired-underlying mapping" is depicted as $H(x)$, where the stacked-nonlinear layers fits a different mapping of $F(x) := H(x) - x$ [6]. This means that original mapping can be reformulated to $F(x) + x$, thus it is simpler to optimize the residual mapping function compared to optimizing the original and un-referenced mapping. The formulation of residual mapping $F(x) + x$ is realizable through the feed-forward-networks with "shortcut-connections" Fig. 2. The shortcut-connections [21,41] skips one or more network layers. Their purpose is to achieve "*identity*" mapping while their outputs are included in the stacked layer's outputs, see Fig. 2. The advantage of identity shortcut-connections mappings is that they add no extra computational cost or parameter.

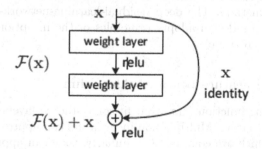

Fig. 2. A building block for residual learning [6]

2.6   Machine Learning, Remote Sensing, Computer Vision, and Networking

Machine learning [7] enables computer systems to process data and make deductions based on knowledge learned. Pattern recognition problems that leverage machine learning include, classification, clustering, and regression [7]. The softmax logistic regression [9], and support vector machine [8] are the popular machine learning techniques utilized in scene classification of remote sensing images. The goal of machine learning in remote sensing is to learn image feature patterns extracted by computer vision techniques for scene classification tasks. Of recent years, machine learning applications which utilize remote sensing data are on the surge, this include, smart farming [10], crop diseases/pests [11], aerial surveillance and security [12], among others. These applications require to process satellite image information effectively and reliably for appropriate responses. Convolutional neural networks extract image features effectively [2], and this information should be transmittable to various monitoring centers reliably through computer networks. This work focuses on effective feature processing of satellite images which can be utilizable in a networked environment for appropriate decision making processes.

3   Materials and Methods

This section presents and discusses the operation mechanism of the deep residual feature learning method. Further, the section discusses properties of remote sensing datasets which apply in evaluating the effectiveness of the proposed deep residual feature learning technique. Furthermore, details on the experimental procedures are given in this section.

3.1   Residual Feature Learning Network

This work develops and presents an algorithm and network architecture that utilizes the principle of residue learning [6] in remote sensing images scene classification. Figure 3 shows the architecture and algorithm of the proposed deep residual learning method. The deep residual-learning-network-architecture has 20 layers, further, its design adopts principles of the inception module [21,24] (depicted in Fig. 1; b and c).

3.2   A Residual Feature Learning Algorithm

Consider a mapping function $H(x)$ that fit some stacked layers, where x depicts the input into first layers. Multiple nonlinear layers can approximate asymptotically functions which are complex [6], similarly, they can approximate asymptotically the residual functions, that is, $H(x) - x$ (for same input and output dimensions). Therefore, instead of stacked functions approximating $H(x)$, they approximate a residual function $F(X) := H(x) - x$, this can be reformulated to

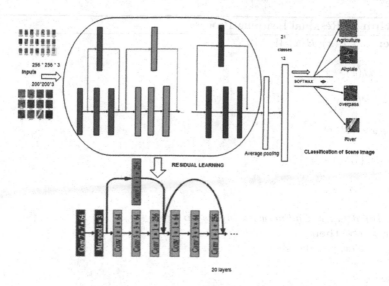

Fig. 3. Deep residual learning network architecture for remote sensing image scene classification

$F(x) + x$ (a form that is equivalent to the original function $H(x)$). The reformulation $F(x) + x$ addresses the network degradation problem. This means that, with added network layers configured as identity-mappings, more deep models should contain same training errors as the shallower counterpart network models. The residual learning algorithm learns the residual mapping building blocks Fig. 2 which as formulated as:

$$y = F(x, \{W_i\}) + x. \tag{4}$$

where x and y are vectors for the input and output of the considered layers. Whereas $F(x\{W_i\})$ is residue mapping function that is learned; for stacked layers (e.g. Fig. 2) where they are two, $F = W_2\sigma(W_1x)$, in this case σ depict a nonlinear learning function ReLU [42]. The residual mapping operation $F + x$ is conducted via short-cut connection with element wise addition. And for subsequent stacked layers with shortcut connection, where the dimensions of F and x are different, a linear projection W_s is performed with the shortcut-connections so as to achieve dimension matching:

$$y = F(x, \{W_i\}) + W_s x. \tag{5}$$

The function $F(x, \{W_i\})$ represents many convolutional layers.

Algorithm 1. Residual Learning Process
___
Require: $x, y, F(x), H(x), W_i, W_s$

$F(x) \leftarrow H(x) - x$

$H(x) \leftarrow F(x) + x$

$F(x) + x \leftarrow H(x)$

$y \leftarrow F(x, \{W_i\}) + x$

$F(X) \leftarrow F(x) + x$

while $input\_layer < no\_network\_layers$ **do**
   if $x \neq y >$ **then**
      $y \leftarrow F(x, \{W_i\}) + W_s x$

   else
      $y \leftarrow F(x, \{W_i\}) + x$

      $F(X) \leftarrow F(x) + x$

   end if

end while
___

3.3 Deep Residual Feature Learning Architecture

Figure 3 depicts the residual network learning architecture which is motivated by the VGG net [44] philosophy and deep residual learning [6] works. The convolutional layer comprises 3×3 filters that apply two design principles: 1) for outputs of feature maps with the same size, the network layers comprise the same filter numbers. 2) when sizes of the feature maps are halved, the filter numbers are doubled to maintain the time complexity in every layer. Downsampling is done on convolutional layers with strides of 2. Shortcut connections are placed in Fig. 2 to the network layers turning it to a residual learning network; Eq. (4) achieves this. Equation (5) performs shortcut projections for dimension matching (normally done with 1×1 convolutions).

3.4 Remote Sensing Datasets

This section discusses the properties of remote sensing datasets which apply in evaluating the effectiveness of the deep residual feature learning method that is proposed in this study.

Fig. 4. Sample images of UCMerced dataset [38]

UCMerced Dataset. UCMerced dataset [38] consist of 21 classes (Fig. 4) and each class contain 100 images with three color channels. Each image dimension is 256×256 pixels and they have a spatial resolution of 1-ft. The classes are highly overlapped e.g. (agricultural and forest are differ by vegetation cover; dense residence and medium residence differ by the number of units), this diverse image content pattern is a challenge for effective feature representations. Further, the images of ucmerced dataset have many common low-level features with multipurpose visible images hence they are suitable candidates for fine-tuning with pre-trained convolutional neural networks.

WHU-SIRI Dataset. WHU-SIRI Dataset [49] comprises of 12 classes (Fig. 5) with a total of 2400 aerial images with three color channels. Each class contains 200 image samples with a size of 200×200 pixels and 2 m of spatial resolution. WHUR-SIRI images are in different scales, orientations under differing lighting conditions. These dataset properties are diverse and challenging thus requiring effective feature learning techniques for accurate scene analysis and interpretation.

3.5 Experiment Setups

The experiments evaluate the effectiveness and efficiency of the deep residual learning network on two publicly available datasets Ucmerced [38] and WHU-SIRI [49] whose properties are described under the Sect. 3.4 remote sensing datasets above. The datasets are split into different train, validation and test ratios (i.e. 70:20:10, for UCMerced) and (150:40:10, WHU-SIRI) with 30, 20, 30 epochs; in batches of 32. Five experiments are conducted and optimal results are reported in both cases.

The training experiment is set up on TensorFlow [43] and implementation adopts the practices in [6, 44]. Input images are sampled at [256/200] on scale augmentation of 0.2, and shear of 0.2 is applied (refer to the source-code: a link provided in the appendix). This research uses batch normalization (BN) [28] immediately after every convolution prior to applying the ReLU [42] activation

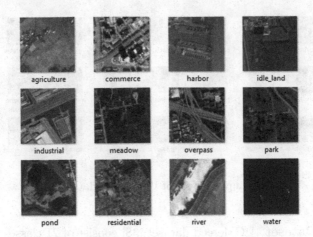

Fig. 5. Sample images of WHU-SIRI dataset [49]

function. The weight initialization follows [29] while the residual feature learning network is trained from scratch. This research applies the Adam optimizer [45] that utilizes stochastic gradient descent to attain faster convergence. The deep residual-feature learning-network is trained with CHPC-GPU, on the lengau cluster (This a high performance GPU owned by the ministry of education, South Africa).

4   Results, Analysis, and Discussion

The experimental results are shown in Tables 1 and 2. For performance evaluation, overall accuracy [3] metrics and confusion matrix [3] are applied.

$$overall\ accuracy = \frac{number\ Of\ Correctly\ Classified\ Images}{Total\ Images} \times 100 \qquad (6)$$

Table 1. Classification results on WHU-SIRI dataset

Feature learning method	Classification accuracy
Structural satellite image indexing [49]	81.67 80
CNN (6conv+2fc) [50]	77. 28
Deep residual feature learning network	**92.56** ± 3.06

Table 2. Classification results on UCMerced dataset

Feature learning method	Classification accuracy
Unsupervised feature learning [47]	81.67
Bag of visual words [38]	71.68
Spatial pyramid matching kernel [48]	74
Adaptive deep pyramid matching [37]	94.75
caffeNet [4]	95.02 ± 0.81
GoogLeNet [4]	94.31 ± 0.60
VGG-16 [4]	95.21 ± 1.20
Deep salient feature based anti-noise transfer network [50]	98.20
Deep residual feature learning network	**94.30** ± 2.54

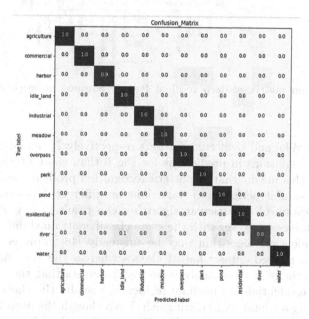

Fig. 6. Confusion matrix of deep residual feature learning network on Whu dataset

4.1 Analysis and Discussion of Results

From Tables (1 and 2), it is observed that the deep networks image feature learning techniques [4,47,50], attain superior classification result with the two remote sensing datasets as compared to low-level feature learning methods [38,48,49]. This is consistent with the fact that deep learning [16] provides an architecture for models that are made up of multiple processing layers to learn feature representations of data with many levels of abstractions to uncovers complex structures in large datasets. Figures 6 and 7 gives the confusion matrices with

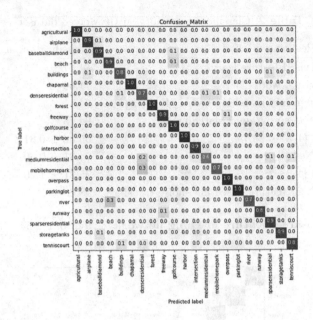

Fig. 7. Confusion matrix of deep residual feature learning network on UCMerced dataset

test images on Whu and Ucmerced datasets respectively. It can be observed that there are few confusions in predictions for Fig. 6 than those of Fig. 7. A possible reason for this maybe that in pattern recognition problems feature learning algorithms to exhibit performance variations on different datasets. This means that the utilization of a particular algorithm can result in powerful classifiers with some databases however with the same classifiers trained different datasets utilizing the identical algorithm may be unsteady [46]. This is consistent to the fact that ucmerced dataset has 21 classes whereas the Whu-dataset has 12 classes. Furthermore, a close analysis on Fig. 7 depicts that there is confusion among dense residential and medium residence, visually the classes are similar even challenging for humans to distinguish. Even though the deep feature-based learning techniques attain good results, there are those deep learning techniques that attain superior classification results. This fact can be attributed to the different design and operation mechanisms that determine how optimal the deep learning technique can be in feature extraction. For instance, the VGG-16 [44] and GoogleNet [23] have 16 and 22 layers respectively. This paper utilizes (a 20-layer network) an intermediate number of layers to the aforementioned network architectures with different design principles, (i.e. residual-connections). Results of Tables 1 and 2 shows the deep residual feature learning network is of significant improvement as compared to the other deep feature learning methods on the same datasets. These show the effectiveness of feature learning of the residual learning network. Owing to these findings, it can be observed that the effectiveness of a deep feature learning network is dependent on the design principles which are taken into consideration when building a deep learning model.

5 Conclusion

This research proposes a deep residual feature learning network that demonstrates to be effective in image feature extraction as evaluated with the ucmerced and whu-rs datasets in Sect. 4. On analysis of the results in Tables 1 and 2, it is evident that the residual learning network achieves better than most deep learning methods in the literature. Additionally, it can be observed from Tables 1 and 2 that the proposed method classification accuracy surpasses the results of other non-deep learning computer vision methods in the literature. Although the results attained by the proposed method are close to those of the other deep learning feature extraction methods in literature; this works gives alternative insights on the novel design approach which utilizes recent literature principles in the context of remote sensing.

Future research work will investigate combining of residual learning networks multiple classifier-fusion in enhancing class prediction accuracies with remote sensing datasets that have more diverse features instead of utilizing only one classifier such as the softmax or the SVM. The recent literature [46] indicate promising trends in this direction.

References

1. Ma, L., Liu, Y., Zhang, X., Ye, Y., Yin, G., Johnson, B.A.: Deep learning in remote sensing applications: a meta-analysis and review. ISPRS J. Photogrammetry Remote Sens. **152**, 166–177 (2019)
2. Bazi, Y., Al Rahhal, M.M., Alhichri, H., Alajlan, N.: Simple yet effective fine-tuning of deep CNNs using an auxiliary classification loss for remote sensing scene classification. Remote Sens. **11**(24), 2908 (2019)
3. Cheng, G., Han, J., Lu, X.: Remote sensing image scene classification: benchmark and state of the art. Proc. IEEE **105**(10), 1865–1883 (2017)
4. Xia, G.S., et al.: AID: a benchmark data set for performance evaluation of aerial scene classification. IEEE Trans. Geosci. Remote Sens. **55**(7), 3965–3981 (2017)
5. Zhou, Q., Zheng, B., Zhu, W., Latecki, L.J.: Multi-scale context for scene labeling via flexible segmentation graph. Pattern Recogn. **59**, 312–324 (2016)
6. He, K., Zhang, X., Ren, S., Sun, J.: Deep residual learning for image recognition. In: Proceedings of the IEEE Conference on Computer Vision and Pattern Recognition, pp. 770–778 (2016)
7. Boutaba, R., et al.: A comprehensive survey on machine learning for networking: evolution, applications and research opportunities. J. Internet Serv. Appl. **9**(1), 16 (2018)
8. Cortes, C., Vapnik, V.: Support-vector networks. Mach. Learn. **20**(3), 273–297 (1995)
9. Bishop, C.M.: Pattern Recognition and Machine Learning. Springer, New York (2006)

10. Tombe, R.: Computer vision for smart farming and sustainable agriculture. In: 2020 IST-Africa Conference (IST-Africa), pp. 1–8. IEEE, May 2020

11. Wójtowicz, M., Wójtowicz, A., Piekarczyk, J.: Application of remote sensing methods in agriculture. Commun. Biometry Crop Sci. **11**(1), 31–50 (2016)

12. Ghassemian, H.: A review of remote sensing image fusion methods. Inf. Fusion **32**, 75–89 (2016)

13. Bu, S., Han, P., Liu, Z., Han, J.: Scene parsing using inference embedded deep networks. Pattern Recogn. **59**, 188–198 (2016)

14. Zhou, B., Khosla, A., Lapedriza, A., Oliva, A., Torralba, A.: Learning deep features for discriminative localization. In: Proceedings of the IEEE Conference on Computer Vision and Pattern Recognition, pp. 2921–2929 (2016)

15. Pohlen, T., Hermans, A., Mathias, M., Leibe, B.: Full-resolution residual networks for semantic segmentation in street scenes. In: Proceedings of the IEEE Conference on Computer Vision and Pattern Recognition, pp. 4151–4160 (2017)

16. LeCun, Y., Bengio, Y., Hinton, G.: Deep learning. Nature **521**(7553), 436–444 (2015)

17. Choromanska, A., Henaff, M., Mathieu, M., Arous, G.B., LeCun, Y.: The loss surfaces of multilayer networks. In: Artificial Intelligence and Statistics, pp. 192–204, February 2015

18. Mishkin, D., Matas, J.: All you need is a good init. arXiv preprint arXiv:1511.06422 (2015)

19. Sutskever, I., Martens, J., Dahl, G., Hinton, G.: On the importance of initialization and momentum in deep learning. In: International Conference on Machine Learning. pp. 1139–1147, February 2013

20. Jia, Y., et al.: Caffe: convolutional architecture for fast feature embedding. In: Proceedings of the 22nd ACM International Conference on Multimedia, pp. 675–678. ACM, November 2014

21. Szegedy, C., Ioffe, S., Vanhoucke, V., Alemi, A.A.: Inception-v4, Inception-ResNet and the impact of residual connections on learning. In: Thirty-First AAAI Conference on Artificial Intelligence, February 2017

22. Russakovsky, O., et al.: ImageNet large scale visual recognition challenge. Int. J. Comput. Vis. **115**(3), 211–252 (2015)

23. Szegedy, C., et al.: Going deeper with convolutions. In: Proceedings of the IEEE Conference on Computer Vision and Pattern Recognition, pp. 1–9 (2015)

24. Szegedy, C., Vanhoucke, V., Ioffe, S., Shlens, J., Wojna, Z.: Rethinking the inception architecture for computer vision. In: Proceedings of the IEEE Conference on Computer Vision and Pattern Recognition, pp. 2818–2826 (2016)

25. Bengio, Y., Simard, P., Frasconi, P.: Learning long-term dependencies with gradient descent is difficult. IEEE Trans. Neural Netw. **5**(2), 157–166 (1994)

26. Glorot, X., Bengio, Y.: Understanding the difficulty of training deep feedforward neural networks. In: Proceedings of the Thirteenth International Conference on Artificial Intelligence and Statistics, pp. 249–256, March 2010

27. Martinez-Covarrubias, J.: Algorithms for large-scale multi-codebook quantization. Doctoral dissertation, University of British Columbia (2018)

28. Ioffe, S., Szegedy, C.: Batch normalization: accelerating deep network training by reducing internal covariate shift. arXiv preprint arXiv:1502.03167 (2015)

29. He, K., Zhang, X., Ren, S., Sun, J.: Delving deep into rectifiers: surpassing human-level performance on ImageNet classification. In: Proceedings of the IEEE International Conference on Computer Vision, pp. 1026–1034 (2015)

30. He, K., Sun, J.: Convolutional neural networks at constrained time cost. In: Proceedings of the IEEE Conference on Computer Vision and Pattern Recognition, pp. 5353–5360 (2015)
31. Schroff, F., Kalenichenko, D., Philbin, J.: FaceNet: a unified embedding for face recognition and clustering. In: Proceedings of the IEEE Conference on Computer Vision and Pattern Recognition, pp. 815–823 (2015)
32. Movshovitz-Attias, Y., Yu, Q., Stumpe, M.C., Shet, V., Arnoud, S., Yatziv, L.: Ontological supervision for fine grained classification of street view storefronts. In: Proceedings of the IEEE Conference on Computer Vision and Pattern Recognition, pp. 1693–1702 (2015)
33. Srivastava, R.K., Greff, K., Schmidhuber, J.: Highway networks. arXiv preprint arXiv:1505.00387 (2015)
34. Arora, S., Bhaskara, A., Ge, R., Ma, T.: Provable bounds for learning some deep representations. In: International Conference on Machine Learning, pp. 584–592, January 2014
35. Wadhwa, A., Madhow, U.: Bottom-up deep learning using the Hebbian principle (2016)
36. Jegou, H., Perronnin, F., Douze, M., Sánchez, J., Perez, P., Schmid, C.: Aggregating local image descriptors into compact codes. IEEE Trans. Pattern Anal. Mach. Intell. **34**(9), 1704–1716 (2011)
37. Liu, Q., Hang, R., Song, H., Zhu, F., Plaza, J., Plaza, A.: Adaptive deep pyramid matching for remote sensing scene classification. arXiv preprint arXiv:1611.03589 (2016)
38. Yang, Y., Newsam, S.: Bag-of-visual-words and spatial extensions for land-use classification. In: Proceedings of the 18th SIGSPATIAL International Conference on Advances in Geographic Information Systems, pp. 270–279, November 2010
39. Perronnin, F., Dance, C.: Fisher kernels on visual vocabularies for image categorization. In: 2007 IEEE Conference on Computer Vision and Pattern Recognition, pp. 1–8. IEEE, June 2007
40. Chatfield, K., Lempitsky, V.S., Vedaldi, A., Zisserman, A.: The devil is in the details: an evaluation of recent feature encoding methods. In: BMVC, vol. 2, no. 4, p. 8, September 2011
41. Ripley, B.D.: Pattern Recognition and Neural Networks. Cambridge University Press, Cambridge (2007)
42. Nair, V., Hinton, G.E.: Rectified linear units improve restricted Boltzmann machines. In: Proceedings of the 27th International Conference on Machine Learning, ICML 2010, pp. 807–814 (2010)
43. Abadi, M., et al.: TensorFlow: large-scale machine learning on heterogeneous distributed systems. arXiv preprint arXiv:1603.04467 (2016)
44. Simonyan, K., Zisserman, A.: Very deep convolutional networks for large-scale image recognition. arXiv preprint arXiv:1409.1556 (2014)
45. Kingma, D.P., Ba, J.: Adam: a method for stochastic optimization. arXiv preprint arXiv:1412.6980 (2014)
46. Zhao, H.H., Liu, H.: Multiple classifiers fusion and CNN feature extraction for handwritten digits recognition. Granular Comput. **5**(3), 411–418 (2020)
47. Cheriyadat, A.M.: Unsupervised feature learning for aerial scene classification. IEEE Trans. Geosci. Remote Sens. **52**(1), 439–451 (2013)
48. Lazebnik, S., Schmid, C., Ponce, J.: Beyond bags of features: spatial pyramid matching for recognizing natural scene categories. In: 2006 IEEE Computer Society Conference on Computer Vision and Pattern Recognition, CVPR 2006, vol. 2, pp. 2169–2178. IEEE, June 2006

49. Xia, G.S., Yang, W., Delon, J., Gousseau, Y., Sun, H., Maître, H.: Structural high-resolution satellite image indexing, July 2010
50. Gong, X., Xie, Z., Liu, Y., Shi, X., Zheng, Z.: Deep salient feature based anti-noise transfer network for scene classification of remote sensing imagery. Remote Sens. 10(3), 410 (2018)

Identifying Device Types for Anomaly Detection in IoT

Chin-Wei Tien[1], Tse-Yung Huang[1], Ping Chun Chen[1], and Jenq-Haur Wang[2(⊠)]

[1] Cybersecurity Technology Institute, Institute for Information Industry, Taipei, Taiwan
{jakarence,tseyunghuang,pcchen}@iii.org.tw
[2] National Taipei University of Technology, Taipei, Taiwan
jhwang@csie.ntut.edu.tw

Abstract. With the advances in Internet of Things (IoT) technologies, more and more smart sensors and devices are connected to the Internet. Since the original idea of smart devices is better connection with each other, very limited security mechanism has been designed. Due to the diverse behaviors for various types of devices, it would be costly to manually design separate security mechanism. To prevent these devices from potential threats, It would be helpful if we could learn the characteristics of diverse device types based on the network packets generated. In this paper, we propose a machine learning approach to device type identification through network traffic analysis for anomaly detection in IoT. First, characteristics of different types of IoT devices are extracted from the generated network packets and learned using unsupervised and supervised learning methods. Second, we apply feature selection methods to the model learned from device type identification module to improve the performance of classification. In our experiments, the performance of device type identification on real data in a smart factory using supervised learning is better than unsupervised learning. The best performance can be achieved by XGBoost with an accuracy of 97.6% and micro-averaging F1 score of 97.6%. This shows the potential of the proposed approach for automatically identifying devices for anomaly detection in smart factories. Further investigation is needed to verify the proposed approach using more types of devices.

Keywords: Anomaly detection · IoT security · Device identification · Machine learning

1 Introduction

With the rapid growth of Internet of Things (IoT) technologies, more and more diverse devices and smart sensors are being deployed in the smart manufacturing scenario, which are connected to the Internet through IoT gateways or local routers. Increasing amount of data could be generated from huge amount of devices and sensors in the manufacturing process. In smart factories, these data need to be automatically collected and transmitted to the server to facilitate predictive analytics. When storing and transmitting data, there will be very high risk of information security if they cannot be effectively protected.

© Springer Nature Switzerland AG 2021
É. Renault et al. (Eds.): MLN 2020, LNCS 12629, pp. 337–348, 2021.
https://doi.org/10.1007/978-3-030-70866-5_22

Since they usually contain critical knowledge of manufacturing and domain know-how for the company, it would further threaten the competitiveness of the company.

The major issue of IoT devices is their lack of security. When all devices and sensors are connected to the Internet, they might be vulnerable to potential attacks. Most conventional devices and sensors are simply designed for better connection with each other, the security standards are neither complete nor mandatory on these devices. Their security level cannot match ordinary computers. Since devices in smart factories are usually controlled and managed in a more automatic way, it's under huge risk of attacks. On the other hand, different types of devices and sensors have different functions. When various malicious operations are enforced on them, they might have different responses than usual. Given many possible types of devices and sensors, it's challenging to detect if they have abnormal behaviors or not.

In this paper, we propose a machine learning approach to device type identification through network traffic analysis for anomaly detection in smart factories. First, identification model for each device type is built from the corresponding generated network packets using both unsupervised and supervised learning techniques. Second, based on the device identification models learned, we use feature selection techniques to keep only important features. This helps to reduce the dimensionality while improving the classification performance. Then, the learned models and selected features can be easily deployed in our embedded sensor. These are used to improve the performance of anomaly detection.

In our experiments, real data were collected in a smart factory with 12 production lines and three types of smart devices. From the experimental results, we can observe better performance of device identification using supervised learning than unsupervised learning. The best accuracy of 97.6% and micro-averaging F1 score of 97.6% can be achieved using XGBoost, an implementation of gradient boosted decision trees, for feature selection.

Also, to test the effects of device identification for anomaly detection, we simulated the normal and abnormal behaviors on real devices in a smart factory by their responses to one example type of malicious network patterns: MS17010. It involves the vulnerability of remote code execution if an attacker sends specially crafted messages to a Microsoft Server Message Block 1.0 (SMBv1) server. From the experimental results on more than 666 thousand simulated messages and responses, feature selection was found useful in improving the classification performance. The F1 score can be improved from 82.62% to 91.06%, which shows the effectiveness of feature selection in classification. This verifies the potential of our proposed approach to automatic device identification for anomaly detection using machine learning methods. Further investigation is needed to evaluate our proposed method with more diverse types of devices and network attacks.

2   Related Work

Internet of Things (IoT) technologies allow various types of smart devices and sensors to be connected to the Internet. This is useful for many scenarios such as smart homes, and smart factories. Due to the high cost and limited hardware resources, very few security mechanisms are enforced in IoT devices. Without suitable protection, this brings huge risk of being compromised, controlled and tempered with for these devices. To enhance the security of IoT devices, different solutions have been proposed. Hardware-based solution such as Physical Unclonable Functions (PUF) [1] was proposed to generate a physically-defined "digital fingerprint" output that serves as a unique identifier for a given input and conditions. For example, PUF-based authentication mechanism [2] was proposed to verify identities for IoT devices that usually have energy and computational constraints. But hardware-based solutions are usually unable to deal with new attacks when they are deployed. In contrast, software-based solutions are usually more flexible if cryptographic functions could be implemented on these devices. As in Internet security, anomaly detection has been a popular research topic in IoT security. For example, Alrashdi et al. [3] proposed an intelligent anomaly detection system in smart cities called AD-IoT based on random forest. They obtain an accuracy of 99.34%. Hasan et al. [4] compared performances of several machine learning methods for anomaly prediction in IoT systems. The best accuracy of 99.4% can be obtained for decision trees, random forests, and artificial neural networks (ANN). Diro and Chilamkurti [5] proposed a distributed deep learning approach to network attack detection in social IoTs. When testing on NSL-KDD dataset for 2-class and 4-class classification, deep learning models outperform shallow models, with the best F1 score of more than 99% and 98%, respectively. To train a good model, a representative dataset is critical. Previous datasets such as NSL-KDD were not useful for IoT environments, since the major characteristics of IoT cannot be well represented. Thus, Koroniotis et al. [6] developed a new dataset called Bot-IoT which incorporates legitimate and simulated IoT network traffics with various types of attacks. Feature selection techniques including correlation coefficient and joint entropy were used. The best classification accuracy of 99.98% can be achieved for SVM.

In Industrial Internet of Things (IIoT) scenarios, Anton et al. [7] employed machine learning-based anomaly detection algorithms to find malicious traffic in a synthetic data set of Modbus/TCP communication of a fictitious industrial scenario. Zolanvari et al. [8] studied the most common used SCADA protocols in IIoT systems. They also developed a machine learning based intrusion detection system where random forest shows the best performance with Matthews Correlation Coefficient (MCC) of 96.81%. However, since IoT devices usually cannot afford the complex deep learning models due to the limited computational power, Ngo et al. [9] proposed an adaptive anomaly detection approach for hierarchical edge computing (HEC) systems. When using deep learning models of different complexity in different layers of the hierarchy, the detection delay can be greatly reduced by 84% while maintaining almost the same accuracy.

Since there are various types of IoT devices in a typical smart factory, the format of network packets generated could also be very different. IOT Sentinel [10] is one of the major works for automatic device type identification for security enforcement in IoT. By monitoring the communication behaviors of devices during the setup process, the generated device-specific fingerprints are mapped to device types using Random Forest classification algorithm and edit distance. Shahid et al. [11] used a machine learning approach to the recognition of IoT devices based on the analysis of packets sent and received. In their experiments, Random Forest classifier can obtain an accuracy of 99.9%. Since different devices might respond with different network packets in different protocol, to better detect attacks in IoT devices, the idea of Chiba et al. [12] was modified to detect malicious websites using the characteristics of IP addresses. Since IP addresses are more reliable than URLs, they might help to detect attacks from unknown sources which masquerade their identities with various URLs. In this paper, we compared the performance of device identification using supervised and unsupervised learning on real IoT devices in smart factories. Given different responses in diverse devices and sensors, we want to improve the performance of anomaly detection based on the normal and abnormal behaviors for the device type that can be learned from their characteristic in responses. Similar to the idea of adaptive anomaly detection in hierarchical edge computing systems [9], we design our solution in the edge. By learning and selecting the features to be deployed in the embedded sensor, the efficiency of our proposed anomaly detection system can be greatly improved.

3   The Proposed Method

In this paper, we propose a machine learning approach to device type identification through network traffic analysis for anomaly detection in IoT. To capture all network packets generated from various smart devices, we design a system architecture consisting of an embedded sensor on the LAN of IoT devices, with three modules: packet collection, device type identification, and packet filtering, as shown in Fig. 1.

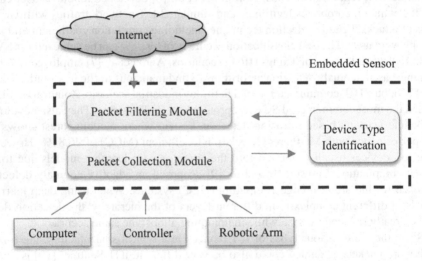

Fig. 1. System architecture of device type identification for anomaly detection.

As shown in Fig. 1, we can see three typical types of devices in a smart factory scenario, including computers, controllers, and robotic arms. First, we collect network packets generated from existing devices by our packet collection module. Specifically, an open source Intrusion Detection system called Suricata is used for packet collection since the detection rules can be customized and deployed on embedded systems. The device identification module then utilizes supervised and unsupervised learning methods to learn the most important characteristics in network traffics to distinguish between different devices. Second, based on the models learned from network analysis results, we use feature selection techniques to keep only important features. Since the device characteristics are learned from their corresponding network packets, the selected features can be automatically deployed into the packet filtering module. It could help anomaly detection when enough training data are collected for training classifiers. This would be particularly useful when new devices are first installed. After a few packet exchanges during the device setup process, the corresponding features can be deployed in the embedded sensor.

3.1 Supervised Learning for Device Type Identification

First, we utilize supervised learning for device type identification. One reason to adopt supervised learning is the potentially higher classification accuracy, with the extra cost of manually labelling the device types in the training data. Since there are usually more than two device types in IoT, we train multiclass classification models for each type. Then, feature selection methods are further used to reduce feature dimensions for improving classification performance.

To extract the most important features for distinguishing between network packets generated from different types of devices, selected fields are extracted from network packets as the features for supervised learning: 'dest_IP', 'src_port', 'dest_port', 'proto', 'pkts_toserver', 'bytes_toserver', 'time_spent', and 'age'. 'dest_IP' denotes the destination IP address for the packet, and 'src_port' and 'dest_port' denotes the source and destination ports, respectively. 'proto' denotes the protocol used in the packet. 'pkts_toserver' and 'bytes_toserver' represents the number of packets and number of bytes to be sent to the server, respectively. Finally, 'time_spent' is calculated from the 'start' and 'end' from the 'flow' field, where 'age' denotes the duration in seconds. Note that flows in Suricata are similar to the idea of sessions which consists of bidirectional packets with the same 5 tuples (protocol, source IP, destination IP, source port, destination port). All these fields are numeric except for the 'proto' field, which contains textual names of the protocol that might have variable lengths. To simplify the processing required for real-time feature extraction, word embedding models are applied to convert texts into a vector of fixed dimension. In this paper, we adopt Word2Vec as the word embedding model with the vector dimension of 100. This makes the total feature dimension as 110.

For the classification model, we adopt an ordinary artificial neural networks (ANN) with multiple dense layers. One advantage of ANN is that it's incremental. No retraining is needed when we have more training data. Specifically, we utilize ELU (Exponential Linear Unit) activation function, and Adam optimizer [13] which is a gradient-based optimization of objective functions. We further apply XGBoost as the feature selection method which is an implementation of gradient boosted decision trees to estimate the relative importance of individual features. Specifically, gradient boosting is a method where new models are created that predict the residuals or errors of prior models and then added together to make the final prediction. By utilizing tree ensembles, gradient tree boosting tries to minimize the sum of loss functions for the residuals of the previous round for each tree. Also, importance score is calculated for each single decision tree, and weighted by the number of observations in the node. The feature importance is then calculated by averaging across all decision trees within the model.

3.2 Unsupervised Learning for Device Type Identification

Due to the time-consuming manual labeling efforts in supervised learning, to facilitate easier device configuration and deployment of features for device identification, we also compare with an unsupervised learning-based identification model. Specifically, we use clustering techniques for unsupervised device type identification based on similarities among packet requests transmitted to these devices and packet responses from them. The idea is that different device types should demonstrate different patterns in packet responses. It would be helpful if the devices could be automatically clustered by similarity among their packet responses since no manual labeling would then be needed. Since there are various devices in IoT, we cannot forecast new device types yet to be developed. If classification methods were used, we have to define very broad categories such as computers, robotic arms, or CNC machines. These general categories might be useful in simple scenarios, but they could not help when devices are only slightly different in their functions.

To group devices with similar behaviors into the same clusters, we adopt K-Means clustering algorithm for all devices based on the content of packet responses. Based on our previous experience, we slightly modify the feature representations from those used in supervised learning for each network packet as follows: 'src_IP', 'dest_IP', 'proto', 'bytes_toserver', and 'time_spent'. They have the same meanings as those used in Sect. 3.1. These features are converted into the corresponding vector representation with the fixed dimension of 110, in which word embedding is also applied for texts in a similar way. Then, with a pre-determined number of clusters K, the goal of K-means clustering is to determine a partition $C = \{C_1, C_2, ..., C_K\}$ that minimizes the sum of squared distances to the cluster centroids.

Since there's no ground truth for the clusters, to verify the clustering results, we accumulated the "IP vectors" by IP address as follows: $[<IP_i>, <F_i>, <f_{i1}>, ..., <f_{id}>]$, where $<IP_i>$ denotes the source IP address, $<F_i>$ denotes the frequency of IP_i, which is the total number of packets that contains IP_i, and $<f_{ij}>$ denotes the frequency of IP_i in class C_j, and d denotes the number of device types. The total number of dimensions is $4 + 1 + d = d + 5$. In this paper, since we only consider the cases when the number of device types is 3, the total number of dimensions is 8, as shown in Fig. 2.

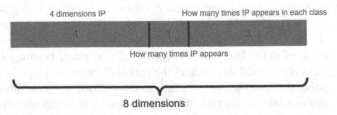

Fig. 2. The "IP vector" for verifying clustering performance.

The assumption is that: for devices of the same type, they usually generate packets originating from similar IP addresses. Under this assumption, we expect to observe similar distribution of IP addresses in each device type.

3.3 Deployment of Features for Anomaly Detection

After device identification model has been trained with the selected important features, they can be automatically deployed in our packet filtering module. Then, classification algorithms can be used to train the anomaly detection model.

For the deployment of features, when new devices are first installed, they will be configured to communicate with the embedded sensor. This involves a few packet exchanges during the device setup process, when our device identification module learns the corresponding features. After the setup process completes, the corresponding features can be automatically deployed to the packet filtering module which configures the corresponding filtering rules for identifying packets with the features.

Since our goal is to distinguish between normal and abnormal behaviors through network packets for devices, in this paper, we propose to simulate potential attacks and intrusions by sending predefined malicious network patterns to devices, and monitoring the resulting packet responses. Specifically, the malicious packet requests from simulated attackers are abnormal, while the corresponding packet responses to malicious requests are considered normal since these are the behaviors of real devices.

Unsupervised learning involves calculating the nearest centroid among all clusters, while supervised learning needs to classify using gradient boosted decision-tree based XGBoost algorithm based on the selected features. Based on our experiences in device identification module, we selected the following 7 features, including: 'src_port', 'dest_port', 'proto', 'age', 'bytes', 'max_ttl', and 'pkts'. In our preliminary experiments, word embedding for texts in the 'proto' field only has marginal improvement in performance. To reduce the dimensionality of features to be matched as well as the efficiency, we chose not to use word embedding when detecting abnormal behaviors. This gives the total number of dimensions as 7.

4 Experiments

In our experiments, we evaluated the performance of our proposed method for device type identification and anomaly detection.

First, to capture network traffic from our embedded sensor, we used the open source IDS and IPS tool called Suricata. We collected the network packets in 12 production lines in our target factory. Then, three types of devices were manually labeled: computers in 17,296 events, robotic arms in 738 events, and controllers in 3,413 events, with a total of 21,447 records as our first dataset for device identification. Note that in Suricata, events represent different types of traffic that can be logged associated with different protocols or other fields.

Second, to evaluate the effects of device identification on anomaly detection, we deployed the rules learned by our device identification module using Suricata on the embedded sensor. Since real attacking data is not available in IoT, we further simulated one example type of malicious network patterns, MS17010 attack, which includes the vulnerability of remote code execution if an attacker sends specially crafted messages to a Microsoft Server Message Block 1.0 (SMBv1) server. Specifically, there are 62,345 and 604,044 events in normal and simulated malicious behaviors, respectively. This gives a total of 666,385 records as our second dataset for anomaly detection.

4.1 Performance of Device Type Identification

In this experiment, we want to compare the performance of device identification using supervised and unsupervised learning. To test the effects of unsupervised learning on device identification, we selected 2,000 records from one production line in the first dataset. We used the 110-dimensional feature when we conducted K-means clustering for K = 3 since there are three types of devices.

After converting the clustering results into "IP vectors", we calculate their pairwise Euclidean distance. When we set the similarity threshold = 0.2, the number of clusters of IP vectors is also 3. For the dataset with 47 different IP addresses, after K-means clustering, the largest cluster includes 41 IP addresses. The clustering result is not as expected. This means that we cannot successfully distinguish between different types of devices using unsupervised learning. The reason is that the behaviors of the three device types are similar in terms of "IP vectors". In order to distinguish the differences between device types, we need to learn more features from more data.

Next, we compared the performance of supervised learning for device identification. We utilized all 12 production lines in the first dataset, which consists of 21,447 packets in total. In order to do supervised learning, we manually labelled the data into three classes: PC, device controller, and robotic arm. With 10% for validation, the dataset is divided into 18,000 events as the training set, 2,000 as the validation set, and 1,447 as the test set. To better learn the classifiers, we used 110-dimensional features. The parameters of the neural network include: ELU (Exponential Linear Unit) as the activation function, Adam optimizer, 150 epochs, with the batch size of 32. In our experiment with three dense layers of neural networks, the final test accuracy of the neural network is 92.8%. This shows the potential of using supervised learning for device identification.

Next, we further improved the performance of our device identification model by utilizing feature selection methods for dimensionality reduction. Specifically, after applying XGBoost which is based on decision trees for feature selection, we obtained the top-10 important features as in Fig. 3.

	src_port	dest_port	proto	app_proto	tcp	age	bytes	max_ttl	pkts	spend_time	label
0	55822	5355	0	1	1	0	128	1	2	98048	2
1	49518	5355	0	1	1	0	128	1	2	104372	2
2	61895	5355	0	1	1	0	128	1	2	96474	2
3	51705	5355	0	1	1	0	128	1	2	97464	2
4	61462	1947	0	1	1	0	82	128	1	0	2

Fig. 3. The top-10 features for device identification selected by XGBoost.

As shown in Fig. 3, among the top-10 selected features, 'proto' is the most important feature, followed by 'bytes' and 'dest_port'. This is reasonable since the same IoT device type usually uses the same protocol for communication. Also, the number of bytes determines the amount of payload, and the 'dest_port' determines the destination service. After applying feature selection, the confusion matrix of device identification performance can be obtained as follows:

Table 1. The confusion matrix of device identification performance.

Actual	Predicted		
	Controller	Arm	PC
Controller	340	0	2
Arm	11	45	18
PC	3	17	1,710

As shown in Table 1, we can see a high accuracy of 97.6%. The accuracy greatly improved by 4.8% comparing to the case without feature selection. This shows the effectiveness of using feature selection methods for improving classification performance. If we look closer into the multiclass classification performance, the micro-averaging and macro-averaging F1 scores are 97.6% and 87.6%, respectively. We found the reason why the macro-averaging F1 score is not as good to be the lower performance for the class of robotic arms, which showed a precision of 72.6%, recall of 60.8%, and F1 score of 66.2%. Some network packets from robotic arms were mistaken as coming from either controllers or PCs. This is reasonable since the distribution of three classes are imbalanced, where robotic arm is the smallest class. In order to prevent bias from imbalanced data distribution, we suggest to adopt the All-versus-All (AVA) approach where a binary classifier for each pair of device types is trained. This can improve the performance of multiclass classification, especially for minority classes.

4.2 Effects of Feature Selection on Anomaly Detection

Next, we deployed the neural network classifier learned from the first part for device identification in the packet filtering module. To evaluate the performance of our proposed method for anomaly detection, we performed 10-fold cross-validation on the second dataset. From the evaluation results of linear SVM classifier, the F1 score is 82.62% and 91.06% for using all features and the 7 selected features, respectively. This shows the effectiveness of feature selection in classification, and potential of the proposed approach for anomaly detection. When we further checked the feature importance, we found the most important feature as 'bytes', which means the number of bytes in the packets to the server. After careful inspection into the data, we found the number of bytes 74 and 62 accounts for 97.6% of all malicious packets. This is consistent with the single type of malicious patterns for the respective 'bytes' field in the packet format. Since they are simple malicious patterns, binary classification of normal and malicious classes can achieve very good performance. More types of malicious patterns are needed to justify the performance of the proposed method in anomaly detection.

5 Discussions

In our experiments of device identification on real data collected from smart factories, we can effectively distinguish between three most commonly used types of devices: computers, controllers, and robotic arms. The proposed approach achieved the best multiclass classification accuracy of 97.6% and micro-averaging F1 score of 97.6% when we use an ordinary artificial neural network as the classification model and XGBoost for feature selection. Due to the imbalanced data distribution in three classes is, the robotic arms class has a lower F1 score which reduced the macro-averaging F1 score to 87.6%. To be practically useful for identifying devices in real word, we suggest a "All-versus-All" approach to train a binary classifier for each pair of device types to improve classification performance for minority classes. With the help of neural networks as the classification model, it will be easier to extend to more device types since the model is incremental. There's no need for re-training all models. When a new device type is added, we simply need to manually label and train the new model with the d existing types. This makes the solution more feasible in practice.

To detect potential anomalies, we can easily deploy the model trained for device identification on our embedded sensor. When adding new malicious patterns, we simply need to collect the response packets from the devices being monitored in the same domain using the same embedded sensor. From our experimental results of simulated attacks on real devices in smart factories, we can effectively distinguish between normal and abnormal behaviors using responses to one example malicious pattern: MS17010 attack. The classification performance in terms of F1 score can be improved from 82.62% to 91.06% by feature selection methods. This shows the potential of our proposed approach.

In addition to smart factories, our proposed approach can be easily deployed in any IoT environments without complex configurations. In our implementation, we integrate device identification module and packet filtering module on the same embedded sensor, which can be simply connected to the same local area network to begin monitoring and detection. It facilitates easy deployment of anomaly detection system in existing IoT. In future, we are going to extend to more types of devices and attacks in different IoT scenarios. Instead of binary classification of normal and malicious behaviors, multiclass classification of different types of attacks will be more challenging. Also, we are working on the unsupervised learning for establishing the baseline for devices so that the efforts of deployment to new sites can be further reduced.

6 Conclusion

In this paper, we have proposed a machine learning approach to device type identification for anomaly detection in IoT. First, with real data collected from smart factories, we can effectively detect various device types by their network responses using multiclass classification models of neural networks. The performance can be further improved with the help of feature selection techniques by gradient boosted decision trees. Second, from the features learned for device identification, we can easily deploy in our embedded sensor. By using the simulated malicious attacks on real devices in smart factories, we can effectively distinguish between normal and abnormal behaviors based on packet responses using SVM algorithm with feature selection. This shows the potential of our proposed machine learning approach to automatic device identification for anomaly detection. Further investigation is needed to test the performance for more device types and malicious behaviors in different IoT scenarios.

Acknowledgement. This study is conducted under the "Artificial Intelligence Oriented for Cyber Security Technology Collaboration Project (1/4)" of the Institute of Information Industry which is subsidized by the Ministry of Economic Affairs of the Republic of China.

References

1. Pappu, R.S.: Physical one-way functions. Ph.D. dissertation, Massachusetts Institute of Technology (2001)
2. Huang, Z., Wang, Q.: A PUF-based unified identity verification framework for secure IoT hardware via device authentication. World Wide Web **23**(2), 1057–1088 (2019). https://doi.org/10.1007/s11280-019-00677-x
3. Alrashdi, I., Alqazzaz, A., Aloufi, E., Alharthi, R., Zohdy, M., Ming, H.: AD-IoT: anomaly detection of IoT cyberattacks in smart city using machine learning. In: Proceedings of IEEE 9th Annual Computing and Communication Workshop and Conference (CCWC), pp. 305–310 (2019)
4. Hasan, M., Islam, M., Zarif, I.I., Hashem, M.M.A.: Attack and anomaly detection in IoT sensors in IoT sites using machine learning approaches. Internet Things **7**, 100059 (2019)
5. Diro, A.A., Chilamkurti, N.: Distributed attack detection scheme using deep learning approach for internet of things. Future Gener. Comput. Syst. **82**, 761–768 (2018)
6. Koroniotis, N., Moustafa, N., Sitnikova, E., Turnbull, B.: Towards the development of realistic botnet dataset in the Internet of Things for network forensic analytics: Bot-IoT dataset. Future Gener. Comput. Syst. **100**, 779–796 (2019)
7. Duque Anton, S.D., Kanoor, S., Fraunholz, D., Schotten, H.D.: Evaluation of machine learning-based anomaly detection algorithms on an industrial Modbus/TCP data set. In: Proceedings of the 13th International Conference on Availability, Reliability and Security (ARES 2018), pp. 41:1–41:9. ACM (2018)
8. Zolanvari, M., Teixeira, M.A., Gupta, L., Khan, K.M., Jain, R.: Machine learning-based network vulnerability analysis of industrial internet of things. IEEE Internet Things J. **6**(4), 6822–6834 (2019)
9. Ngo, M.V., Chaouchi, H., Luo, T., Quek, T.Q.S.: Adaptive anomaly detection for IoT data in hierarchical edge computing. In: Proceedings of the AAAI Workshop on Artificial Intelligence of Things (AIoT) (2020)
10. Miettinen, M., Marchal, S., Hafeez, I., Sadeghi, A., Asokan, N., Tarkoma, S.: IoT sentinel: automated device-type identification for security enforcement in IoT. In: Proceedings of 2017 IEEE 37th International Conference on Distributed Computing Systems (ICDCS 2017), pp. 2177–2184 (2017)
11. Shahid, M.R., Blanc, G., Zhang, Z., Debar, H.: IoT devices recognition through network traffic analysis. In: Proceedings of the IEEE International Conference on Big Data (BigData 2018), pp. 5187–5192 (2018)

12. Chiba, D., Tobe, K., Moriy, T., Goto, S.: Detecting malicious websites by learning IP address features. In: Proceedings of the 2012 IEEE/IPSJ 12th International Symposium on Applications and the Internet (SAINT 2012), pp. 29–39 (2012)
13. Kingma, D.P., Ba, J.: Adam: a method for stochastic optimization. In: Proceedings of ICLR 2015 (2015)

A Novel Heuristic Optimization Algorithm for Solving the Delay-Constrained Least-Cost Problem

Amina Boudjelida[✉] and Ali Lemouari[✉]

Laboratory of Mathematics and Applied Mathematics (LMAM), University of Jijel, Jijel, Algeria
boudjelidaamina@gmail.com, lemouari_ali@yahoo.fr

Abstract. The quality of the multicast routing service (QoS) is an NP multi-objective optimization problem. It is one of the main issues of transmission in communication networks that consist of concurrently sending the same information from a source to a subset of all possible destinations in a computer network. Thus, it becomes an important technology communication. To solve the problem, a current approach for efficiently supporting a multicast session in a network is establishing a multicast tree that covers the source and all terminal nodes. This problem can be reduced to a minimal Steiner tree problem (MST), which aims to look for a tree that covers a set of nodes with a minimum total cost that has been proven to be NP-complete. In this paper, we propose a novel algorithm based on the greedy randomized search procedure (GRASP) for the Delay-Constrained Least-Cost problem. Constrained with the construction and improvement phase, the proposed algorithm makes the difference in the construction phase through using a new method called EB heuristic. The procedure was first applied to improve the KMB heuristic in order to solve the Steiner tree problem. Obtained solutions were improved by using the tabu search method in the enhancement process. Computational experiments on medium-sized problems (50–100 nodes) from literature show that the proposed metaheuristic gives competitive results in terms of cost and delay compared to the proposed results in the literature.

Keywords: Multicast routing · DCLC · QOS · Optimization · Heuristic · Metaheuristic

1 Introduction

Depending on the number of destinations, network routing can be categorized into three basic types: unicast (one-to-one), broadcast (one-to-all) and multicast (one-to-many). The multicast communication model has been defined firstly in [1]. Multicast, or selective streaming, is a communication approach for the

© Springer Nature Switzerland AG 2021
E. Renault et al. (Eds.): MLN 2020, LNCS 12629, pp. 349–363, 2021.
https://doi.org/10.1007/978-3-030-70866-5_23

transmission of datagrams to a group of zero or more hosts identified by a single destination group address [1]. The notion of a group is often associated with multicast communications. A host group is a set of network entities sharing a common identifying multicast address, each member of this group received any data packets addressed to this multicast address by senders (sources) that may or may not be members of the same group and have no knowledge of the groups' membership [2]. This definition implies that, from the sender's point of view, this model reduces the multicast service interface to a unicast one, moreover, multicast routing can utilize network resources more efficiently, as a data packet traverses each link only once, and some of the links are shared [2,3].

In underlying computer networks, the rapid development of real-time multimedia technologies such as videoconferencing, distance learning and coordination require strict quality of service conditions such as cost, bandwidth, packet loss rate, delay, and delay jitter. The stringent delay constraint is enforced on multimedia traffic to guarantee that audio and video data are transformed smoothly to the audience [22]. Besides, multicast is assigned to a condition in which the sender wants to send its data packets to a group of networks or receivers that actually form a multicast group. It is obvious that the benefits of this task include less wastage of bandwidth and network resources, parallelism in the network, transmitter load and reduced network traffic.

Multicast routing based on QoS is an serious issue for network research as well as for high-efficiency and performance networks in future generations. Multicast routing based on the quality of service, therefore, it aims to find an optimized multicast routing tree in order to meet the service quality limitations, it is an NP-complete problem. Establishing a multicast tree could solve multicast routing problems. One of the most important issues of implementing multicast services is the type of tree structure designed to ensure increased quality and efficiency of the multicast tree. There are abundant methods for solving problems related to multicast routing. Among those methods are the exact methods that seek to find optimal solutions for the problems. Other methods are the approximate methods, where one was satisfied with the good quality solutions, without guaranteeing optimality to the port of reduced computation time. In return, the disadvantage is having no information on the quality of the solutions obtained. Thus, hybrid methods combine exact methods and/or approximate methods to create new methods that have given rise to a pseudo-class of methods. There are two ways to solve this problem: An optimal solution at the final moment and An optimal close solution by a heuristic algorithm. The first solution is an optimal solution, but the complexity of the NP-Complete problems makes it impracticable. On the other hand, the second method is a possible way. Thus, the routing algorithm has a significant impact on the development and performance of computer communication networks.

In this paper, our objective is to find solutions to the problems related to the Qos multicast routing, particularly the Delay-Constrained Least-Cost routing problem. Mathematically, the problems can be considered as Steiner tree problems (PST) in a graph. Hence, we propose a novel approach based on Randomized Search Procedure (GRASP) metaheuristic, using the EB heuristic for construction phase, and the Tabu Search algorithm (TS) as a local search procedure used to improve the solution. The remainder of this paper is structured as follows. Section 2 outlines the multicast routing and reviews the existing related work. Section 3 discusses the QoS multicast routing problem. Section 4 defines the Delay-Constrained Least-Cost routing problem. Section 5 describes the proposed method for solving DCLC problem. Section 6 shows simulation and tested experiments. Section 7 summarizes the main contributions and results of this paper.

2 Multicast Routing: Related Work

Since the 1990s, the rapid evolution of numerous real-time multimedia applications has been stimulating the demand for QoS based multicast routing in the underlying computer communication networks. The main goals of QoS multicast routing are to efficiently allocate network resources, balance the network load, reduce congestion hot spots and provide adequate QoS guarantees for end-users of multimedia applications [3].

The traditional unicast model is extremely inefficient for the group-based applications (videoconferencing, shared workspaces, distributed interactive simulations (DIS), software upgrading, and resource location) since the same data is unnecessarily transmitted across the network to each receiver, these applications require the underlying network to satisfy certain quality of service (QoS) multicast communication [2,4]. These QoS requirements include the cost, delay, delay variation, lost and hop count, etc. [4]. The multicast model was proposed to reduce the many unicast connections into a multicast tree for a group of receivers [2]. The phenomenal growth of group communications and quality of service (QoS) aware applications over the Internet has accelerated the need for scalable and efficient network support [2].

Multicasting has emerged as one of the most focused areas in the field of networking. As the technology and popularity of the Internet have grown, applications that require multicasting are becoming more widespread, where information needs to be sent to multiple end-users at the same time in the underlying computer networks [5]. Another interesting recent development has been the emergence of dynamically reconfigurable wireless ad hoc networks to interconnect mobile users for applications ranging from disaster recovery to distributed collaborative computing [5]. In this context, self-organized wireless mobile nodes that share a common wireless channel can work without the support of fixed infrastructure or centralized administration [6]. Multicasting is more complex than in wired networks, the main constraints in these networks are bandwidth limitation and unpredictable host mobility. The challenge is to propose multi-hop routes for multicast routing protocols [7], multi-hopping is usually required due to limited transmission power where each node participates in the network as both host and a router [6].

Many works were carried in the last decade for QoS multicast routing problem such as a method based on genetic algorithms (GA) proposed by Haghighat et al. [8]. In this algorithm, the connectivity matrix of edges is used for genotype representation the matrix where it tells whether or not a specific edge connects the pair of nodes. The initial population is based on the randomized depth-first search algorithm. Also, different heuristics are proposed for reproduction process. The proposed GA-based algorithm overcomes existing algorithms such as BSMA heuristic [9] is the best deterministic Delay Constraint Low-Cost, Wang-GA [10] used for solving the Bandwidth-Delay-Constrained Least-Cost problem which accepted non-uniform and real-valued delay bounds and used the mutation operations to convert the algorithm from a local optimal point into the global solution, moreover this algorithm used a pre-processing mechanism to decrease the search space. In addition, several heuristics have been developed with the GRASP heuristic by N. Skorin-Kapov et al. [11] to solve the Delay-Constrained Multicast Routing (DCMR), however, the neighborhood of the TS-CST algorithm used in the local search procedure proved too restrictive and take both the cost and the time frame into consideration. Zhang et al. [12] proposed a method for least-cost QoS multicast routing based on genetic simulated annealing algorithm NGSA, the genetic algorithm and simulated annealing algorithm are combined to improve the computing performance in this method.

In the last recent years, as in [13], a multi-objective differential evolution algorithm named as MOMR-DE proposed using the modified crossover and mutation operators to build the shortest-path multicast tree. Constraint handling scheme is used to handle QoS constraints. Furthermore, ranking technique, fast non-dominated sorting process and crowding distance sorting process were combined together in order to select the elitism and preserve the diversity of the solutions. The last year, Zhang et al. [14] Combine the solution generation process of Ant Colony Optimization (ACO) algorithm with the cloud model (CM). The cloud model enhances the performance of the ACO algorithm by adjusting the pheromone trail on the edges.

Recently, Askari et al. [23] proposed an algorithm named EMSC to resolve the problem of jointing multicast routing, scheduling, and calling admission control in MRMC-WMNs. Accordingly, this is an efficient cross-layer algorithm that jointly considers the minimum-cost minimum-interference path selection, the QoS requirement of each path in tree construction, and the minimum number of occupied time slots. In a recent paper by Hassan et al. [27], and based on ant colonies, a multi-objective algorithm is developed to construct a multicast tree for data transmission in a computer network. This algorithm simultaneously optimizes the cost, delay and hop (total weight) of the multicast tree. additionally, a novel encoding-free non-dominated sorting genetic algorithm (EF-NSGA) has recently been presented in [25] in order to optimise the Application Layer Multicast (ALM) routing problem as a multi-objective optimization problem to minimize the tree. For achieving encoding-free, genotypes are directly represented as tree-like phenotypes in this algorithm. Accordingly, the genetic operators acting on genotypes, like crossover and mutation, need to be redesigned to adapt the tree-like genotypes. Furthermore, in [26], the authors proposed a discrete artificial fish school algorithm (DAFSA) to optimize multiple co-existing multicast routing problems in a link-capacitated network as a whole rather than sequentially optimize them each in isolation in order to avoid the link-congestion and minimizing their overall tree cost as well.

3 QoS Multicast Routing Problem

The QoS multicast routing problem concerns the search of optimal routing trees in the distributed network, where messages or information are sent from the source node to all destination nodes while meeting all QoS requirements. A common approach constructs a multicast tree structure which covers the source and all terminals nodes, is to bring the problem towards the minimal Steiner tree problem (MST), which aims to look for a tree that covers a set of nodes with a minimum total cost. The Constrained Steiner tree is a well-known NP-complete problem.

Let us represent a network with a graph $G = (V, E)$, where V is a set of n nodes and E a set of m edges. $e(i, j) \in E$ a link that associates a weighting function $f_W(e)$ between nodes i and j. The link is bidirectional, i.e. the existence of a link $e(i, j)$ from node i to node j implies the existence of another link $e'(j, i)$. Due to the asymmetric nature of computer networks, it is possible that $f_W(e) \neq f_W(e')$.

Let us define s a node called the destination node, R a set of nodes where $R \subseteq V - \{s\}$ is the destination nodes that receive a data stream from source node s. The set R called multicast groups or terminal nodes, relay nodes which are intermediate hops on the paths from the source to destinations. The rest of the paper uses the following notations [4]:

- $(i, j) \in E$ the link from node i to node j, i, j $\in$ V.
- $c_{ij} \in \Re^+$ the cost of link (i, j).
- $d_{ij} \in \Re^+$ the delay of link (i, j).
- $z_{ij} \in \Re^+$ the capacity of link (i, j), measured in Mbps.
- $t_{ij} \in \Re^+$ the current traffic of link (i, j), measured in Mbps.
- $s \in V$ the source node of a multicast group.
- $R \subseteq V$-s the set of destinations of a multicast group.
- $r_d \in R$ the destinations in a multicast group.
- $\phi \in \Re^+$ the traffic demand (bandwidth requirement) of a multicast request, measured in Mbps.
- $T(s, R)$ the multicast tree with the source node s spanning all destinations $r_d \in R$.
- $p_T(s, r_d) \subseteq T(s, R)$ the path connecting the source node s and a destination $r_d \in R$.
- $d(p_T(s, r_d))$ the delay of path $p_T(s, r_d)$.

4 Delay-Constrained Least-Cost Multicast Routing Problem (DCLC)

A variety of Quality of Service (QoS) constraints have been established in real-life applications. That is, cost, packet loss ratio, use of links, bandwidth, delay variation, etc. Multicast Routing Problems (MRPs) based on QoS become much more complicated multi-objective problems when various conflicting objectives are considered simultaneously, these problems are named a Multi-Objective Multicast Routing Problem (MMRP). The problems consist of finding a multicast tree noted T that minimiz the following objectives:

$$min\ Z = C(T) + DM(T) + \alpha(T) + DA(T) \tag{1}$$

where: C(T): the cost of the multicast tree:

$$C(T) = \phi \sum_{(i,j)\in T} c_{i,j} \tag{2}$$

DM(T): the maximal end-to-end delay:

$$DM(T) = Max\left\{d(p_T(s, r_d))\right\}, r_d \in \Re \tag{3}$$

$$d(p_T(s, r_d)) = \sum_{(i,j)\in p_T(s,r_d)} d_{ij}, r_d \in \Re \tag{4}$$

α(T): the maximal link utilization:

$$\alpha(T) = Max\left\{\frac{\varnothing+t_{ij}}{z_{ij}}\right\}, (i, j) \in T \tag{5}$$

DA(T): the average delay:

$$DA(T) = \frac{1}{|\Re|} \sum_{r_d\in R} d(p_T(s, r_d)) \tag{6}$$

The MMRP is subject to a link capacity constraint as follows:

$$\varnothing + t_{ij} \leqslant z_{ij}, \forall(i, j) \in T(s, \Re) \tag{7}$$

The DCLC problem is a particular case of MMRP problem. It concerns only the two QoS requirements the cost and delay of the multicast tree. The DCLC multicast routing problem is equivalent to the Delay-Constrained Steiner tree (DCST) problem, which is also NP-complete [15]. The objective of the Delay-Constrained Steiner Tree (DCST) Problem is constructing a multicast tree T such that the tree delay is within the delay bound, and the tree cost is minimized.

The end-to-end delay from the source to each destination is the sum of delays along the path, it plays a key role in obtaining feasible solutions in search algorithms, the smaller the delay bound is, the tighter the problem is constrained [3,4]. The objective function of the DCLC can be rewritten as follow:

$$Min\ C(T)\ |\ Delay(T) \leqslant \Delta, T \in T(s, \Re) \tag{8}$$

The inhibiting factor in the Delay-Constrained MRP is the value of Δ, which is the delay bound. This means that the smaller the delay bound is, the stronger is the constraint [11]. We note that in the majority of the literature papers and this one, the same delay bound is applied to all destinations. Otherwise different applications may have different upper bound for each destination.

5   Resolution Methodologies

Based on Greedy Randomized Adaptive Search Procedure (GRASP) combined with the EB heuristic as the initial solution, and the tabu search procedure as the local search method, an algorithm for solving the multicast routing problem have been proposed in this paper. The proposed algorithm is dedicated to the problem of single-objective multicast routing, is reduced to the MStTG problem, where the only constraint is the cost. Furthermore, this algorithm was adapted for resolving the problem of multi-objective multicast routing DCLC, where both constraints: cost and delay are considered.

In order to resolve the multicast routing problem (MRP), we propose in this paper a novel one-off optimization approach based on the GRASP metaheuristic. The greedy randomized adaptive search procedure (GRASP) is a metaheuristic proposed by Feo et al. [24], it consists of an iterative randomized sampling technique in which each iteration provides a solution to the problem. The best incumbent solution overall GRASP iterations is kept as the final result. The procedure builds a solution based upon two phases: a construction phase and a local search phase.

5.1   The Construction Phase

The KMB heuristic [16], proposed by Kou, Markovsky and Bermann, is one of the most efficient approximate method that used to construct the Steiner tree covering all terminals nodes, the problem is well known as an NP-hard. The method based on two main algorithms allowing the development of the shortest paths and minimal spanning tree. There is a practical interest in this heuristic due to its simple implementation, however, it is rarely enough to apply an algorithm such as KMB directly to a multicast routing problem [17]. The produced results by the KMB are not necessarily minimal.

In 2016, M. Fujita et al. have proposed a Steiner tree construction heuristic to improve the KMB algorithm [18]. The heuristic is called Edge Betweenness (EB) (Algorithm 1.1) based on information derived from the edges of the graph and has good performance for various types of Steiner tree problems.

Algorithme 1.1 : EB heuristic.
1 **Input:** an undirected distance graph G;
2 Step1. Construct the G' graph from a given network using Eq. 10;
3 Step2. Construct the complete undirected graph G_1;
4 Step3. Find the minimal spanning tree (MST_1) according to Prim;
5 Step4. The paths in MST_1 are replaced by those from the original network to construct the Steiner tree;
6 Step5. Remove from the tree all branches that contain only non-terminals;
7 **Output:** a Steiner tree for G;

The EB heuristic is based on maximizing the use of low weight edges. This reduces the cost of the Steiner tree result. Therefore, in [18], the authors introduced a parameter allowing to calculate the rate of use of the edges; this parameter is called Edges Betweenness and it represents the centrality of the edges in the network. If an edge is used by many paths between nodes, then that edge has high betweenness [18]. The edge betweenness of the edge e is defined as follows:

$$eb(e) = \frac{\sum_{s=1}^{|V|} \sum_{g=1,g \neq s}^{|V|-1} \frac{I(s,g)}{n(s,g)}}{(|V|-1)(|V|-2)/2} \tag{9}$$

where $|V|$ is the number of nodes, s is a starting node of the shortest paths, g is a terminating node of the shortest paths, $I(s, g)$ is the number of shortest paths between s and g through the edge e and $n(s, g)$ is the number of shortest paths between s and g. M. Fujita et al. [18] define a new cost for the edge as follows, by using the edge betweenness:

$$C_{new} = C(e) - \alpha * eb(e) \tag{10}$$

In Eq. 10, C (e) is the given cost of the edge e, eb (e) is the edge betweenness of the edge e, and α is a controlling parameter that determines the priority between cost and betweenness of the given edge. By using the new cost, an edge is susceptible to be included in the Steiner tree, if it has a low cost and a high edge betweenness. The heuristic gives better results than the KMB algorithm.

The construction phase constructs a solution in two steps. First, in our proposed algorithm (Algorithm 1.4), the EB heuristic is employed to create a so-called restricted candidate list (RCL) of elements that can be added to a partial solution. Second, the elements are randomly selected according to the P factor to create a feasible solution while $P \in \;]0,1[$ is the percentage of components included in the candidate list (RCL) at each iteration, which the factor allows to diversify the initial solution generated.

5.2 The Local Search Phase

Once a solution is obtained in the construction phase, our TS algorithm is applied to improve the current solution with a best solution reached in neighborhood. The metaheuristic Tabu search (TS) is a global optimization method guides a local heuristic search procedure to explore the solution space beyond local optima using intensification and diversification strategy. The heuristic tries to avoid trapping into local optima by using a special memory called tabu list. Any solution which has been recently selected from the best in neighborhood is put into a tabu list so that it becomes (taboo) for a short period of time, depending on the length of the list. The process minimizes the chance of cycling in the same solution and therefore creates more chances of improvement by moving into other space region.

Our procedure TS starts using the EB heuristic as the initial solution, it is done after a fixed number of iterations or a maximum number of continuous iterations without improvement of the best-known solution. The current solution is improved using the best in neighborhood. Furthermore, three movements are used to construct a neighborhood: random movement, path movement, and node movement:

a. Random move (random position - random path): this movement consists of randomly removing a path between two terminals from its position and replacing it with another path that does not belong to the tree.

b. Steiner node move: a no-Steiner node is a node that does not belong to the multicast tree. Moving the Steiner node is a basic transformation that switches the status of one of the elements of the current solution from a Steiner node to a no-Steiner node or vice versa [19]. This movement is an exchange between two non-terminal nodes, the first node is a worse node (a non-Steiner node), but the second is a better node that minimizes the cost of the movement (a Steiner node). This movement was used in the TS-based algorithm for the problem of multicast routing at Delay Constrained and Least-Cost proposed by Skorin Kapov et al. [20].

c. Path move (random position - selected path): this movement is a special case of random movement, it was used in the TS-based algorithm for the problem of multicast routing at DCLC proposed by Youssef et al. [21]. In this paper, to find the alternative path, Dijkstra algorithm is used to calculate the path length between nodes, then Prim algorithm is used for Steiner tree construction. The neighborhood structure based on "Delete and Add" operations inspired from [21] have been chosen for generating neighbors. This movement is a path change operation using two data structures, the solution is encoded as an array of $|M|$ elements, the first represents the current solution, while the second represents a secondary solution. Furthermore, this move consists of inserting the selected path in a random position, the path is defined between a source s and the destination d terminal nodes. The choice of the source s at each iteration is random, and the destination d is one of the other multicast group nodes. At each iteration, we randomly delete one superpath from the encoding of the current solution and then generate different feasible solutions by adding superpaths from the secondary solution which can be at a low cost.

In the context of our paper, the proposed algorithm (Algorithm 1.4) take into account the end-to-end delays in order to solve the DCLC problem.

Algorithme 1.4 : The proposed algorithm.

1 **Input:** Graph G;

2 $C_T \leftarrow$ the set of edges in the complete graph of the terminal nodes; // the costs of these edges are constructed using the Edge Betweenness method

3 P $\leftarrow$ percentage of components to include each iteration; // P $\in$]0,1[

4 n $\leftarrow$ length of time to do TS;

5 Best $\leftarrow$ initially the best solution is the resulting tree of the C_T ;

6 **repeat**

7   /* Construction of the initial solution */

8   S {}; // Our candidate solution

9   **repeat**

10   Sol_C_T $\leftarrow$ components in C_T $-$ S arranged in ascending order which could be added to S;

11   **if** *(Sol_C_T is empty)* **then**

12   S {}; //Try again

13   **else**

14   RCL $\leftarrow$ the P % lowest cost components in Sol_C_T; // restricted candidate list

15   S $\leftarrow$ S $\cup$ s; // s is component chosen uniformly at random from RCL

16   Sol_C_T $\leftarrow$ Sol_C_T $-$ s;

17   **until** *S is a complete solution (Includes all terminal nodes)*;

18   /* Local search */

19   **for** *(a period of time n)* **do**

20   S' = TS (G, S);

21   **if** *(((cost (S')< cost (Best))& &(Delay (S')< Δ)) || ((cost (S')= cost (Best))& &(Delay (S')< Delay(Best))))* **then**

22   Best=S';

23 **until** *Best is the ideal solution or we have run out of time*;

24 **Output:** Best solution found;

At each iteration of our algorithm, we generate a new feasible solution, so the initial solution used for local search is changing at each iteration in order to avoid being trapped into a local optimum. Moreover, to prevent the proposed algorithm stuck into local optima, Edge Betweenness heuristic with α factor is incorporated to our algorithm. Concerning the EB heuristic, the control parameter α, whose optimal value depends on the topological characteristics of the considered network, determines the priority between the cost of the edge and the cost edge betweenness. The value of α can change the cost of the edges, and this changed the Steiner tree result. It is our mechanism for resetting the initial solution. When creating the RCL, a function can be used to evaluate the quality of the elements. According to the percentage factor P, only qualified elements are included in the RCL. Looking for a best solution of the exploration of space continues until a criterion is met, the best solution obtained returned as the final one of our proposed algorithm. This procedure is repeated several times until the ideal solution is achieved or following a maximum number of iterations. Beside that, the ideal solution is the optimal one of benchmark instances from the Steiner library.

6   Experimental Results

To properly test and evaluate the implemented algorithms, we used the SteinLib instances library [29], which is a library of test instances of different sizes for the Minimum Steiner Tree Problem in Graphs. We generate class B instances and transform them into non-oriented graphs. Thus, we will notice that the difficulty of an instance of the problem does not depend only on the size of the problem, there are relatively small instances that are still difficult to resolve.

Table 1. Results of proposed algorithms.

N	OPT	KMB	EB	MRP	DCLCΔ1	DCLCΔ2
B01	82	82*	82*	**82***	82*	**82***
B02	83	90	85	**83***	–	–
B03	138	140	139	**138***	138*	**138***
B04	59	64	59*	**59***	59*	**59***
B05	61	62	62	**61***	61*	**61***
B06	122	128	127	**123**	123	**122***
B07	111	111*	111*	**111***	111*	**111***
B08	104	104*	104*	**104***	104*	**104***
B09	220	223	221	**220***	220*	**220***
B10	86	98	98	**86***	86*	**86***
B11	88	92	90	**88***	88*	**89**
B12	174	174*	174*	**174***	174*	**174***
B13	165	175	175	**170**	170	**170**
B14	235	238	238	**238**	238	**237**
B15	318	325	322	**318***	318*	**318***
B16	127	137	137	**127***	127*	**127***
B17	131	134	133	**132**	132	**133**
B18	218	223	223	**220**	220	**221**

(with DCLC $\Delta_1 = \infty$ and $\Delta_2 = 1.1 * $ Delay(OPT)).

Based on proposed algorithm, combined with the EB and TS methods, an approach is proposed in this paper for the first time to solve the DCLC problem. In Table 1, we presente the final results for the proposed algorithms. Computational results demonstrate the effectiveness of our proposed algorithm for this problem. OPT, denotes the cost of the optimal solution for each unconstrained Steiner tree instance, and values marked with * indicate optimal solutions.

From Table 1, we can observe that the KMB heuristic has given good results (B01, B07, B08, B12), but it does not ensure optimal solutions for all instances. It is found that the EB method has obtained improved solutions than the KMB algorithm, which implies that several edges are shared by the paths in the Steiner tree obtained using the new cost calculated from the cost and the edge betweenness. Concerning the EB heuristic, the control parameter α, in which optimal value depends on the topological characteristics of the considered network, determines the priority between the cost of the edge and the cost edge betweenness. Also, we consider that the maximum cost of the edge is an important factor in determining the optimal value of α. It was necessary to determine a good balance between parameter α and the number of iterations of the proposed algorithm, for at each iteration, the value of α can change the cost of the edges, and thus changed the Steiner tree result. Recall that candidates in the restricted candidate list (RCL) are chosen according to the cost of adding them to the existing tree in ascending order.

In the column MRP, we represent the results based on our proposed algorithm for solving the single-objective multicast routing problem (reduced to the MStTG problem). The proposed algorithm gave the optimal solution to 13 test problems for the MStTG problem. Therefore, the results provide a good indication of the competency of the proposed algorithm compared to the KMB and EB. However, some instances take a bit of time to run. The initial solution is based on EB heuristic and the choice of position in this neighborhood that minimizes cost. The mechanism consists of prohibiting the return to the last positions explored. The "Size of the taboo list" parameter can range between [10, 50], and the number of iterations in our first tests can go up to 1000 iterations. Through the random part of the initial solution, based Edge Betweenness heuristic allows to differ the solutions generated, but they are still of good quality since the random choice is made among a set of good candidates. The movements using in the TS metaheuristic, applies to the realizable solution to see if it is still possible to improve this solution.

Regarding QoS multicasting with a bounded end-to-end delay, to evaluate the performance of the proposed algorithm for solving the DCLC problem, a large number of simulations have been conducted on the benchmark Steinb instances where end-to-end delay is added. The results obtained are presented in the columns DCLC in Table 1. This algorithm have been executed with a sufficiently high value of the delay bound Δ so as not to act as a constraint (The link delay cannot be set to ∞). That means simulating the MStTG problem. Also, we set the delay bound $\Delta_2 = 1.1 * \mathrm{Delay(OPT)}$, where OPT is the multicast tree of the optimal solution with the minimal cost and delay.

Lastly, to evaluate the performance of the proposed algorithm, we compare the results obtained with other existing in the literature which are implemented GRASP algorithm. In Table 2, we compare the obtained results with two algorithms: GRASP−CST (Greedy Randomized Adaptive Search

Procedure–Constrained Steiner Tree) proposed in the literature [11], and GRASP–VND (Greedy Random Adaptive Search Procedure–Variable Neighborhood Search) mentioned in [28] in terms of the average tree cost for this set of benchmark problems. All experiment results demonstrate that the proposed algorithm has the overall best performance compared with other existing results in the literature through the use of EB heuristic combined with the movements of tabu searh in the implementation of our proposed algorithm.

Table 2. Ccomparison of the proposed algorithm with GRASP–CST and GRASP–VNS.

N	DCLC$\Delta2$	GRASP–CST	GRASP–VNS
B01	82*	82	82
B02	–	83	83
B03	138*	138	138
B04	59*	59	59
B05	61*	62	61
B06	122*	124	122.2
B07	111*	111	111
B08	104*	104	104
B09	220*	221	220
B10	86*	86	86.5
B11	89	88	88
B12	174*	174	174
B13	170	165	169.5
B14	237	235	235
B15	318*	318	319.5
B16	127*	132	127
B17	133	131	131.5
B18	221	219	218.5

(with DCLC $\Delta_2 = 1.1 *$ Delay(OPT)).

7   Conclusion and Future Work

In this paper, we try to find optimal solutions for single-objective multicast routing problem and multi-objective multicast routing problem, where both constraints: cost and delay are considered. To achieve this goal, we investigate algorithms such as heuristic EB and Tabu Search. Our proposition is based on the Greedy Randomized Search Procedure using the EB heuristic to construct the initial solution, then we used the movements of Tabu Srearch to diversify the research space. The chosen algorithms are adapted, implemented, and tested via an extensive set of experiments on a number of benchmark instances from the Steiner library. The results obtained show that the EB algorithm gives improvements to the KMB algorithm. Thus, the proposed method gives good results by contributing to EB heuristic and the movements of TS. After, we compared these algorithms to the optimal solutions from the Steiner library in order to study the efficiency of these algorithms to optimize the total cost of the constructed Steiner tree. Finally, we test our proposed algorithm on the DCLC problem and compare the results obtained with other existing one in the literature. The results provide a good indication of the competency of our proposed algorithm. In our future work, we intend to investigate the influence of our algorithm for solving MRPs with a wider range of real-world features. It is also interesting to extend our algorithms to solve other multi-objective optimization problems with reduced execution time.

References

1. Diot, C., Levine, B.N., Lyles, B., Kassem, H., Balensiefen, D.: Deployment issues for the IP multicast service and architecture. IEEE Netw. **14**(1), 78–88 (2000)
2. Striegel, A., Manimaran, G.: A survey of QoS multicasting issues. IEEE Commun. Mag. **40**(6), 82–87 (2002)
3. Xu, Y.: Metaheuristic approaches for QoS multicast routing problems. Ph.D. thesis, University of Nottingham Nottingham (2011)
4. Qu, R., Xu, Y., Kendall, G.: A variable neighborhood descent search algorithm for delay-constrained least-cost multicast routing. In: Stützle, T. (ed.) LION 2009. LNCS, vol. 5851, pp. 15–29. Springer, Heidelberg (2009). https://doi.org/10.1007/978-3-642-11169-3_2
5. Lee, S.J., Gerla, M., Chiang, C.C.: On-demand multicast routing protocol. In: 1999 IEEE Wireless Communications and Networking Conference, WCNC (Cat. no. 99TH8466), vol. 3, pp. 1298–1302. IEEE (1999)
6. Boppana, A.: A scalable simplified multicast forwarding for mobile ad-hoc networks (2011)
7. Chelius, G., Fleury, É.: Performance evaluation of multicast trees in adhoc networks (2002)
8. Haghighat, A.T., Faez, K., Dehghan, M., Mowlaei, A., Ghahremani, Y.: GA-based heuristic algorithms for QoS based multicast routing. Knowl.-Based Syst. **16**(5–6), 305–312 (2003)
9. Parsa, M., Zhu, Q., Garcia-Luna-Aceves, J.J.: An iterative algorithm for delay-constrained minimum-cost multicasting. IEEE/ACM Trans. Netw. **6**(4), 461–474 (1998)
10. Zhengying, W., Bingxin, S., Erdun, Z.: Bandwidth-delay-constrained least-cost multicast routing based on heuristic genetic algorithm. Comput. Commun. **24**(7–8), 685–692 (2001)
11. Skorin-Kapov, N., Kos, M.: A grasp heuristic for the delay-constrained multicast routing problem. Telecommun. Syst. **32**(1), 55–69 (2006). https://doi.org/10.1007/s11235-006-8202-2
12. Zhang, L., Cai, L.B., Li, M., Wang, F.H.: A method for least-cost QoS multicast routing based on genetic simulated annealing algorithm. Comput. Commun. **32**(1), 105–110 (2009)

13. Wei, W., Qin, Y., Cai, Z.: A multi-objective multicast routing optimization based on differential evolution in MANET. Int. J. Intell. Comput. Cybern. **11**(1), 121–140 (2018)

14. Zhang, X., Shen, X., Yu, Z.: A novel hybrid ant colony optimization for a multicast routing problem. Algorithms **12**(1), 18 (2019)

15. Matta, I., Guo, L.: QDMR: an efficient QoS dependent multicast routing algorithm. J. Commun. Netw. **2**(2), 168–176 (2000)

16. Kou, L., Markowsky, G., Berman, L.: A fast algorithm for Steiner trees. Acta Informatica **15**(2), 141–145 (1981). https://doi.org/10.1007/BF00288961

17. Oliveira, C.A., Pardalos, P.M.: Mathematical Aspects of Network Routing Optimization. Springer, New York (2011). https://doi.org/10.1007/978-1-4614-0311-1

18. Fujita, M., Kimura, T., Jin'no, K.: An effective construction algorithm for the Steiner tree problem based on edge betweenness. J. Sig. Process. **20**(4), 145–148 (2016)

19. Ghaboosi, N., Haghighat, A.T.: Tabu search based algorithms for bandwidth-delay-constrained least-cost multicast routing. Telecommun. Syst. **34**(3–4), 147–166 (2007). https://doi.org/10.1007/s11235-007-9031-7

20. Skorin-Kapov, N., Kos, M.: The application of Steiner trees to delay constrained multicast routing: a tabu search approach. In: 2003 Proceedings of the 7th International Conference on Telecommunications, ConTEL 2003, vol. 2, pp. 443–448. IEEE (2003)

21. Youssef, H., Al-Mulhem, A., Sait, S.M., Tahir, M.A.: QoS-driven multicast tree generation using tabu search. Comput. Commun. **25**(11–12), 1140–1149 (2002)

22. Zhang, K., Wang, H., Liu, F.: Multicast routing for delay and delay variation bounded Steiner tree using simulated annealing. In: Proceedings of the 2005 IEEE Networking, Sensing and Control, pp. 682–687. IEEE (2005)

23. Askari, Z., Avokh, A.: EMSC: a joint multicast routing, scheduling, and call admission control in multi-radio multi-channel WMNs. Front. Comput. Sci. **14**(5), 1–16 (2020). https://doi.org/10.1007/s11704-019-8199-9

24. Feo, T.A., Resende, M.G.: Greedy randomized adaptive search procedures. J. Global Optim. **6**(2), 109–133 (1995). https://doi.org/10.1007/BF01096763

25. Liu, Q., Tang, R., Ren, H., Pei, Y.: Optimizing multicast routing tree on application layer via an encoding-free non-dominated sorting genetic algorithm. Appl. Intell. **50**(3), 759–777 (2019). https://doi.org/10.1007/s10489-019-01547-9

26. Liu, Q., Ren, H.P., Tang, R.J., Yao, J.L.: Optimizing co-existing multicast routing trees in IP network via discrete artificial fish school algorithm. Knowl.-Based Syst. **191**, 105276 (2020)

27. Hassan, M., Hamid, A., Alkinani, M.: Ant colony optimization for multi-objective multicast routing. Comput. Mater. Continua **63**(3), 1159–1173 (2020)

28. Xu, Y., Qu, R.: A GRASP approach for the delayconstrained multicast routing problem. In: Proceedings of the 4th Multidisplinary International Scheduling Conference (MISTA4), Dublin, Ireland, pp. 93–104 (2009)

29. Koch, T., Martin, A., Voß, S.: SteinLib: an updated library on steiner tree problems in graphs. In: Cheng, X.Z., Du, D.Z. (eds.) Steiner Trees in Industry. COOP, vol. 11, pp. 285–325. Springer, Boston (2001). https://doi.org/10.1007/978-1-4613-0255-1_9

Terms Extraction from Clustered Web Search Results

Chouaib Bourahla[1], Ramdane Maamri[1], Zaidi Sahnoun[1], and Nardjes Bouchemal[2(✉)]

[1] Abdelhamid Mehri Constantine 2 University, Constantine, Algeria
{chouaib.bourahla,ramdane.maamri,
zaidi.sahnoun}@univ-constantine2.dz
[2] Mila University Center, Mila, Algeria
n.bouchemal.dz@ieee.org

Abstract. With a significant increase in web content, users cannot find easily the exact information that needs, so users are forced to read many search results sequentially until they reach what they want. In this paper, we proposed an approach to extract terms that represent sets of search results. These terms help users to access what they are looking for quickly and easily. Experimental results show that our proposed approach efficiently extract terms that represent these sets.

Keywords: Search result organization · Keywords extraction · Unsupervised method · Information retrieval · World Wide Web

1 Introduction

In the face of the huge data, the user cannot find easily the information that he needs on the web. Several approaches proposed personalized search systems based on the user profile (user's interests). However, the user still cannot find what he needs easily because he has multiple interests in different domains, and he does not use the exact query. Thus, He is forced to read many search results sequentially until he reaches what he wants, even they are very similar and they are not the desired. In addition, sometimes the user needs to enrich his query to get results that are more accurate.

As well, there are several approaches that extract keywords from text documents to represent these documents. However, these approaches are not appropriate for extracting highly relevant keywords from clustered text documents.

In this work, we focus on extracting terms or keywords from clustered search results given by a search engine like Google, Bing and Yahoo. These terms represent different subjects (clusters) and can help the user to determine which subject he is looking for. These terms are extracted from the search results clusters by our proposed scoring function.

The rest of the paper has been organized as follows. Section 2 is giving a brief review of the clustering and terms extraction methods. In Sect. 3, our approach has been detailed. Section 4 shows the experimental work. In Sect. 5, we discuss the results. The Sect. 6 gives the conclusion and the extensions possible for this work in the near future.

É. Renault et al. (Eds.): MLN 2020, LNCS 12629, pp. 364–373, 2021.
https://doi.org/10.1007/978-3-030-70866-5_24

2   Related Work

Given the variety of search results, this requires clustering these results according to their similarity. There are several clustering algorithms [1, 2], some of them use hard clustering approach and others use soft clustering approach. Among these algorithms, the Agglomerative Hierarchical Clustering Algorithms [3, 4], where it can be used to cluster the text documents based on their similarity. This similarity can be calculated using cosine similarity, which is one of the most popular similarity measures applied to text documents [5].

There are many previous works about clustering web search results, including [6–8] which they depend on clustering the search results according to their content similarity.

The work proposed in [9] clusters the search results then ranked the documents in these clusters. These clusters are not titled or described. So, we have to read some documents in the cluster to know the topic of this cluster.

In the other part, many approaches exist for extracting keywords or terms from text documents. All these approaches follow one of these four methods, Rule Based Linguistic approaches, Machine-learning approaches, Domain specific approaches or Statistical approaches [10].

The Rule Based Linguistic approaches are generally based on the linguistic knowledge/features, like the lexical analysis, syntactic analysis, and discourse analysis. Machine-learning approaches generally use supervised methods, they use a model that builds it from training documents to extract keywords from new ones. The Do-main specific approaches use the appropriate methods for each specific domain corpus, such as ontology, to extract keywords [10].

The Statistical approaches generally use a statistical feature on the terms of the document. The important advantages of these approaches that can use them with different languages in the same way and they can benefit the availability of a large number of datasets to perform analysis and get good results [10].

One of the Statistical approaches is the TF-IDF (term frequency-inverse document frequency) [9, 11, 12] which is the most widely used and considered as one of the most appropriate term weighting schemes. The TF-IDF scores words by their frequency in one document compared to their frequency in other documents of the corpus. This score can determine how important this word in a given document. The TF-IDF split in two parts; TF (Term Frequency) which is simply the number of occurrences of this term in the given document, and the second part is the IDF (Inverse document frequency) which is the measure of the importance of the term in the corpus. It gives a high value to the least frequent terms, which are considered more discriminating.

The TF-IDF can be used to score terms that represent a document in a corpus. Nevertheless, it is not appropriate to score terms that represent a cluster of documents in the corpus, because the approach does not take into consideration the notion of clusters.

Our approach falls within this context, which focuses on extracting the most important terms from the search results clusters.

3 Our Proposed Approach

The main objective of terms extraction approach is to help users to find the exact information they are looking for, by extracting different terms from different search results clusters. This allows them to access what they need accurately and easily, without having to read all the search results.

3.1 The Steps of the Approach

Our approach is based on the results (titles and snippets) given by a search engine (or multiple search engines). The process of the proposed approach is shown in (Fig. 1), and detailed in the following steps:

Getting the Search Results
After the user submits his/her query, we forward this query to the search engine. The search engine returns a set of results; each result is represented in a title and a snippet (short text). Then we display these results to the user.

Preprocessing the Search Results
After getting the search results, we preprocess them, by removing all punctuation and numbers, tokenizing, stop words removing and stemming [13] to get the bag of words.

Similarity Calculation
Next, we calculate the number of occurrences of each word in each search results item. We represent that in a Vector Space Model [14], where the number of dimensions is the number of terms in the bag of words, and the ith index of the vector contains the number of occurrences of the ith term of the bag of words in this search results item. Thus, we calculate the similarity of each pair of search results items using the cosine similarity as shown in the (Eq. 1).

$$sim(d_i, d_j) = \frac{d_i d_j}{|d_i||d_j|} \tag{1}$$

Where d is referred to a vector that contains the number of occurrences of terms in this document. This document contains the title and the snippet of the search results item.

Clustering Search Results Items
Therefore, we cluster these search results items using the agglomerative hierarchical clustering algorithm based on the similarity between them. We use this algorithm because it is one pass algorithm, that mean a fast algorithm [15]. This algorithm allows us to select different sub-clusters and different numbers of clusters without re-execute the algorithm. However, we can use also any other clustering methods.

Terms Extraction from Clusters
Then, we select K clusters from the top-levels clusters in the hierarchy. The number of clusters K is selected by the user. After that, from the previous bag of words, and in each

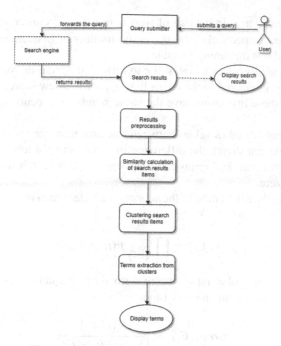

Fig. 1. The process of the proposed approach.

cluster, we select the terms that frequent a lot in this cluster and frequent a less in the other clusters. This selection of terms is done by giving a score to each term. The terms with the highest score are selected to distinguish the cluster.

Display the Terms
Finally, we display the selected terms to the user. The user selects one of these terms to filter the search results items. These two last steps can be repeated after each time the user chooses a term, by selecting the sub-clusters of the selected cluster, until he reaches what he wants.

3.2 Score Calculation

We proposed our own method to calculate the score of the terms instead of using the TF-IDF method. This score is calculated by multiplying the number of frequencies of the term in each search results item included in a cluster, after adding a tuning variable. Then, divide by the multiplication of the number of frequencies of the term in each search results item in the other clusters, after adding the same tuning variable. The equation is defined in (Eq. 2).

$$score(t_i, C_j) = \frac{\prod_{d_k \in C_j} f(t_i, d_k) + a}{\prod_{d_l \notin C_j} f(t_i, d_l) + a} \tag{2}$$

Where t_i is the ith terms in the bag of words, c_j is the jth cluster, d_k and d_l are the kth and l^{th} document respectively, $f(t_i, d_k)$ is the number of frequencies of a term t_i in a document d_k, and a is the tuning variable.

The (Eq. 2) generally, gives a high score to term that appear in many search results items of the cluster, compared to the terms that appear in a few search results items of this cluster, when these two terms have the same number of occurrences in the entire cluster.

The tuning variable a takes values between one and zero, zero is not included (a $\in$] 0, 1]). This variable can stretch the difference in score between terms those appear in many search results items, and terms those appear in a few search results items, if its value is closed to zero.

We can perform the calculation of the score by calculate each term's frequency score in each cluster as shown in (Eq. 3).

$$fc(t_i, C_j) = \prod_{d_k \in C_j} f(t_i, d_k) + a \tag{3}$$

Then, calculate the global term's score with these frequency scores as shown in (Eq. 4), without recalculate them every time.

$$score(t_i, C_j) = \frac{fc(t_i, C_j)}{\prod_{C_k \neq C_j} fc(t_i, C_k)} \tag{4}$$

Where $fc(t_i, C_j)$ is the frequency of the term t_i in the cluster c_j, d_k is the kth document, $f(t_i, d_k)$ is the number of frequencies of a term t_i in a document d_k, and a is the tuning variable.

4   Experimental Evaluation

The goal of our approach is to help users to find the exact information they are looking for quickly and easily. This is done by extracting terms that represent different subjects (clusters), and users select one of these terms to filter the search results, instead of reading all the search results including many similar unwanted results.

We present in this section a sample example that explains the working of this approach and the results it shows. We have done two experiments, one using search results items, and the second, using a labeled dataset.

In the example that explains the work of our approach, we assume that the user submitted the query "Olympics", he also chose the number of clusters K = 3 and the tuning variable a = 1. Regarding the search engine, we used Google search engine, and for simplicity, we only use the first 10 results provided by the search engine, which are represented in a title and a snippet as shown in (Table 1).

After preprocessing the search results and generating the vector space model, we calculate the similarity between the search results items [16] using the cosine similarity. The generated clusters and the terms that extracted from each cluster are shown in (Table 2).

Table 1. Search results items.

Search results items	Title
	Snippet
Item 1	Olympics	Olympic Games, Medals, Results, News	IOC
	Official website of the Olympic Games. Find all past and future Olympics, Youth Olympics, sports, athletes, medals, results, IOC news, photos and videos
Item 2	Olympic Games	Winter Summer Past and Future Olympics
	Access official videos, photos and news from all summer, winter, past and future Olympic Games - London 2012, Sochi 2014, Rio 2016, Pyeongchang 2018...
Item 3	Olympic Games - Wikipedia
	The modern Olympic Games or Olympics are leading international sporting events featuring summer and winter sports competitions in which thousands of...
Item 4	Olympic - YouTube
	Welcome to the Olympic Channel! With new videos every day, we are the place where the Games never end! From the best musical moments at the Olympic...
Item 5	NBC Olympics
	Olympic Talk. Follow the latest news from around the Olympic sports world. Djangbaev is sixth world weightlifting medalist to fail drug test · Japan Olympic...
Item 6	Home	The Tokyo Organising Committee of the Olympic and...
	The official website for the Olympic and Paralympic Games Tokyo 2020, providing the latest news, event information, Games Vision, and venue plans
Item 7	Olympics (@Olympics)	Twitter
	The latest Tweets from Olympics (@Olympics). The Olympic Games. Lausanne, Switzerland
Item 8	Olympic - Home	Facebook
	Olympic. 19112786 likes. Excellence Friendship Respect, excellence amitié respect
Item 9	The Olympic Games (@olympics) • Instagram photos and videos
	2.1m Followers, 2316 Following, 2901 Posts - See Instagram photos and videos from The Olympic Games (@olympics)
Item 10	Olympics - Apps on Google Play
	The official app of the Olympic Games View historical results from past Olympic Games, watch highlights, and stay up to date with the latest Olympic News and...

Table 2. Clusters and terms extracted.

Cluster	Search results items in the cluster	Terms extracted from the cluster
Cluster 1	1, 2, 3, 7, 9, 10	Games, photo
Cluster 2	8	Respect, excellence, friendship
Cluster 3	4, 5, 6	Tokyo, world

In the first experiment, we use a large collection of a search result items obtained from a search engine. These search results contain a title and snippet. We took the first (10, 20, 30, 40 and 50) documents of every search results of 20 keyworks. After, we mixed these search results and apply our approach for c = 10.

(Figure 2) shows the precision p of our approach. The precision p is calculated as shown in (Eq. 5):

$$p = \frac{\sum_{i=1}^{i=NC} \frac{1}{ti_{rank}}}{NC} \tag{5}$$

Which ti_{rank} is the rank of the terms given by the approach compared to the wanted term. In addition, NC is the number of clusters. The max value of p is 1, p equals 1 when the first keyword.

In the second experiment, we used a labeled dataset. This dataset is a collection of approximately 20,000 newsgroup documents, partitioned (nearly) evenly across 20 different newsgroups. To the best of our knowledge, it was originally collected by Ken Lang [17]. The data is organized into 20 different newsgroups, each corresponding to a different topic. Some of the newsgroups are very closely related to each other (e.g. comp.sys.ibm.pc.hardware/comp.sys.mac.hardware), while others are highly unrelated (e.g. misc.forsale/soc.religion.christian).

We select a collection of documents (5, 10, 20, 30, 50 and 200 documents) in each one of these three newsgroups: comp.os.ms-windows.misc, talk.politics.guns and sci.space. Then, we mixed these documents together. In addition, to apply our approach, we chose the number of clusters K = 3 and the tuning variable a = 1. (Table 3) shows the first three highest scored terms given by our approach that extracted from 30 documents in each newsgroup of the previous dataset.

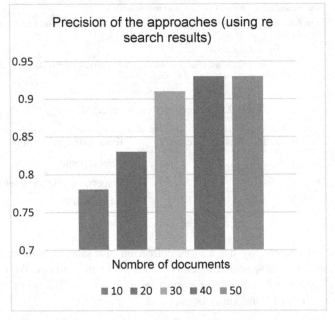

Fig. 2. Precision of the approaches (using real search results).

(Table 4) shows the precision p of our approach. In this case ti_{rank} is the rank of the terms given by the approach compared to these three terms (one terms by cluster): Windows, Space, Guns (or Handguns) which they are the names of the newsgroups (e.g. if the term "space" is the second highest scored term then, p = ½).

Table 3. The first highest scored terms.

Cluster	Terms extracted from the cluster with our approach
Cluster 1	Space, nasa, mission
Cluster 2	Windows, edu, writes
Cluster 3	Homicide, gun, handgun

Table 4. Precision p (using the labeled dataset).

Number of documents	5	10	20	30	50	200
Precision (p)	0.52	0.66	0.66	0.83	0.83	1.00

5 Discussion

In this section, we discuss the results given by our approach according to the example and the two experiments of the previous section.

After selecting the number of clusters k = 3, we see that the search results item 1,2,3,7,9 and 10 are in the first cluster, item 8 is in the second cluster and item 4,5 and 6 are in the third cluster. These clusters items are suitable to be with each other, and this is evident from (Table 1).

(Table 3) shows the extracted terms from each cluster. As we see, the first cluster is about photos and videos of the Olympic Games. The second cluster contains one item that represent the Olympic values, which are excellence, friendship and respect. The last cluster includes items that talk about the Olympic Games world and the Olympic Games that will be organized in Tokyo - Japan in 2020.

The inconvenient of using the TF-IDF approach to extract the terms from clusters in a corpus is that not take into a consideration the occurrences of terms over the clusters (the distribution of the terms over the clusters). As an example, if we have a cluster containing three search result items, and two terms, each one appears three times in the cluster, but the first appears three times in one search result item, and the second one time in each search result item. In this case, this method gives the same score to the two terms, which is not accurate. Because, the second term represents more the cluster, for that it should has a high score compared to the first.

In the first experiment that we have done. We see in (Fig. 2) that our approach gives an increasing precision **p** every time we increase the number of search results. This means that our approach becomes more precise as the number of search results increases.

In the second experiment, as the (Table 4) shows, our approach generally gives good accuracy. Our approach gives better results every time we increase the number of documents. It gives acceptable results when the number of documents used is 30 documents (Table 4). As we see also, it gives best results when we use 200 documents from each one of the three newsgroups. This best result means that our approach gives the first term (the highest scored one) that represent each cluster the same as the name of the newsgroup.

6 Conclusion and Future Works

In this paper, in order to facilitate user search, we proposed an approach to extract terms that represent different sets of search results. These terms are obtained from different clusters that contain similar search results.

The experiment results show that our approach can extract terms from clusters of search results to represents these clusters, so it helps the user to find in which cluster he finds the document he is looking for.

Our future work will be about improving the clustering method, especially the automatic selection of the number of clusters, because the extracted terms depend on the clustered search results.

References

1. Aggarwal, C.C.: Mining text data. In: Data Mining, pp. 429–455. Springer, Cham (2015)
2. Reddy, C.K., Vinzamuri, B.: A survey of partitional and hierarchical clustering algorithms. In: Data Clustering, pp. 87–110. Chapman and Hall/CRC (2018)
3. Jain, A.K., Dubes, R.C.: Algorithms for Clustering Data. Englewood Cliffs: Prentice Hall, New Jersey (1988)
4. Kaufman, L., Rousseeuw, P.J.: Finding Groups in Data: An Introduction to Cluster Analysis, vol. 344. John Wiley & Sons, Hoboken (2009)
5. Huang, A.: Similarity measures for text document clustering. In: Proceedings of the Sixth New Zealand Computer Science Research Student Conference (NZCSRSC2008), Christchurch, New Zealand, vol. 4, pp. 9–56, April 2008
6. Zamir, O., Etzioni, O.: Grouper: a dynamic clustering interface to Web search results. Comput. Netw. **31**(11–16), 1361–1374 (1999)
7. Zamir, O., Etzioni, O.: Web document clustering: a feasibility demonstration. In: SIGIR, vol. 98, pp. 46–54, August 1998
8. Hammouda, K.M., Kamel, M.S.: Efficient phrase-based document indexing for web document clustering. IEEE Trans. Knowl. Data Eng. **16**(10), 1279–1296 (2004)
9. Chaudhari, H.C., Wagh, K.P., Chatur, P.N.: Search engine results clustering using tF-IDF based Apriori approach. Int. J. Eng. Comput. Sci. **4**(05) (2015)
10. Siddiqi, S., Sharan, A.: Keyword and keyphrase extraction techniques: a literature review. Int. J. Comput. Appl. **109**(2) (2015)
11. Soucy, P., Mineau, G.W.: Beyond TFIDF weighting for text categorization in the vector space model. In: IJCAI, vol. 5, pp. 1130–1135, July 2005
12. Robertson, S.E., Jones, K.S.: Relevance weighting of search terms. J. Am. Soc. Inf. Sci. **27**(3), 129–146 (1976)

13. Gurusamy, V., Kannan, S.: Preprocessing techniques for text mining. Int. J. Comput. Sci. Commun. Netw. **5**(1), 7–16 (2014)
14. Salton, G., Wong, A., Yang, C.S.: A vector space model for automatic indexing. Commun. ACM **18**(11), 613–620 (1975)
15. Nguyen, H.-L., Woon, Y.-K., Ng, W.-K.: A survey on data stream clustering and classification. Knowl. Inf. Syst. **45**(3), 535–569 (2014). https://doi.org/10.1007/s10115-014-0808-1
16. Han, J., Kamber, M., Pei, J.: 10-cluster analysis: basic concepts and methods. Data Mining (Third Edition), The Morgan Kaufmann Series in Data Management Systems (third edition ed.), pp. 443–495 (2012)
17. Lang, K.: NewsWeeder: learning to filter netnews. In: Machine Learning Proceedings 1995, pp. 331–339. Morgan Kaufmann (1995)

Author Index

Printed in the United States
By Bookmasters